Modelling with Differential
and Difference Equations

AUSTRALIAN MATHEMATICAL SOCIETY LECTURE SERIES

Editor-in-chief: Professor J.H. Loxton, School of Mathematics, Physics, Computing and Electronics, Macquarie University, NSW 2109, Australia

1 Introduction to Linear and Convex Programming, N. CAMERON
2 Manifolds and Mechanics, A. JONES, A. GRAY & R. HUTTON
3 Introduction to the Analysis of Metric Spaces, J. R. GILES
4 An Introduction to Mathematical Physiology and Biology, J. MAZUMDAR
5 2-Knots and their Groups, J. HILLMAN
6 The Mathematics of Projectiles in Sport, N. DE MESTRE
7 The Peterson Graph, D. A. HOLTON & J. SHEEHAN
8 Low Rank Representations and Graphs for Sporadic Groups,
                          C. PRAEGAR & L. SOICHER
9 Algebraic Groups and Lie Groups, G. LEHRER (ed)

Modelling with Differential and Difference Equations

GLENN FULFORD
Department of Mathematics,
University College ADFA, Canberra

PETER FORRESTER
Department of Mathematics,
Melbourne University

ARTHUR JONES
Department of Mathematics,
Latrobe University

CAMBRIDGE
UNIVERSITY PRESS

CAMBRIDGE UNIVERSITY PRESS
Cambridge, New York, Melbourne, Madrid, Cape Town, Singapore, São Paulo

Cambridge University Press
The Edinburgh Building, Cambridge CB2 2RU, UK

Published in the United States of America by Cambridge University Press, New York

www.cambridge.org
Information on this title: www.cambridge.org/9780521440691

First published 1997
Reprinted 2001

A catalogue record for this publication is available from the British Library

ISBN-13 978-0-521-44069-1 hardback
ISBN-10 0-521-44069-6 hardback

ISBN-13 978-0-521-44618-1 paperback
ISBN-10 0-521-44618-X paperback

Transferred to digital printing 2006

Contents

Preface		*page* ix
Introduction to the student		1
	Part one: Simple Models in Mechanics	5
1	Newtonian mechanics	7
1.1	Mechanics before Newton	7
1.2	Kinematics and dynamics	10
1.3	Newton's laws	13
1.4	Gravity near the Earth	16
1.5	Units and dimensions	18
2	Kinematics on a line	21
2.1	Displacement and velocity	22
2.2	Acceleration	28
2.3	Derivatives as slopes	33
2.4	Differential equations and antiderivatives	37
3	Ropes and pulleys	41
3.1	Tension in the rope	41
3.2	Solving pulley problems	44
3.3	Further pulley systems	49
3.4	Symmetry	57
4	Friction	60
4.1	Coefficients of friction	60
4.2	Further applications	64
4.3	Why does the wheel work?	68
5	Differential equations: linearity and SHM	71
5.1	Guessing solutions	71
5.2	How many solutions?	74
5.3	Linearity	77
5.4	The SHM equation	81

6	Springs and oscillations	85
6.1	Force in a spring	85
6.2	A basic example	88
6.3	Further spring problems	94
	Part two: Models with Difference Equations	103
7	Difference equations	105
7.1	Introductory example	105
7.2	Difference equations — basic ideas	109
7.3	Constant solutions and fixed points	114
7.4	Iteration and cobweb diagrams	118
8	Linear difference equations in finance and economics	126
8.1	Linearity	127
8.2	Interest and loan repayment	133
8.3	The cobweb model of supply and demand	138
8.4	National income: 'acceleration models'	142
9	Non-linear difference equations and population growth	146
9.1	Linear models for population growth	146
9.2	Restricted growth — non-linear models	152
9.3	A computer experiment	157
9.4	A coupled model of a measles epidemic	164
9.5	Linearizing non-linear equations	170
10	Models for population genetics	177
10.1	Some background genetics	177
10.2	Random mating with equal survival	185
10.3	Lethal recessives, selection and mutation	193
	Part three: Models with Differential Equations	201
11	Continuous growth and decay models	203
11.1	First-order differential equations	203
11.2	Exponential growth	212
11.3	Restricted growth	218
11.4	Exponential decay	227
12	Modelling heat flow	232
12.1	Newton's model of heating and cooling	232
12.2	More physics in the model	237
12.3	Conduction and insulation	241
12.4	Insulating a pipe	249
13	Compartment models of mixing	257
13.1	A mixing problem	257
13.2	Modelling pollution in a lake	265
13.3	Modelling heat loss from a hot water tank	270

	Part four: Further Mechanics	275
14	Motion in a fluid medium	277
14.1	Some basic fluid mechanics	277
14.2	Archimedes' Principle	282
14.3	Falling sphere with Stokes' resistance	286
14.4	Falling sphere with velocity-squared drag	290
15	Damped and forced oscillations	295
15.1	Constant-coefficient differential equations	295
15.2	Damped oscillations	302
15.3	Forced harmonic motion	311
16	Motion in a plane	318
16.1	Kinematics in a plane	318
16.2	Motion down an inclined plane	326
16.3	Projectiles	331
17	Motion on a circle	336
17.1	Kinematics on a circle	336
17.2	Uniform circular motion	343
17.3	The pendulum and linearization	348
	Part five: Coupled Models	353
18	Models with linear interactions	355
18.1	Two-compartment mixing	355
18.2	Solving constant-coefficient equations	360
18.3	A model for detecting diabetes	366
18.4	Nutrient exchange in the placenta	373
19	Non-linear coupled models	379
19.1	Predator–prey interactions	379
19.2	Phase-plane analysis	384
19.3	Models of combat	389
19.4	Epidemics	394
References		399
Index		403

Preface

This book provides an introduction to modelling with both differential and difference equations. Our approach to mathematical modelling is to emphasize what is involved by looking at specific examples from a variety of disciplines. From each discipline enough background is provided to enable students to understand both the assumptions and the predictions of the models. Exercises have been included at the end of each section. They are intended to provide a balanced development of some of the main skills used in mathematical modelling, and hence they are an essential part of the book.

The main mathematical tools used in the book are differential and difference equations. *Differential equations* have their origins in mechanics: Newton's laws of motion lead to differential equations whose solutions can be used to predict the position of a body at some later time. Differential equations have been closely associated with the rise of physical science in previous centuries and they are now being used as models for real world problems in a variety of other disciplines. *Difference equations* are the discrete analogues of differential equations. They have risen to prominence in the last decade, during which it has become generally known that solutions of even very simple difference equations can exhibit complex chaotic behaviour.

To allow time for the development of other modelling skills besides solving the equations arising from the models, we have selected only models involving differential equations which are relatively easy to solve. Although our treatment of differential equations is intended to be self-contained, it is only fair to point out that our students were taking concurrently a first course in mathematical methods (beginning with the elements of differentiation and ending with some practice at solving separable and linear constant-coefficient differential equations, towards the end of the year). Some chapters of the book assume a knowledge of linear equations, complex numbers, vector algebra, or the elements of probability theory. The mathematical prerequisites are listed at the start of each chapter.

This book grew out of notes prepared for a first-year course given at La Trobe University for each of the last six years. The total time allotted to the course was 65 hours, including lectures and practice sessions. Not all chapters were covered in the same year and some choice is possible. A longer course could be organized by covering all the material in the book and including some computer work where relevant.

Each year we refined the material and its presentation, based on our experience in teaching the course during the previous years. Some curious incidental difficulties were faced each year by some students. These included (a) correct use of minus signs in setting up equations (b) sketching diagrams to illustrate the choice of a particular coordinate in a mechanics problem (c) distinguishing between the parameters of a problem and its unknowns. We have attempted to address these and other difficulties in both the text and the exercises. We have also analysed the steps involved in solving various types of problem, at least in the early part of the book, and we find this helps students to present their solutions to exercises clearly.

We wish to thank Sid Morris and Ed Smith for assistance with the overall planning of the original course and Alan Andrew, Jeff Brooks, Peter Stacey and John Strantzen for improvements in certain sections. One of the authors (G.F.) also wishes to thank Colin Pask for his encouragement. We also thank Dorothy Berridge and Annabelle Lippiatt for assisting with the typing.

G.F., P.F., A.J.
1996

Introduction to the student

Scientists, engineers and economists, working on a wide variety of problems, nowadays find it useful to set up mathematical models of the systems which they are investigating. To do this, they give a simplified description of the problem which allows equations to be set up and then solved to make predictions.

The following sample of the problems treated in this book shows the wide-ranging areas in which the modelling process is currently being applied.

- Mechanics. *We model the tension in a rope passing over pulleys and then use the model to explain how it is possible for a worker pulling on a rope to raise a heavy load many times his own weight.*
- Genetics. *In a genetics problem we describe a model which allows us to predict how many generations it takes for a recessive gene resulting in defective offspring to effectively disappear.*
- Thermal physics. *By modelling the loss of heat through the insulation surrounding a hot water pipe, we derive the paradoxical result that, in certain circumstances, it is possible for more heat to be lost from the pipe with insulation than without it.*
- Engineering. *We study the phenomenon of resonance, which was responsible for a famous bridge collapsing due only to wind fluctuations.*
- Fisheries management. *We study a model of interacting populations of certain types of fish and sharks and use it to explain how increasing the amount of fishing can actually increase the fish population.*

Some other areas in which models are constructed in this book are finance, economics, population studies, and physiology.

Enough background is given in each of these subjects to make the problems intelligible and to enable you to make suitable modifications to

1

the models when the problems are changed slightly. Carrying out these modifications will help you to develop many of the skills which are used in mathematical modelling.

Setting up a mathematical model of a problem involves the following steps.

- *Identifying the quantities most relevant to the problem and then making assumptions about the way in which the quantities are related. This usually involves simplifying the original problem so as to emphasize the features which are likely to be most important.*
- *Introducing symbols to denote the various quantities, and then writing the assumptions as mathematical equations.*
- *Solving the equations and interpreting their solutions as statements about the original problem.*
- *Checking the results obtained to see whether they seem reasonable and, if possible, whether they are in agreement with experimental data.*

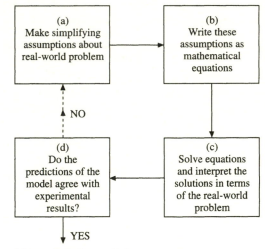

Use model to make further predictions,
and for control, in the real world

The modelling cycle

The modelling process is illustrated in the above figure, which brings out its cyclic nature. The process may fail at stage (c) if the equations are too complicated to be solved. One then returns to stage (a) of the process and tries to simplify the modelling assumptions to produce equations which are easier to solve. At stage (d), moreover, there may be insufficient agreement between the actual experimental results and the results predicted from the model. If this happens, one again returns to stage (a) to see whether the assumptions can be made more realistic. The process of returning to (a) may be repeated many times until a satisfactory model is obtained.

The problems discussed in this book will illustrate various stages of the modelling cycle. While stage (a) of the process is in some ways the most creative, it is often also the most difficult, involving the intuition and experience of specialists in the various areas. Hence, although we describe lots of models in full detail, the examples we set for practice are aimed more at the later stages of the process:

- *introducing mathematical symbols and writing the assumptions as equations;*
- *solving the equations and interpreting their solutions;*
- *checking to see if the answers seem reasonable.*

The book is divided into parts, each of which corresponds to the areas in which the problems arise and to the types of mathematical equations to which they lead.

Thus in *Part one* we model some basic problems in mechanics. The equations which arise in these models express the rate of change of one quantity in terms of other quantities. They are known as *differential equations*. Elementary mechanics provides a lot of interesting models in which the differential equations can be solved easily.

In *Part two* we turn to some models in which the equations express the value of a quantity at the end of the year, say, in terms of the values of the quantity at the end of previous years. Such equations are known as *difference equations*.

In *Parts three, four* and *five* we consider some models from a variety of areas which lead to progressively more advanced types of differential equations.

By studying the material in this book you will appreciate the extent to which mathematical modelling is currently being used in the various sciences and why mathematics is so useful in helping to solve scientific problems. By working through the exercises, moreover, you will develop many of the basic skills used by scientists engaged in mathematical modelling. In addition, the book provides practice and consolidation of the parts of mathematics associated with differential and difference equations. Stimulated by their potential applications, these topics now form one of the most flourishing areas of mathematical research.

Part one
Simple Models in Mechanics

1

Newtonian mechanics

The aim of mechanics is to explain and predict the motion of bodies. Early in the history of mankind the motion of celestial objects — sun, moon, stars, planets, comets — became a source of curiosity and wonder. At the terrestrial level, questions arising from the motion of falling bodies and of projectiles — whether stones or arrows — attracted the interest of some of the greatest thinkers of antiquity.

The system of mechanics devised by Newton in the seventeenth century made it possible to explain, for the first time, motion of both celestial and terrestrial objects with the one set of postulates, or laws. Newtonian mechanics proved to be one of the most successful mathematical models ever devised and it showed conclusively the value of mathematics in understanding nature. For these reasons, its advent is generally regarded as one of the turning points in the history of human thought.

This chapter introduces Newton's laws and sets them against the background of the set of postulates for mechanics due to the ancient Greek philosopher Aristotle. It also refers to the results discovered by Galileo and Kepler, which showed the inadequacy of the model proposed by Aristotle and thereby set the stage for Newton.

The outstanding success of Newtonian mechanics should not be allowed to blind us to the fact that it is only a model of what really happens. In the present century it has been found inadequate to explain motion at the subatomic level and at speeds close to that of light, which occur in astronomy. More useful mathematical models at these two extremes are provided by quantum mechanics and relativity respectively.

1.1 Mechanics before Newton

The first attempt to construct a mathematical model for the motion of bodies was made by ARISTOTLE (384–322 BC). During the Middle

Ages, European scholars regarded him as the authority on all scientific matters, including mechanics. By contrasting the assumptions made by Aristotle with the laws of motion given by Newton, we can gain a better appreciation of Newton's achievement.

The mechanics of Aristotle, like that of Newton, involves the idea of a *force* as an explanation of why bodies move. The basic idea of a force is familiar from everyday experience. If we want to move an object — be it a book, a chair or a wardrobe — we must apply a force to the object. Everyday experience also suggests, in a rudimentary way, how we might compare forces. Thus, most of us would agree that two people pushing together can exert, on average, twice the force exerted by one person pushing alone.

Relationships between force and motion are stated explicitly in the first two of the following assumptions introduced by Aristotle:

1. *Force is necessary to maintain a body in motion and once the force is removed the body will come to rest.*
2. *The force acting on a body is proportional to the velocity that it produces.*
3. *With equal bulk, heavier bodies fall faster than lighter ones.*

Aristotle claimed even more than is stated in the third of these assumptions, namely, *the speed at which bodies fall is proportional to their weight.* Thus if a ball is twice as heavy as another one, it should reach the ground in half the time (when released simultaneously from the same height).

Although these assumptions did not go unchallenged during the Middle Ages, the decisive break with Aristotle's ideas came during the Renaissance with the work of GALILEO (1564–1642). Part of the folklore surrounding Galileo concerns his dropping unequal weights from the leaning tower of Pisa to disprove Aristotle's claim about the relative speeds at which bodies fall. Less well known but equally deserving of mention is an interesting 'thought experiment' which Galileo suggested to show that this claim of Aristotle led to absurdity. This is explained later in the exercises.

To obtain further details about the motion of falling bodies, Galileo rolled a brass ball down a wooden beam. This had the effect of slowing the falling motion of the ball, making it possible to record, with reasonable accuracy, the times at which the ball passed various points marked on the beam. Whereas Aristotle's assumptions had referred to *velocity*, Galileo expressed the result of this experiment in terms of *acceleration* (or rate of

increase of velocity). His results were consistent with the postulate that
bodies fall with uniform acceleration.

In place of Aristotle's claim that force was necessary to maintain a body in motion, Galileo further postulated that

in the absence of forces a body need not be
at rest but can proceed with uniform speed.

A body moving in the absence of forces is said to be kept going by its own *inertia*. Galileo's views on such *inertial motion*, however, were, by later standards, rather timid. To account for the fact that objects do not fly off the Earth into outer space, Galileo allowed the inertial motion to take place along circles. He did point out, however, that motion along a big enough circle would be indistinguishable from motion along a line.

Meanwhile Johannes KEPLER (1571–1630) had been investigating the orbits of the planets. He believed he could explain the existence of six planets (Neptune, Uranus and Pluto were then undiscovered) and the sizes of their orbits from the geometry of the regular solids. Although none of this is taken seriously today, Kepler devoted his life to elaborating these ideas. Almost as a by-product he discovered the three laws of planetary motion for which he is now famous.

1. *The planetary orbits are ellipses with the sun as a focus.*
2. *The area swept out by a ray from the sun to the planet is proportional to the time taken.*
3. *The squares of the periods of the planets are proportional to the cubes of the major axes of their orbits.*

Kepler's laws were based on the observations of the planets' motion by the astronomer Tycho BRAHE (1546–1601). The basis for Kepler's laws was thus empirical. There was at that time no mathematical model from which these laws could have been deduced.

Anyone interested in learning more about the work of Aristotle, Galileo or Kepler should consult the book by Cohen (1987), the relevant chapters being 2, 5 and 6.

Exercises 1.1

1. Two balls have the same size but one is made of material which makes it five times as heavy as the other. According to Aristotle's model, what is the ratio of the speeds with which they fall?

2. Galileo refuted the idea that, in the absence of air-resistance, heavier bodies fall faster than light ones by the following 'thought experiment': Suppose it were true that *'heavier bodies fall faster in a vacuum'.*

(a) Of the two bodies shown below, which would fall faster, given the above assumption?

(b) In view of (a), what should be the effect on the speed of the 2 kg body if you placed the 1 kg body beneath it, as shown below?

(c) How does your answer to (b) contradict the original assumption?

1.2 Kinematics and dynamics

Two distinct aspects of mechanics, called *kinematics* and *dynamics*, are clearly discernible in the work of Galileo and Kepler.

Kinematics

Kinematics is concerned with the geometry of motion. What paths do moving objects follow? What coordinate systems are best suited to describing their paths? What are their velocities? What are their accelerations? In other words, kinematics is concerned with the most obvious aspects of motion — the things we can see with our eyes. Kepler's laws of planetary motion belong to kinematics since they give a geometric description of the path traced out by a planet. Galileo's law that bodies fall with uniform acceleration also belongs to kinematics.

The idea of a *particle* is basic for the kinematic model we shall be using in this book. A particle is defined as a body which has zero dimensions and hence may be regarded as occupying a single point. Clearly such a body cannot exist in the real world, but the idea is none the less useful in modelling, say, the solar system since the size of a planet is very small compared with the interplanetary distances. On the other hand, once a rocket ship gets close to a planet it is no longer appropriate to regard the planet as a single point, although it may well be still appropriate to think of the rocket ship in this way. In this book the primary concern is with situations in which it is appropriate to model the bodies by particles.

The concepts of *velocity* and *acceleration* will be explained more fully later in the book. For the present, however, it is sufficient to recall that velocity measures how fast a body is moving, while acceleration measures how quickly the velocity is changing. Velocity and acceleration must be measured relative to a *frame of reference*. For example, Parliament House is at rest relative to the Earth, but it is moving with a large velocity relative to the sun. Hence its velocity depends on whether the Earth or the sun is used as the frame of reference.

Strictly speaking, velocity and acceleration should be considered as *vector* quantities since they have direction as well as magnitude. Initially, however, we shall simplify matters by supposing that all motion takes place along a line; hence the number of possible directions is reduced to two (up and down, left and right, etc.).

Dynamics

Dynamics, in contrast to kinematics, is concerned with what *causes* bodies to move in certain ways. It involves concepts like *mass* and *force*, which are so basic that it is difficult to define them in terms of any simpler ideas. Hence, instead of attempting to define these concepts in a general way, we point to some of the everyday situations from which these concepts have arisen.

As to the concept of *force*, it had already been mentioned in the previous section that we exert a force when we push against objects, with a view to moving them. Although we cannot *see* a force directly, we are aware that a force is acting when we push against a table because we *feel* the tension in our limbs. Galileo is generally regarded as the founder of modern dynamics because he told what happens when there is no force acting on the body: it moves with uniform velocity (whereas Aristotle had claimed it would necessarily come to rest).

Like velocity and acceleration, force is a vector quantity, which has direction as well as magnitude. For example, the result of kicking a football depends not only on how big the kick is, but also the direction in which the kick is aimed. It will be assumed, moreover, that if two forces act at the same point of a body, then they will have exactly the same effect as their sum, calculated in accordance with the rules of vector algebra, as illustrated in Figure 1.2.1.

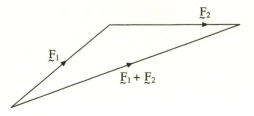

Fig. 1.2.1. Forces combine by vector addition.

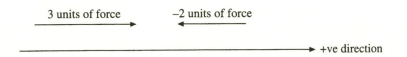

Fig. 1.2.2. Attaching signs to forces.

Initially, all the action will be taking place along a fixed line and so there will be only two directions to worry about. One of the directions will be selected as the *positive direction* and the opposite direction will be called the *negative direction*. Our convention for representing the direction of a force acting along the line is illustrated in Figure 1.2.2.

On the left of the figure is shown a force of magnitude 3 units, acting in the positive direction. On the right, however, the force has magnitude 2 units but acts in the negative direction. (Our convention is thus different from the usual one, in which both arrows would be labelled with positive numbers.)

The other dynamical concept, that of *mass*, arises from our everyday experiences in which a given force does not always produce the same effect. For example, kicking a football full of water produces a markedly different result from kicking the same ball full of air. The motion is smaller in the first case because of the increased mass of the ball.

Newton himself described mass as the 'quantity of matter' in a body. Implicit in this description is the idea that putting two identical bodies together gives double the mass of each of the individual bodies. This idea opens up the possibility of constructing a graded set of standard masses against which the mass of any body (of moderate size!) can be compared on a balance, as in Figure 1.2.3. (Strictly speaking it is their *weights* which are being compared, but at a given point on the Earth's surface, these are proportional to their *masses*.)

The above discussion is intended to provide a general idea of the

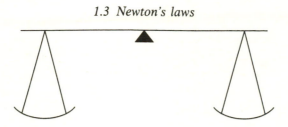

Fig. 1.2.3. Scales are used to determine when masses are equal.

meaning of the concepts of force and mass, but it stops short of giving precise definitions. The status of the concepts of *mass* and *force* in mechanics is analogous to that of a point and a line in Euclidean geometry. These concepts are so basic for geometry that it is not possible to define them in terms of any ideas which are more basic. Instead, there are the axioms or postulates of geometry, which tell us all we need to know in order to make logical deductions about points and lines. In mechanics, the rôle of the axioms is taken over by Newton's laws, discussed in the next section.

Exercises 1.2

1. How many standard masses ('weights') do you need to be able to find the mass, correct to the nearest kilogram, of any mass less than 10 kg?

1.3 Newton's laws

Isaac NEWTON (1642–1727) was born in the year in which Galileo died. His tremendous contribution to mechanics (and hence to the whole of physical science) was to discover a set of postulates or laws for dynamics which would explain the kinematical discoveries of his predecessors. In this way Newton achieved a complete synthesis of the laws of falling bodies near the Earth's surface (discovered by Galileo) and the laws of planetary motion (discovered by Kepler).

In substance, Newton's laws of motion state that when velocities and accelerations are measured in a suitable frame of reference:

1. *Every body continues in a state of rest or uniform motion in a straight line unless compelled to change its state by the action of a force.*

2. *The force produces an acceleration in the direction in which it acts. The magnitude of the force is jointly proportional to the acceleration and the mass of the body.*

3. *To every action there is an equal and opposite reaction; that is, the forces which two bodies exert on each other are always equal and opposite.*

Newton assumed the existence of an ideal frame of reference in which his laws would be true. The Earth itself provides an approximation to this ideal frame of reference, which is often useful for the study of motion near its surface. A better approximation is obtained by regarding the sun or stars as fixed. These latter frames of reference would be more useful for the study of the motion of the planets or of space vehicles. A frame of reference in which Newton's laws are true is called an *inertial frame*. We now make some comments on the significance of the above laws.

Newton's *first law* of motion is sometimes said to express the *principle of inertia*. Newton is here in agreement with Galileo, as opposed to Aristotle, that in the absence of a force, bodies maintain uniform speed. Whereas Galileo believed the motion could be in circles, however, Newton asserts that it must take place along a line.

In the *second law*, Newton asserts that the force depends on the acceleration — rather than on the velocity, as Aristotle had assumed. Nowadays, of course, anyone who has travelled in a jet aircraft will have had direct confirmation of the correctness of Newton's view. You feel the seat thrusting in your back when the jet accelerates from rest to take-off speed, not when it is cruising steadily at near sonic speeds above the clouds.

Given a suitable choice of units, we may write Newton's second law as an equation:

$$\{\text{force}\} = \{\text{mass}\} \times \{\text{acceleration}\}$$

or, more simply, as

$$F = ma.$$

As will be seen later, this is really a *differential equation*. Hence our ability to solve problems in Newtonian mechanics is inextricably linked to our ability to solve differential equations.

Although we have stated Newton's second law in its most ready-to-use form, there is a slightly more general version which equates force with the time rate of change of momentum. This more general version can be applied to a body whose mass changes with time — for example, a body consisting of rocket plus fuel. We shall not use this form of Newton's second law in this book. As to Newton's *third law*, a simple illustration is provided by the propulsion of rocket ships. The rocket ship exerts a force on the material which it expels backwards. This material exerts an equal but opposite reaction on the rocket ship, which causes it to accelerate forwards.

In later chapters, detailed models will be given for the forces which act in various situations, such as those where friction or air-resistance is present. There is, however, one type of force which acts, according to Newton, on every body in the universe. Hence it deserves to be mentioned here, alongside his laws of motion. *Newton's law of universal gravitation* states:

Each pair of bodies in the universe exerts a force of mutual attraction of magnitude

$$\frac{Gm_1m_2}{r^2}$$

where m_1 and m_2 are the masses of the bodies, r is the distance between them, and G is a constant (independent of the bodies).

On the basis of the above laws, Newton was able to derive Kepler's laws of planetary motion — one of the first big successes of Newtonian mechanics.

A very readable account of Newton's work in celestial mechanics is given in Koestler (1958), pages 504–517. A popular biography of Newton is Andrade (1979). A critical discussion of Newton's own statement of his laws of motion is contained in Westfall (1971), chapter 8.

Exercises 1.3

1. To each statement below attach the most appropriate name from the list: Aristotle, Galileo, Newton.

   1. *A force is necessary to maintain a body in motion.*

   2. *The negation of (a).*

   3. *Bodies fall to earth with uniform acceleration.*

   4. *Velocity is proportional to the force acting on the body.*

   5. *Acceleration is proportional to the force acting on the body.*

   6. *In the absence of forces, bodies move along straight lines.*

   7. *In the absence of forces, bodies move with uniform speed but can move around circles.*

Newtonian mechanics

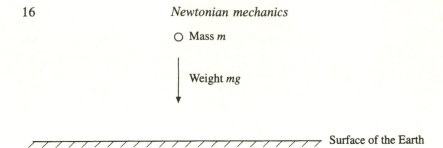

Fig. 1.4.1. Weight is the force *mg* due to gravity which acts on a particle of mass *m*.

1.4 Gravity near the Earth

According to Galileo's model, falling bodies in the absence of air-resistance have a uniform acceleration towards the Earth, which is the same for all bodies and is denoted by *g*. Some aspects of this model will now be discussed from the standpoint of Newton's laws.

First, Newton's second law of motion shows that the force acting on a particle of mass *m* which has an acceleration *g* towards the Earth is *mg* vertically downwards, as shown in Figure 1.4.1. This force, which is due to the gravitational attraction between the particle and the Earth, is called the *weight* of the particle (as distinct from its mass).

Second, Newton's law of universal gravitation, stated in the previous section, suggests that the gravitational attraction between the particle and the Earth diminishes with the height of the particle, and prompts us to discuss the range of validity of Galileo's model. Before this can be done, however, there is a mathematical difficulty to overcome in that Newton's law of universal gravitation applies, strictly speaking, to a pair of *particles*. For motion near the Earth's surface, it is clearly no longer appropriate to regard the Earth as a single point in space.

This difficulty was overcome by Newton himself. By using the integral calculus, which he invented for solving such problems, he was able to show that if a particle is attracted by the gravitational pull of a sphere of homogeneous material it experiences the *same* force as if the entire mass of the sphere is concentrated at its centre. Thus, the Earth being taken as a homogeneous sphere of mass *M*, the force of gravity pulling a particle of mass *m* towards the centre of the Earth is

$$\frac{GMm}{r^2}$$

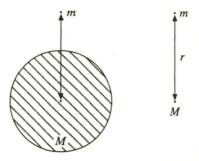

Fig. 1.4.2. Gravitational attraction of a solid sphere equals that of a single point of the same mass.

where r is the distance of the particle from the centre of the Earth, as shown in Figure 1.4.2.

It follows from this formula that, as the altitude of a particle increases, the gravitational attraction of the Earth on the particle decreases. For everyday purposes, however, the variation of gravity with height can be ignored. The radius of the Earth is so large (about 6400 km) that changes in height lead to very small proportional changes in the distance of the particle from the centre of the Earth and hence to correspondingly small changes in gravity. This is illustrated by the following example.

Example 1. *When a jet airliner ascends from take-off to an altitude of* 10 km, *by how much does the gravitational attraction acting on it decrease?*

Solution. *Let* r_0 *and* r_1 *be the distances of the airliner from the centre of the Earth at take-off and at an altitude of 10 km respectively. The ratio of the gravitational force when the distance is* r_1 *to that when it is* r_0 *is*

$$\left(\frac{GMm}{r_1^2}\right) \Big/ \left(\frac{GMm}{r_0^2}\right) = \left(\frac{r_0}{r_1}\right)^2 = \left(\frac{6400}{6410}\right)^2 = 0.99688.$$

Hence the fraction by which gravity decreases is $1 - 0.99688 = 0.00312$ *and hence the percentage decrease is about 0.31%.*

The problem of determining the value of g accurately has an interesting history and it impinges on several topics studied later in this book. The earliest method of determining g, using direct observations of falling bodies, had limited accuracy because of the speed with which bodies fall. In 1664, however, Christiaan HUYGENS accurately determined the value of g from pendulum experiments (using the formula which he had discovered for the period of oscillation of a pendulum). Within a few years it was discovered that g had a smaller value at the equator than it had in Paris. Newton's explanation for this was that the Earth is not a perfect sphere, but is flattened at the poles. The larger distance of a particle from the centre of the Earth when the particle is on the equator accounts for the smaller gravitational attraction there.

Some values of g obtained experimentally at various latitudes are given in Cohen (1987), page 175. They range from about 9.78 metres/s^2 at the equator to about 9.83 metres/s^2 at the poles. Unless stated otherwise we shall ignore the variation in the value of g and use the value of 9.8 metres/s^2. This will give a satisfactory model for the gravitational force on a particle provided it stays near the surface of the Earth and the duration of its motion is not too large.

Exercises 1.4

1. At what altitude is the weight of a given sample of material 1% less than when it is on the ground, directly beneath?

2. What is the percentage increase in the weight of a given sample of material when it is taken from the equator to a pole?

1.5 Units and dimensions

The units for the various quantities used in mechanics are shown in Table 1.5.1 for the international system (SI). Since velocity is obtained by dividing length by time, its units are derived from those for length and time in the obvious way. Similar remarks apply to the unit for acceleration, which comes from dividing velocity by time. The unit for force is chosen to make the constant of proportionality in Newton's second law of motion equal to unity.

How big is a force of 1 newton? To answer this note that, by Newton's second law, if the mass *m* of a body is measured in kilograms and the acceleration g due to gravity is measured in metres/s^2 then its weight is *mg* newtons. In particular if the mass of the body is about 0.1 kg then its weight is approximately 1 newton. This seems rather appropriate in view of the story about Newton and the falling apple, the mass of an apple of moderate size being about 0.1 kg!

Each quantity in mechanics has associated with it an expression involving the letters M, T, L which is called the *dimensions* of the quantity. The dimensions of some common types of quantities are listed in Table 1.5.2.

Table 1.5.1. *SI units for various quantities in mechanics.*

Quantity	Name of unit	Abbreviation
mass	kilogram	kg
time	second	s
length	metre	m
velocity	metres per second	m/s or m s^{-1}
acceleration	metres per second^2	m/s^2 or m s^{-2}
force	newton	N
area	square metres	m^2
volume	cubic metres	m^3

One of the main uses of dimensions is to check formulae involving physical quantities where the formulae are *valid independently of any choice of a system of units*. The check consists in seeing whether both sides of the equation have the same dimensions. The rules which enable us to get the relevant dimensions needed to apply the check are as follows:

1. *Only quantities with the same dimension may be added or subtracted. The result of adding or subtracting these quantities gives a quantity of the same dimensions.*
2. *The dimension of a product or quotient of two quantities is the product or quotient respectively of their dimensions.*
3. *All numerical factors, such as 1/2 and π, do not change the dimensions of a quantity. We therefore say that these factors are dimensionless.*

These rules can be written in symbols. Let *a* and *b* denote two quantities and let the symbol '[ ]' stand for the operating of taking dimensions of a quantity. The rules are:

$$[a + b] = [a] = [b]$$
$$[a - b] = [a] = [b]$$
$$[ab] = [a][b]$$
$$[a/b] = [a][b]^{-1}$$

There is also a rule for the *m*th power of a quantity, where *m* is a fraction:

$$[a^m] = [a]^m.$$

Table 1.5.2. *Dimensions of various quantities in mechanics.*

Type of quantity	Dimensions
mass	M
time	T
length	L
velocity	LT^{-1}
acceleration	LT^{-2}
force	MLT^{-2}
area	L^2
volume	L^3

Note that, from rule 3, $[\pi] = 1$, for example.

Example 1. *The formula for the period τ of small oscillations of a simple pendulum is*

$$\tau = 2\pi\sqrt{\ell/g} \tag{1}$$

where ℓ is the length of the pendulum and g is the acceleration due to gravity. Check that this formula is dimensionally correct.

Solution.

$$[LHS] = [\tau] = T$$

$$[RHS] = [2\pi\sqrt{\ell/g}] = [\ell/g]^{\frac{1}{2}} = \left(L/(LT^{-2})\right)^{\frac{1}{2}} = T.$$

Thus each side of the formula has the same dimensions, as required. Note that the multiplicative numerical factor 2π in (1) is dimensionless.

Exercises 1.5

1. The kinetic energy of a particle of mass m which is moving with a velocity v is defined to be

$$\frac{1}{2}mv^2.$$

What is the unit for kinetic energy when mass and velocity are measured in *SI* units? What are the dimensions of kinetic energy?

2. Let F be a force, let m_1 and m_2 be masses and let ℓ be a length. Is the following formula correct dimensionally?

$$F = \frac{m_1}{m_1 + m_2}\ell.$$

2

Kinematics on a line

This chapter contains a careful discussion of the ideas of velocity and acceleration for a particle moving along a line. Dealing first with this special case enables the use of vectors to be postponed. Familiarity is assumed, however, with the elements of the differential calculus including the idea of the derivative ϕ' of a function ϕ, its geometrical interpretation as the slope of the tangent to the graph of the function, and the basic rules for differentiating functions.

It will be assumed that the displacement x of a particle moving along a line is given as a function of the time t by

$$x = \phi(t)$$

for some twice differentiable function ϕ. The discussion will then lead us to define the velocity $\dot{x}$ of the particle at time t in terms of the derivative of ϕ by

$$\dot{x} = \phi'(t).$$

The acceleration $\ddot{x}$ of the particle at time t will be defined similarly in terms of the second derivative of ϕ by

$$\ddot{x} = \phi'(t).$$

The velocity and acceleration can then be calculated by using the rules of differentiation.

Just as important for mechanics is the reverse problem: given the velocity or the acceleration as a function of the time, what is the displacement at a given time? This problem will be solved by antidifferentiation and will provide an introduction to the idea of a differential equation.

This chapter provides all that is needed on differential equations for

Chapters 3 and 4, while Chapter 5 introduces some harder types of differential equations, which will be used in Chapter 6.

2.1 Displacement and velocity

The idea of *uniform speed* will be explained first. To illustrate this we consider a car moving between two towns A and B. The car makes the trip from A to B, as in Figure 2.1.1, taking 1/2 hour to do it. For a precise measurement of the distance travelled it is necessary to idealize the towns A and B as points and the car as a particle.

Its *average speed* for the trip is defined as

$$\frac{\text{distance travelled}}{\text{time taken}} = \frac{30}{1/2} = 60\,\text{km/hour}.$$

Along some stretch of the road, it may well happen that the average speed between every pair of points is the same. The speed for this part of the trip is then said to be *uniform* or *constant*. The speed could not be uniform for the whole trip, however, because there is an initial period as the car leaves A when it is picking up speed and a final period as the car approaches B when it is slowing down.

While speed is always ≥ 0, *velocity* may be negative to allow for motion in the opposite direction. Problems of uniform velocity involve only elementary arithmetic, whereas those of non-uniform velocity are much more difficult and involve the use of calculus. Before studying such problems, however, it is necessary to understand how plus and minus signs are used to help describe the position of the particle.

Coordinates and displacement

On the line along which the particle is moving, a point 0 is chosen to act as origin. Points on one side of 0 are assigned positive numbers which measure their distances from 0, in suitable units. Points on the opposite

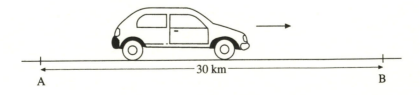

Fig. 2.1.1. A car travelling between two points.

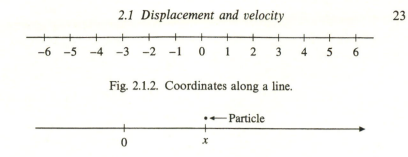

Fig. 2.1.2. Coordinates along a line.

Fig. 2.1.3. A particle with coordinate x.

side of 0 are then assigned negative numbers, as in Figure 2.1.2. The absolute value of the number assigned to a point measures the distance of the point from 0. The numbers are called the *coordinates* of the points on the line.

The choice of the origin is at the discretion of the user. It does not really matter which point is chosen, provided it is fixed in the inertial frame of reference. It is often the case, however, that a suitable choice of origin can simplify subsequent calculations.

Likewise, the choice of the side of the origin to label with positive numbers is entirely up to the user. In Figure 2.1.2, we have chosen to label the points to the right of 0 with positive numbers, but we could have equally well chosen to label the points to the left with them. The line with coordinates attached is called a *coordinate axis*. The direction in which the coordinates increase is called the *positive direction* of the coordinate axis, while the reverse direction is called its *negative direction*. The positive direction of the x-axis is also called the positive direction for the coordinate x itself.

If a particle moves along the line, its coordinate will change as time goes on. For this reason, a letter such as x is used to denote the coordinate of the particle at an arbitrary time t. When showing the significance of the coordinate x on a diagram, it is important to show the particle in a 'typical' or 'general' position, as in Figure 2.1.3, rather than in a special position such as the origin.

The coordinate x is often called the *displacement* of the particle from the origin 0. An alternative way of representing x which goes with this terminology is as an arrow from the origin to the particle, as shown in Figure 2.1.4.

Sometimes the coordinate x for the particle at time t is implicitly specified by the problem, without explicit mention of the origin or the coordinate axis. In such problems the choice of the origin and the

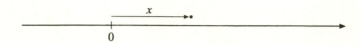

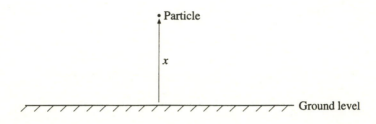

Fig. 2.1.4. The coordinate x measures the displacement of the particle from the origin.

Fig. 2.1.5. Here the coordinate x measures the displacement of the particle above ground level.

positive direction of the axis has been made already. The following example illustrates how the origin of the coordinate axis and its positive direction may be inferred from the description given in the problem.

Example 1. *A particle is moving in a vertical line. A coordinate x is defined for the particle by the statement that x is the height of the particle above ground level at time t. Where is the origin and which is the positive direction for the x-axis?*

Solution. *The displacement x of the particle above ground level at a typical time t is shown in Figure 2.1.5.*
The origin is at ground level since this is where the particle must be when $x = 0$. The positive direction for the coordinate axis, moreover, is upwards since this is the direction in which the particle moves if x increases.

It will be assumed that the displacement x of the particle may be expressed as a function of the time t by

$$x = \phi(t)$$

where ϕ is a twice differentiable function mapping times into displacements. The function ϕ is called the *displacement–time* function since it relates the displacement to the time. It is not necessarily assumed that $x = 0$ when $t = 0$ and negative values of t are not necessarily excluded.

Students often confuse the use of negative values of t with the science-fiction feat of going backwards in time. Negative values of t, however, simply indicate that we have agreed to label some instant as the origin for time; any instant prior to that is then labelled with a negative t value. For example, if we choose today as day 0, then tomorrow is day 1 and yesterday was day -1.

Some books use the same symbol x for the displacement and for the function ϕ. This can lead to confusing notation such as '$x = x(t)$'. We try to avoid such usage in this book, even though it means we have to introduce extra symbols.

With the aid of the above ideas on displacement, we can now complete our discussion of velocity.

Non-uniform velocity

In non-uniform motion, the velocity may change from one instant to another. Hence there is a need to analyze the idea of *velocity at an instant* or *instantaneous velocity*. At first sight, however, there appears to be something paradoxical about this idea. An instant has zero duration, hence during an instant the particle cannot move. How then can it have an instantaneous velocity?

We resolve this paradox by looking at smaller and smaller time intervals surrounding the given instant and taking the limit of average velocities over these intervals. Let the time change from t to $t + \delta t$ and let the displacement change from x to $x + \delta x$. At the start of the change

$$x = \phi(t)$$

and at the end of the change

$$x + \delta x = \phi(t + \delta t).$$

Hence by subtraction

$$\delta x = \phi(t + \delta t) - \phi(t).$$

This gives the change in displacement as the time changes from t to $t + \delta t$ and hence the average velocity of the particle during this period is

$$\frac{\delta x}{\delta t} = \frac{\phi(t + \delta t) - \phi(t)}{\delta t}.$$

Since ϕ was assumed differentiable, this ratio will approach a limit as $\delta t \to 0$ which is just the derivative of ϕ at t. Hence we may define the *instantaneous velocity* of the particle at time t as

$$\lim_{\delta t \to 0} \frac{\delta x}{\delta t} = \phi'(t).$$

Thus we have been led to define the instantaneous velocity (or simply the velocity) at time t as the derivative at t of the displacement–time function. The function ϕ' is called the *velocity–time function*.

Notation

The velocity at time t is written in Leibniz's notation as

$$\frac{dx}{dt}.$$

This notation helps recall the way in which the velocity was defined above as the limit of a *ratio*: a small increase in x divided by a small increase in t. It also reminds us that velocity is a *rate of change*: of displacement with respect to time. For these reasons it is understandably popular with those who are primarily concerned with setting up mathematical models of physical problems.

On the other hand, Leibniz's notation ignores the functions which link the physical quantities. There are many situations in mathematics where it is necessary to recognize these functions quite explicitly and then the notation $\phi'(t)$ has the advantage. Familiarity with both notations, together with the ability to translate from one to the other, is therefore desirable.

An alternative notation for the velocity — the one used by Newton in fact — is

$$\dot{x}.$$

This is logically equivalent to Leibniz's notation in that one thinks of the dot as meaning $\frac{d}{dt}$, and it is quicker to write.

Sign of the velocity

Since

$$\dot{x} = \lim_{\delta t \to 0} \frac{\delta x}{\delta t}$$

it follows that, if $\dot{x} > 0$ at some time t, then for δt sufficiently small

$$\frac{\delta x}{\delta t} > 0.$$

Hence,

$$\text{if} \quad \delta t > 0 \quad \text{then} \quad \delta x > 0.$$

This means that, during the interval when the time increases from t to $t + \delta t$, the net change of displacement of the particle has been in the positive direction of the coordinate x. We may summarize this by saying that

> when $\dot{x} > 0$ *the particle moves in the positive direction for the x-coordinate.*

Similarly it can be shown that

> when $\dot{x} < 0$ *the particle moves in the negative direction for the x-coordinate.*

The particle is said to be *stationary* when $\dot{x} = 0$ because the velocity is then instantaneously zero.

The following example illustrates the ideas that have been introduced in this section.

Example 2. *A particle moves along a horizontal line in such a way that its displacement x metres to the right of an origin 0 at time t seconds is given by*

$$x = t^3 + 2t^2 - 9t + 9.$$

At the instant when $t = 1$, find the displacement and velocity of the particle. State also the side of the origin on which the particle lies, the direction in which it is moving, and its speed at this instant.

Solution. *Here the displacement–time function is defined by $\phi(t) = t^3 + 2t^2 - 9t + 9$. By differentiation, the velocity at time t is $\dot{x} = \phi'(t) = 3t^2 + 4t - 9$. Thus, when $t = 1$,*

$$x = 3 \quad and \quad \dot{x} = -2.$$

Since the particle moves to the right if x increases, this is the positive direction for the coordinate x. Hence, when $t = 1$, the particle is to the right of the origin but is moving left. Its speed is $2\,\text{m/s}$.

Exercises 2.1

1. A particle drops over the edge of a table and falls vertically downwards. The coordinate x is chosen as the distance through which the particle has fallen at an arbitrary time t.

   (a) Which of the four diagrams below best conveys the meaning of x?

   (b) Which is the positive direction for the coordinate x? Why?

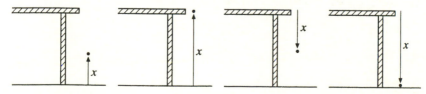

2. Repeat Example 2 in the text, but this time suppose that x metres is the displacement of the particle to the *left* of the origin at time t seconds.

3. Let ϕ be the displacement–time function for Example 2 in the text. Write down each of the numbers $\phi(0), \phi(2), \phi(t+1)$. Write down also $\phi'(0), \phi'(2), \phi'(t+1)$.

4. Two particles P_1 and P_2 are connected by a string (of constant length). P_1 lies on the table while P_2 hangs from the rightmost edge of the table, shown below. Let x be the distance of P_1 from the rightmost edge of the table and let y be the distance of P_2 below this edge at time t.

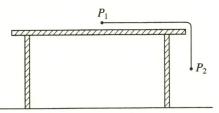

(a) Copy the diagram and mark x and y on it.

(b) Which is the positive direction for the coordinate x? for the coordinate y?

(c) In each of the following cases state the direction of motion for each particle:

    (i) $\dot{x} > 0$ (ii) $\dot{y} > 0$.

(d) Write the length of the string in terms of x and y and then, using the rule for differentiating a sum, deduce a relationship between $\dot{x}$ and $\dot{y}$.

2.2 Acceleration

In everyday usage, a car is said to be *accelerating* when its speed is increasing and *decelerating* when its speed is decreasing. In mechanics, however, the word acceleration is used in a slightly more technical way, being defined in terms of rate of change of velocity rather than speed, and it may be positive or negative.

To frame a definition of acceleration we use the notation of the previous section and we suppose that as the time changes from t to $t+\delta t$ the velocity changes from $\dot{x}$ to $\dot{x}+\delta \dot{x}$ so that

$$\dot{x}+\delta \dot{x} = \phi'(t+\delta t)$$

with

$$\dot{x} = \phi'(t)$$

and so, by subtraction,

$$\delta\dot{x} = \phi'(t + \delta t) - \phi'(t).$$

This gives the change in velocity as the time changes from t to $t + \delta t$ and hence the ratio

$$\frac{\delta\dot{x}}{\delta t} = \frac{\phi'(t + \delta t) - \phi'(t)}{\delta t}$$

gives the change in velocity per unit time. This ratio is called the *average acceleration* during the time interval from t to $t + \delta t$. Since ϕ' is assumed differentiable, the ratio on the right approaches the limit $\phi''(t)$ as $\delta t \to 0$. Hence we are led to define the *instantaneous acceleration* (or simply the *acceleration*) of the particle at time t as

$$\lim_{\delta t \to 0} \frac{\delta\dot{x}}{\delta t} = \phi''(t).$$

Thus the acceleration of the particle at time t is the first derivative of the velocity–time function or the second derivative of the displacement–time function at this instant.

Notation

The acceleration of the particle at time t is written in Leibniz's notation as

$$\frac{d}{dt}\left(\frac{dx}{dt}\right) \quad \text{or, more briefly, as} \quad \frac{d^2x}{dt^2}.$$

As this is a little cumbersome to write, however, we shall mainly use Newton's notation $\ddot{x}$. Thus we may write

$$\ddot{x} = \phi''(t).$$

Sign of the acceleration

Suppose that the acceleration $\ddot{x}$ of the particle is positive at some time t. This implies that a small positive change δt in the time will produce a positive change $\delta\dot{x}$ in the velocity. Thus the velocity increases from $\dot{x}$ to $\dot{x} + \delta\dot{x}$.

This does not necessarily mean, however, that the *speed* increases. For example: if $\dot{x} = 2$ and $\delta\dot{x} = 0.05$ then both the velocity and the speed increase from

2 to 2.05

If, however, $\dot{x} = -2$ and $\delta\dot{x} = 0.05$ then the velocity increases from

$$-2 \text{ to } -1.95$$

whereas the speed *decreases* from

$$2 \text{ to } 1.95.$$

Thus there is some divergence between the everyday usage of the word acceleration and its use in mechanics. Our usage is summarized by saying that:

If $\ddot{x} > 0$ the particle is accelerating in the positive direction of the x-axis; if $\ddot{x} < 0$ it is accelerating in the negative direction of the x-axis.

Example 1. *Suppose that a particle moves along a horizontal line in such a way that its displacement x metres to the right of an origin 0 at time t seconds is given by $x = \phi(t)$ where*

$$\phi(t) = t^2 - t - 2.$$

(a) Sketch the graph of the displacement–time function. At which times is the particle

  (i) at the origin?
  (ii) to the right of the origin?
  (iii) to the left of the origin?

(b) At which times is the particle

  (i) stationary?
  (ii) moving to the right?
  (iii) moving to the left?

(c) Which is the leftmost point reached by the particle?
(d) Find the acceleration. In which direction is the particle accelerating?

Solution. *The coordinate x of the particle at a typical time t is shown in Figure 2.2.1. The positive direction for the x-axis is to the right since this is the direction in which the particle moves if x increases.*

(a) To sketch the graph of $x = \phi(t)$, note that the quadratic $\phi(t)$ factorizes to give

$$x = t^2 - t - 2 = (t+1)(t-2).$$

This leads to the following table of signs

t	large $-ve$	-1	between -1 and 2	2	large $+ve$
x	large $+ve$	0	$-ve$	0	large $+ve$

which in turn helps us to sketch the graph in Figure 2.2.2 — a parabola with vertex at the bottom.

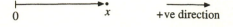

$$0 \qquad\qquad x \qquad\qquad +\text{ve direction}$$

Fig. 2.2.1. Coordinates for Example 1.

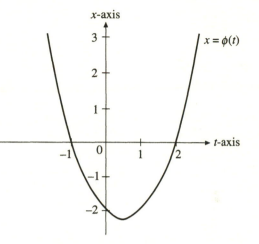

Fig. 2.2.2. Displacement–time graph for Example 1.

Note that in sketching the graph we draw the x-axis vertically upwards even though in the original problem it was horizontal. From the graph it is clear that the particle is

    (i) at the origin *when* $t = -1$ *or* $t = 2$,
    (ii) to the right of the origin *when* $t < -1$ *or* $t > 2$,
    (iii) to the left of the origin *when* $-1 < t < 2$.

(b) Differentiation gives

$$\dot{x} = \phi'(t) = 2t - 1.$$

The particle is thus

    (i) stationary *when* $\dot{x} = 0$, *hence* $t = \frac{1}{2}$,
    (ii) moving right *when* $\dot{x} > 0$, *hence* $t > \frac{1}{2}$,
    (iii) moving left *when* $\dot{x} < 0$, *hence* $t < \frac{1}{2}$.

(c) The particle reaches its leftmost point when $\dot{x} = 0$, *hence* $t = \frac{1}{2}$ *and so* $x = -2\frac{1}{4}$. *The leftmost point is thus* $2\frac{1}{4}$ *metres to the left of* 0.

(d) Differentiation of $\dot{x}$ *with respect to t gives*

$$\ddot{x} = \phi''(t) = 2.$$

Thus the acceleration is 2 m/s². *Since* $\ddot{x} > 0$, *the particle is accelerating towards the right.*

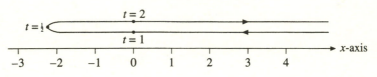

Fig. 2.2.3. Tracking diagram for the motion described in Example 1.

Geometry is a valuable tool in the study of problems in mechanics. The well-known saying that a good picture is worth a thousand words is just as true in mechanics as in any other part of mathematics. In addition to representing the motion of a particle by a displacement–time graph, we often find it informative to have diagrams which relate the motion directly to the line on which it is taking place.

Thus, in Example 1, the displacement–time graph may be used to derive the diagram shown in Figure 2.2.3, which helps us to keep track of what is happening on the horizontal line on which the particle is moving.

There does not seem to be any standard name for these diagrams. We shall call them *tracking diagrams* because they help us keep track of the particle's motion.

Exercises 2.2

1. In Example 1 in the text, how would the solution change if x were defined to be the displacement of the particle to the left of the origin at time t?

2. Suppose that a particle moves along a horizontal line in such a way that its displacement to the right of an origin 0 at time t is given by $x = \phi(t)$ where

$$\phi(t) = t^3 - 6t^2 + 9t.$$

(a) Sketch the graph of the displacement–time function, showing where it crosses the axes.

(b) Express $\dot{x}$ and $\ddot{x}$ as functions of time, and sketch their graphs.

(c) At which times is the particle
  (i) at the origin?
  (ii) to the right of the origin?
  (iii) to the left of the origin?

(d) At which times is the particle
  (i) stationary?
  (ii) moving to the right?
  (iii) moving to the left?

(e) At which times is the particle accelerating towards the right?

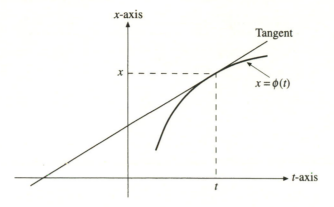

Fig. 2.3.1. The derivative is the slope of the tangent.

(f) Sketch a tracking diagram showing how the particle moves along the line.

3. Repeat the above exercise, but with $\phi(t) = -t^4 + 2t^2 - 1$. What is the rightmost point reached by the particle?

2.3 Derivatives as slopes

Velocity was defined earlier in this chapter as the derivative of the displacement–time function. The geometrical interpretation of the derivative, as the slope of a tangent to the graph of the original function, can therefore be used to predict various features of the velocity–time graph directly from the displacement–time graph. In a similar way, information about the acceleration–time graph can be obtained from the velocity–time graph.

To illustrate the process consider the graph in Figure 2.3.1 showing the displacement x of a certain particle as a function of the time t, given by $x = \phi(t)$. The tangent is shown at a typical point (t, x) on the graph. Its slope is the 'rise over run' between any two of its points. This slope is the derivative $\phi'(t)$ of the function ϕ at time t and hence taking $\dot{x}$ to be this slope gives a point $(t, \dot{x})$ on the velocity–time graph. By plotting points $(t, \dot{x})$ in this way for a number of values of t, we can build up a rough picture of the velocity–time graph. The following example illustrates this procedure.

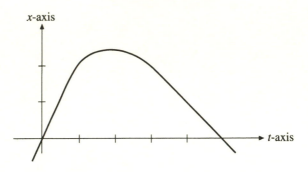

Fig. 2.3.2. Displacement–time graph for Example 1.

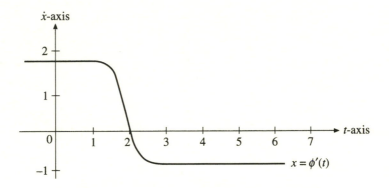

Fig. 2.3.3. Velocity–time graph for Example 1.

Example 1. *Suppose that a particle moves along a line in such a way as to give the graph in Figure 2.3.2 for the displacement x metres as a function of the time t seconds. By examining slopes of tangents at various points obtain a rough sketch of the velocity–time graph. Show also the general shape of the acceleration–time graph.*

Solution. *The part of the graph for which $t \leq 1$ is a line segment of slope 2, while the part of the graph for which $t \geq 3$ is a line segment of slope −1. In between, as t increases from 1 to 3 the slope decreases steadily from 2 to −1. Hence the velocity–time graph contains a horizontal line segment at height 2 for $t < 1$ and another such segment of height −1 for $t > 3$. These are linked by a curved portion which drops smoothly from height 2 to height −1 as in Figure 2.3.3.*

The acceleration–time graph is now obtained by taking slopes on the velocity–time graph. Along the constant parts of the velocity–time graph the slopes are zero. Between $t = 1$ and $t = 3$ the slope drops from zero to some negative minimum value and then comes back up again to zero. Hence the acceleration–time graph has the general shape shown in Figure 2.3.4.

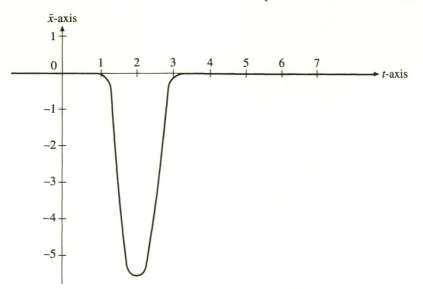

Fig. 2.3.4. Acceleration–time graph for Example 1.

It is difficult to give a precise estimate of the minimum value of the acceleration since the curved portion of the $\ddot{x}$*–t graph is only a rough approximation.*

Exercises 2.3

1. A particle is moving along a horizontal line and its displacement to the right of an origin 0 at time t is x. The displacement–time graph for the motion is shown below.

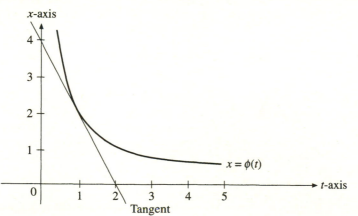

What is the velocity when $t = 1$? In which direction is the particle then moving?

2. Repeat the previous exercise, but this time suppose that x denotes the displacement of the particle to the left of the origin 0 at time t.

3. The graph below shows the velocity–time graph for the motion of a particle whose velocity and acceleration are both positive when $t = 2$.

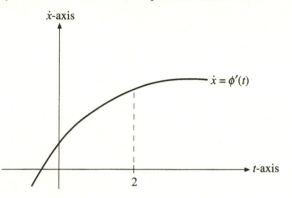

(a) Sketch similar velocity–time graphs to illustrate each of the following possibilities, when $t = 2$:

   (i) velocity positive and acceleration negative,

   (ii) velocity negative and acceleration positive,

   (iii) velocity negative and acceleration negative.

(b) In which of the cases in part (a) is the speed decreasing when $t = 2$?

4. A particle moves on a horizontal line and x is its displacement to the right of the origin at time t. The tracking diagram for the motion is shown below.

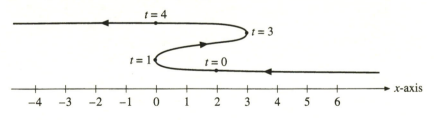

(a) Give a rough sketch of the displacement–time graph.

(b) At which times is the velocity zero?

(c) Between which pair of times must there be an instant when the acceleration is zero? Choose the times as close together as you can and give reasons for your answer.

5. The displacement–time graph for the motion of a certain particle is shown below. Copy the graph and then, directly underneath, sketch

   (a) the velocity–time graph,

   (b) the acceleration–time graph.

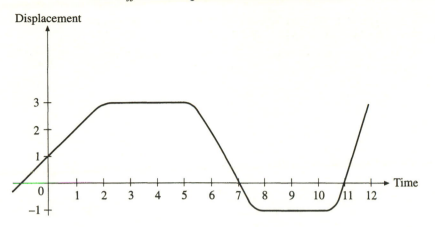

2.4 Differential equations and antiderivatives

In previous sections it was shown how to proceed in the direction

displacement as a function of time

↓

velocity as a function of time

↓

acceleration as a function of time.

When expressed in terms of our definitions, all that is involved here is the process of differentiation (once or twice in succession).

The reverse process is just as important in mechanics. To proceed in the reverse direction, going back from acceleration to displacement, turns out to involve the process of antidifferentiation. How this works will be illustrated by an example.

Example 1. *What can be said about the displacement x of a particle if its acceleration is given as a function of the time t by*

$$\ddot{x} = t? \tag{1}$$

The particle starts from rest at the origin.

Solution. *First, we find all the functions ϕ such that, when $x = \phi(t)$,*

$$\ddot{x} = t$$

or equivalently

$$\phi''(t) = t.$$

The first antidifferentiation gives the derivative ϕ':

$$\phi'(t) = \frac{1}{2}t^2 + c_1$$

for some constant c_1. A further antidifferentiation now gives

$$\phi(t) = \frac{1}{6}t^3 + c_1 t + c_2 \tag{2}$$

for some constant c_2. But since the particle starts from rest, at the origin, the initial values of x and $\dot{x}$ are $x = 0$ and $\dot{x} = 0$ at $t=0$. Hence $\phi(0) = 0$ and $\phi'(0) = 0$ from which using (1) it may be deduced that $c_1 = c_2 = 0$. Hence the required answer is

$$x = \frac{1}{6}t^3.$$

The equation (1) is an example of a simple *differential equation* — that is, an equation involving derivatives. In the above example we solve the differential equation by antidifferentiating both sides of the equation.

Exercises 2.4

The exercises from number 4 onwards provide practice at applying differential equations to constant acceleration problems. In each of these exercises you must set up a differential equation and then solve it — merely quoting a formula is not what is wanted. You may assume the acceleration g due to gravity is 9.8 m/s^2 vertically downwards.

1. Use antidifferentiation to find the solution of the differential equation

$$\dot{x} = t + 1$$

which satisfies the initial condition $x = 2$ when $t = 0$.

2. Find the solution of the differential equation

$$\ddot{x} = e^t + e^{-t}$$

which satisfies the initial conditions $x = 2$ and $\dot{x} = 1$ when $t = 0$.

3. A particle moves along a line and its displacement to one side of an origin at time t is x. The acceleration is given as a function of time by the differential equation

$$\ddot{x} = \sin(2t).$$

Find x as a function of t, given that $x = \dot{x} = 1$ when $t = 0$.

4. Read through the following problem, then answer the questions (a) to (e) below:

A stone was dropped from rest at the top of a cliff and a clunk was heard 3 seconds later when it struck the ground at the foot of the cliff. How high was the cliff? With what speed did the stone hit the ground?

(a) There are two obvious points P_1 and P_2 either of which would be a natural choice for origin. Which points are they?

(b) Show on a diagram how to choose a coordinate x for the particle, with P_1 as the origin for the x-axis.

(c) Write down a differential equation for x as a function of t and state the initial conditions. Hence solve the original problem, stated above.

(d) Repeat parts (b) and (c) but this time choose the coordinate x so that P_2 is the origin for the x-axis.

(e) Which of the two choices of origin gives the neater solution?

5. Consider the following problem; then answer the questions below.

A jet airliner has a constant acceleration of $\frac{1}{3}$g m/s^2 while on the runway, and it becomes airborne at a speed of 300 km/h. Find how long it takes to achieve take-off speed, starting from rest, and find the distance it travels along the runway during this period. How would your answer change if there were a headwind?

(a) Draw a diagram showing how to choose an origin and showing the displacement x of the airliner from the origin at time t.

(b) Write down a differential equation for x as a function of t and state the initial conditions. Hence solve the original problem.

6. Consider the following problem.

A particle moves along a line with uniform acceleration. In the first second of its motion it moves a distance of 1 metre. What is the total distance it has moved after 2 seconds? 3 seconds? 4 seconds? n seconds?

(a) Write down a differential equation for the particle's displacement x metres to one side of the origin 0 at time t seconds.

(b) Hence solve the original problem.

(Galileo used the answers to the above problem in designing his experiment to test the uniformity of acceleration down an inclined plane. Thus his ability as an experimentalist depended on his prior ability as a mathematician!)

7. On a frosty morning, a person of average height accidentally drops a milk bottle. Estimate approximately how far the bottle falls before it hits the ground. Hence estimate the time before the bottle strikes the ground and the speed with which it hits.

8. Assume that, in dry weather, a car travelling at 60 km/h takes a distance of 38 metres to stop once the brakes are applied. What distance would the car need to pull up if initially it had been travelling at 90 km/h.

9. Consider the following problem.

A particle is thrown vertically upwards with an initial velocity of v_0. Find the time which elapses before the particle reaches its maximum height and then express the maximum height as a function of v_0.

(a) Choose a coordinate x to measure the position of the particle at time t and illustrate on a sketch.

(b) Write down a differential equation for x as a function of t and hence solve the problem.

10. A particle moving along a line is displaced a distance x to one side of an origin at time t. It is at rest at the origin when $t = 0$.

(a) Show from the relevant differential equation that if the particle has a constant acceleration a $(a > 0)$ then the velocity and displacement at time t satisfy

$$\dot{x}^2 = 2ax.$$

(b) Verify that the above equation is satisfied by $x = \frac{1}{2}at^2$. Can you spot another function $x = \phi(t)$ which satisfies it?

(c) Write down the *converse* of the result stated in part (a). Is the statement you have written down true?

3

Ropes and pulleys

This chapter is about mechanical systems in which particles are attached to a rope which passes around some pulleys. To keep the mathematical model simple, we shall ignore friction, together with the masses of the rope and the pulleys. Newton's laws can then be used to find the tension in the rope and the acceleration of the particles.

On the basis of this mathematical model, it is possible to explain the operation of the 'block and tackle', which is often used in factories to raise heavy loads. It will be shown how a small force exerted by a workman pulling on one end of a rope can be converted by the pulleys into a large force acting on the heavy load.

The mathematical model also provides the theory underlying Atwood's machine, a contraption which is sometimes used to measure the acceleration due to gravity.

The main skill which this chapter aims to develop is that of using Newton's laws to derive equations of motion. The chapter also provides incidental practice at solving systems of simultaneous linear equations, solving differential equations by antidifferentiation, and manipulation of inequalities.

First, however, a model will be constructed for the forces acting in the rope.

3.1 Tension in the rope

The dynamical rôle of a piece of rope may be illustrated by what happens in a tug of war. By pulling on the rope, one team is able to exert a force on the other team, even though the teams are some distance apart. Thus the rope enables force to be transmitted over a distance.

Does the magnitude of the force diminish along the length of the rope?

41

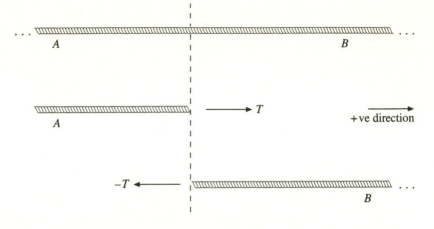

Fig. 3.1.1. The idea of tension in a rope.

Most people who have thought about it at all would probably answer NO! The pull exerted by the team at one end of the rope will be the same as the force felt by the team at the other end. A mathematical model for the rope will be set up in which this can actually be proved from Newton's laws of motion. First, however, it is necessary to analyse the idea of tension in a rope.

Consider a length of rope which is being pulled at either end. Figure 3.1.1. shows a section through a rope at a point along its length. We imagine this section as determining two separate pieces of rope, say A and B.

The piece of rope A will have a force acting on it due to B of magnitude say $T > 0$. Since the piece of rope B can pull, but cannot push (it would go slack), the direction of this force on A must be in the direction shown in Figure 3.1.1. By Newton's third law, the piece of rope B will have a force acting on it due to A of the same magnitude $T > 0$, but with opposite direction.

We define the *tension* in the rope at the point of section to be the common magnitude $T > 0$ of the above forces.

A rope is said to be *light* if it has zero mass. Although such ropes cannot occur in practice, we assume their existence as part of our idealized mathematical model. They should be closely approximated in practice by ropes whose masses are small compared with the masses of the objects at either end. We also assume Newton's laws are applicable even when

$-T_1$ ←——————— ⫿⫿⫿⫿⫿⫿⫿⫿⫿⫿⫿⫿⫿⫿⫿⫿⫿⫿⫿⫿ ——————→ T_2

——————→
+ve direction

Fig. 3.1.2. Diagram used in the proof of Proposition 1.

the mass is zero. The assumption of a light rope gives the following proposition, which is the key to modelling many practical problems.

Proposition 1 *For a light rope, stretched taut by forces at either end, the tension stays constant along the length of the rope.*

Proof Let T_1 and T_2 be the tensions at any two points along the rope, so that the forces acting on the piece of rope between these points are as in Figure 3.1.2.

The net force acting on the rope is $T_2 - T_1$ and so, by Newton's second law,

$$T_2 - T_1 = \{\text{mass of rope}\} \times \{\text{acceleration}\}$$
$$= 0 \times \{\text{acceleration}\}$$
$$= 0.$$

Thus

$$T_2 = T_1.$$

This shows the tension is the same at the two points. As these two points were chosen arbitrarily, the tension remains constant along the entire length of the rope. ☐

The tension in a rope may change as it passes around a pulley. For example, if the axle of a pulley is not properly lubricated, a large part of the force exerted by the worker in Figure 3.1.3 might be expended in getting the pulley to turn, instead of in raising the load.

A similar result might ensue if the pulley were large and massive — a lot of the force would be wasted in getting the pulley to spin.

The contrary case, in which there is little friction at the centre of the pulley and the mass of the pulley is small compared with that of the load, will be modelled by an idealized pulley in which both friction and mass are zero. Such a pulley is said to be *smooth* and *light*. It is possible to argue — although we shall not go into detail here — that such a pulley does not change the tension in the rope.

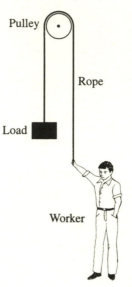

Fig. 3.1.3. Using a pulley to lift a load.

The upshot of this section is therefore that, *if a light rope passes around one or more smooth light pulleys, the tension will be the same at either end of the rope.*

Exercises 3.1

1. By modifying the proof of Proposition 1 show that for a rope not necessarily light, stretched taut by forces at either end, the tension stays constant along the length of the rope provided its acceleration is zero.

2.   A light rope has a particle of positive mass firmly attached at a point somewhere between its two ends, as shown below. Are the two tensions at either end of the rope always equal when the rope is stretched taut? Give reasons for your answer.

$$-T_1 \longleftarrow \text{———} \quad \text{⬤} \quad \text{———} \longrightarrow T_2$$

3.2 Solving pulley problems

In a typical pulley problem, particles will be suspended by a rope passing over pulleys. The aim will be to find the tension in the rope and the accelerations of the particles. Although the central step in the solution is to apply Newton's second law, some preliminary steps and some concluding steps are also necessary. In outline, the procedure for solving these problems is as follows.

STEP 1: *Draw a diagram. Show the particles in typical positions. Introduce a coordinate for each particle and a letter for the tension in the rope.*

STEP 2: *Express the length of the rope in terms of these coordinates. Hence obtain a relationship between the accelerations of the particles.*

STEP 3: *For each particle in turn, draw a diagram showing all the forces acting on that particle.*

STEP 4: *Apply Newton's second law to each particle in turn.*

STEP 5: *Solve the resulting equations for the unknowns — the accelerations of the particle and the tensions in the rope.*

STEP 6: *Look carefully at your answers, think about what they mean physically, and check whether they seem reasonable.*

While some students may wish to refer to the above list as a guide to solving the problem, others may use it merely as a checklist at the end to ensure nothing essential has been omitted from their solutions. The following example illustrates how these steps are carried out in detail.

Example 1. *Two particles, of mass 1 kg and 2 kg respectively, are attached to the ends of a light rope which passes over a smooth light pulley suspended at a fixed distance below the ceiling. Find the accelerations of the particles and the tension in the rope.*

Solution.
STEP 1: Draw a diagram and set up notation. *The particles are shown in typical positions in Figure 3.2.1 at time t.*
  Let x_1 and x_2 be the coordinates of the particles as in the diagram. Thus

$$x_1 = \text{distance of first particle below centre of pulley}$$

and similarly for x_2.
  Now note that, for either particle, if its coordinate increases *then it moves* downwards. *Hence* downwards *is the* positive direction *for each coordinate.*
STEP 2: Relate the coordinates. *Note from the diagram that*

$$x_1 + x_2 = \text{length of the rope} - \frac{1}{2} \times \text{circumference of pulley.}$$

Ropes and pulleys

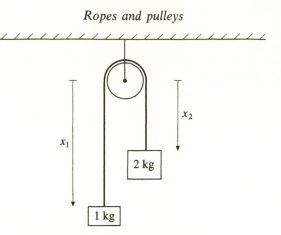

Fig. 3.2.1. Setting up coordinates for Example 1.

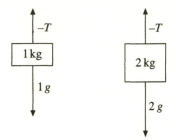

Fig. 3.2.2. Force diagrams for each mass.

But, since this does not change with time, $x_1 + x_2$ must be constant and hence its derivative with respect to t is 0. By the rule for differentiating a sum, applied twice in succession, it therefore follows that

$$\ddot{x}_1 + \ddot{x}_2 = 0. \qquad (1)$$

STEP 3: Show forces on each particle. *Since the rope can only* pull, *the forces it exerts on the particles must be in the* upwards *direction, which is negative for each coordinate. In addition, each particle has its weight acting on it in the downwards direction. Hence the forces on each particle are as in Figure 3.2.2.*

STEP 4: Apply Newton's second law. *For each particle, the mass times the acceleration equals the net force, which can be read off Figure 3.2.2. Hence*

$$1\ddot{x}_1 = 1g - T \qquad (2)$$

$$2\ddot{x}_2 = 2g - T \qquad (3)$$

STEP 5: Solve the equations. *The equations (1), (2) and (3) form a system of three simultaneous linear equations in three unknowns $\ddot{x}_1, \ddot{x}_2$ and T. The standard method for solving such systems is the Gaussian elimination algorithm. To apply this procedure, first bring all the unknowns to the left-hand side of each equation* *to get*

$$\ddot{x}_1 + \ddot{x}_2 \qquad = 0$$
$$\ddot{x}_1 \qquad + T = g$$
$$2\ddot{x}_2 + T = 2g$$

Next, the coefficients of the unknowns are read off and placed in a matrix alongside the column of right-hand sides to give

$$\left[\begin{array}{ccc|c} 1 & 1 & 0 & 0 \\ 1 & 0 & 1 & g \\ 0 & 2 & 1 & 2g \end{array} \right]$$

The aim now is to reduce the coefficient matrix to the unit matrix by using operations on the rows. This gives the following matrices.

$$\left[\begin{array}{ccc|c} 1 & 1 & 0 & 0 \\ 0 & -1 & 1 & g \\ 0 & 2 & 1 & 2g \end{array} \right] \quad \text{new row 2} = \text{row 2} - \text{row 1}$$

$$\left[\begin{array}{ccc|c} 1 & 1 & 0 & 0 \\ 0 & 1 & -1 & -g \\ 0 & 0 & 3 & 4g \end{array} \right] \quad \begin{array}{l} \text{new row 2} = -\text{row 2} \\ \text{new row 3} = \text{row 3} + 2 \times \text{row 2} \end{array}$$

$$\left[\begin{array}{ccc|c} 1 & 1 & 0 & 0 \\ 0 & 1 & -1 & -g \\ 0 & 0 & 1 & \frac{4}{3}g \end{array} \right] \quad \text{new row 3} = \tfrac{1}{3} \times \text{row 3}$$

$$\left[\begin{array}{ccc|c} 1 & 1 & 0 & 0 \\ 0 & 1 & 0 & \frac{1}{3}g \\ 0 & 0 & 1 & \frac{4}{3}g \end{array} \right] \quad \text{new row 2} = \text{row 2} + \text{row 3}$$

$$\left[\begin{array}{ccc|c} 1 & 0 & 0 & -\frac{1}{3}g \\ 0 & 1 & 0 & \frac{1}{3}g \\ 0 & 0 & 1 & \frac{4}{3}g \end{array} \right] \quad \text{new row 1} = \text{row 1} - \text{row 2}$$

The final matrix is called the row echelon *form of the original matrix. Reinsertion of the unknowns in the row echelon form makes the solutions obvious:*

$$\ddot{x}_1 = -\frac{1}{3}g \qquad (4a)$$

$$\ddot{x}_2 = \frac{1}{3}g \qquad (4b)$$

$$T = \frac{4}{3}g \qquad (4c)$$

Thus the first particle has an acceleration vertically upwards of magnitude $\frac{1}{3}g$ m/s^2, the second particle has a downwards acceleration of the same magnitude, and the tension in the rope is $\frac{4}{3}g$ newton.

STEP 6: *Check the answers.* Certain features of the answers (4) could have been predicted from Figure 3.2.1. For example $\ddot{x}_1$ should be negative and $\ddot{x}_2$ should be positive because the lighter particle will accelerate upwards and the heavier one downwards. The magnitude of the accelerations should be less than g, moreover, because the fall of the heavier particle is impeded by the upwards pull of the rope.

The answer for the tension should (and does) lie between g and 2g since the tension must be larger than the weight of the lighter particle (to accelerate it upwards) and less than the weight of the heavier particle (to allow it to accelerate downwards).

The answers obtained for the accelerations provide differential equations, which may be solved (subject to suitable initial conditions) to give a complete description of the motion of the particles. The steps involved in applying the Gaussian elimination algorithm are routine and can easily be implemented on a computer.

Example 2. *Suppose that in Example 1 the particles are released from rest when the heavier particle is at a height of 2 metres above the floor. Find how long it takes to reach the floor (given that the rope is long enough for this to happen before the lighter particle hits the pulley).*

Solution. *By Example 1, the distance x_2 of the heavier particle below the centre of the pulley, when considered as a function of time, satisfies the differential equation*

$$\ddot{x}_2 = \frac{1}{3}g.$$

Suppose that initially the particle is a distance a metres below the centre of the pulley. The initial conditions may be written as

$$x_2 = a \qquad and \qquad \dot{x}_2 = 0 \qquad when \quad t = 0.$$

The problem is to find t such that $x_2 = a + 2$, since the particle has to drop a further 2 metres.

Antidifferentiation applied twice to the differential equation and use of the initial conditions gives the solution

$$x_2 = \frac{1}{6}gt^2 + a.$$

Hence $x_2 = a + 2$ when $2 = \frac{1}{6}gt^2$ and hence when $t = \sqrt{12/g}$, as $t > 0$. Thus the particle hits the ground after about 1.1 seconds.

Exercises 3.2

1. Two particles, of mass 2 kg and 3 kg respectively, are connected by a light rope passing over a smooth light pulley. The coordinates for the particles are to be taken as their respective heights above the floor, say y_1 and y_2 metres. The particles are shown in typical positions in the diagram below.

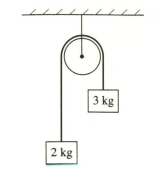

(a) Show the coordinates y_1 and y_2 of the particles on the above diagram.

(b) Suppose now that the system is set in motion. What does your physical intuition tell you about

   (i) the sign of $\ddot{y}_1$,

   (ii) the sign of $\ddot{y}_2$,

   (iii) the range in which the tension in the rope must lie.

(c) Suppose the particles are initially released from rest when the second particle is at a height of 2 metres above the floor. What are the values of y_2 and $\dot{y}_2$ when $t = 0$?

2. Repeat the solution of Example 1 in the text, but this time choose the coordinates x_1 and x_2 to be the heights above the ground of the respective particles, at time t.

How have the answers changed with the new choice of coordinates? Does the first particle still accelerate upwards?

3. Two particles, of mass 2 kg and 3 kg respectively, are attached to the ends of a light rope which passes over a smooth light pulley, which is suspended at a fixed distance below the ceiling. Find the accelerations of the particles and the tension in the rope, by following the steps explained in the text.

4. Suppose that in the preceding exercise the particles are released from rest when the heavier particle is at a height of 1 metre above the floor. Find how long it takes to reach the floor (given that the rope is sufficiently long for this to occur before the lighter particle reaches the pulley).

3.3 Further pulley systems

Each of the following examples exhibits some new feature which was not present in Example 1 of Section 3.2. Since the same general procedure for solving pulley problems is still applicable, we will not give the complete solutions but will concentrate instead on the steps which need modification.

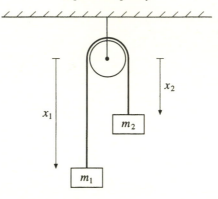

Fig. 3.3.1. Setting up the coordinates for Example 1.

The first example generalizes Example 1 of Section 3.2 to allow for arbitrary masses m_1 and m_2 for the attached particles. Thus m_1 and m_2 come with the problem and are regarded as 'knowns'. The aim in solving the problem is to express the 'unknowns' in terms of them. Letters like m_1 and m_2 which allow us to consider a whole range of problems simultaneously are called *parameters* of the problem.

Example 1. *Two particles of mass $m_1 > 0$ and $m_2 > 0$ respectively are attached to the ends of a light rope which passes over a smooth light pulley suspended a fixed distance below the ceiling. Find the accelerations of the particles and the tension in the rope.*

Discussion   STEPS 1–5: These may be followed much as in Example 1 of Section 3.2. As always the first step is to draw a diagram to indicate which coordinates are to be used, as in Figure 3.3.1.

The remaining steps then lead to the following answers for the accelerations and the tension:

$$\ddot{x}_1 = \frac{m_1 - m_2}{m_1 + m_2} g \tag{1a}$$

$$\ddot{x}_2 = \frac{m_2 - m_1}{m_1 + m_2} g \tag{1b}$$

$$T = \frac{2m_1 m_2}{m_1 + m_2} g \tag{1c}$$

STEP 6: Because the answers (1) contain parameters, it is possible to use them to make a wide range of predictions about the behaviour of the system in Figure 3.3.1 and to perform a variety of checks on the answers.

The simplest of the checks is just to substitute $m_1 = 1$ and $m_2 = 2$ into the answers (1) and observe that these are then the same as the answers (4) found for Example 1 of Section 3.2. Also, it is easily verified that $\ddot{x}_1 + \ddot{x}_2 = 0$ which must be true since the length of the rope stays fixed. Another check is that $T > 0$ which must hold by the definition of tension.

Further checks are as follows.

(a) *Dimensional checks.* Recall that $[g] = LT^2$. The answer (1a) gives for the dimensions of $\ddot{x}_1$

$$[\ddot{x}_1] = \left[\frac{m_1 - m_2}{m_1 + m_2}g\right] = MM^{-1}[g] = LT^{-2}$$

which are the correct dimensions for acceleration. The answer (1a) gives

$$[T] = \left[\frac{2m_1 m_2}{m_1 + m_2}g\right] = M^2 M^{-1}[g] = MLT^{-2}$$

which are the correct dimensions for a force.

(b) *Equilibrium cases.* If the two particles have equal masses, then their weights should exactly balance. Hence the particles should stay at rest or move with uniform velocity. The answers predict this will happen, since putting $m_1 = m_2$ in (1a) and (1b) gives

$$\ddot{x}_1 = \ddot{x}_2 = 0$$

while (1c) shows the tension is then the common weight of the particles.

(c) *Limiting values of the parameters.* If we keep the first mass fixed and allow the other to approach zero, we expect to obtain answers appropriate to free fall of the first particle. This is what happens since

$$\ddot{x}_1 = \frac{m_1 - m_2}{m_1 + m_2}g \qquad \text{by (1a)}$$

$$\rightarrow \frac{m_1 - 0}{m_1 + 0}g = g \quad \text{as} \quad m_2 \rightarrow 0$$

while

$$T = \frac{2m_1 m_2}{m_1 + m_2}g \qquad \text{by (1c)}$$

$$\rightarrow \frac{0}{m_1}g = 0 \quad \text{as} \quad m_2 \rightarrow 0.$$

(d) *Further inequality checks.* In the case $m_1 < m_2$, it is shown in one of the exercises how to derive from the answers the following inequalities:

$$0 < \ddot{x}_2 < g \tag{2}$$

$$m_1 g < T < m_2 g. \tag{3}$$

These inequalities say that the answers must lie within certain ranges which are very plausible on physical grounds. Thus the inequalities (2) say that the heavier particle must accelerate downwards and that the magnitude of the acceleration must be less than that for free fall. The inequalities (3), on the other hand, say that the tension in the rope must be larger than the weight of the lighter particle (to make it accelerate upwards) and less than the weight of the heavier particle (to allow it to accelerate downwards). A similar discussion applies in the case where $m_1 > m_2$.

The solution to the above example was based on assumptions about the lightness of the rope and the pulley, and the smoothness of the pulley. Examples encountered in practice only approximate our idealized model, without satisfying our assumptions exactly. Hence it is desirable to consider ways to test the accuracy with which our model approximates the real world.

To this end, recall that the answers for the accelerations provide very simple differential equations which can be solved to give the heights of the particles as functions of the time. By adjusting the relative masses of the particles, we could achieve small accelerations and hence, with the aid of a stop-watch, get empirical plots of height against time. These results could then be compared with those predicted by our model.

Alternatively, assuming the validity of the model, we could use the measurements to determine the acceleration g due to gravity. When put to this use, the mechanical system shown in Figure 3.3.1 is called *Atwood's machine*.

The next example introduces some of the complexities that arise when more than one pulley is involved.

Example 2. *A light rope is attached to the ceiling at one end. It then*

  (a) passes under a smooth light pulley from which a particle of mass $m_1 > 0$ is suspended,
  (b) passes over a smooth light pulley suspended from the ceiling, and finally
  (c) has a particle of mass $m_2 > 0$ attached to its other end.

The portions of the rope not directly in contact with the pulleys lie in the vertical direction. Find the accelerations of the particles and the tension in the rope.

Discussion   A suitable choice of coordinates for the particles is shown in Figure 3.3.2.

In this example, it is convenient to choose the coordinate of the first particle to be the distance between the centre of the first pulley and the particle. Since the distance of the first particle below the ceiling differs from x_1 by a constant, its acceleration will be $\ddot{x}_1$.

In deriving a relation between the coordinates, you should note that the x_1-coordinate contributes *twice* to the length of the rope. As a result, the relation between $\ddot{x}_1$ and $\ddot{x}_2$ will be different from that obtained in previous examples.

On the dynamical side, some thought is needed to model the forces acting on the first particle. Let T be the tension in the rope. Since the rope is pulling upwards on both sides of the pulley, as in Figure 3.3.3, a plausible assumption is that it exerts a net upwards force of magnitude $2T$ on the pulley. Since the pulley is light, this force will be transmitted unchanged to the first particle.

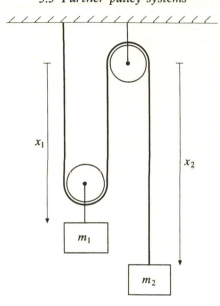

Fig. 3.3.2. Coordinates for Example 2.

Fig. 3.3.3. Forces on a light pulley.

Pulleys find practical application in the use of 'block and tackle' to lift heavy loads. A small force exerted by a worker can thereby be converted into a large force acting on the load. Further details are given in Exercises 7 and 8 below.

Exercises 3.3

1. Complete the solution of Example 1 in the text by carrying out Steps 2–5 (explained in Section 3.2).

To solve the linear equations you may use the fact that the matrix below on the left has the row echelon form shown on the right, where $\mu = (m_1 + m_2)^{-1}g$.

$$\left[\begin{array}{ccc|c} m_1 & 0 & 1 & m_1 g \\ 0 & m_2 & 1 & m_2 g \\ 1 & 1 & 0 & 0 \end{array} \right] \qquad \left[\begin{array}{ccc|c} 1 & 0 & 0 & (m_1 - m_2)\mu \\ 0 & 1 & 0 & -(m_1 - m_2)\mu \\ 0 & 0 & 1 & 2m_1 m_2 \mu \end{array} \right]$$

2. (a) Show that the answers obtained for $\ddot{x}_1$ and $\ddot{x}_2$ in Example 1 in the text can be written so that they involve the masses m_1 and m_2 only in the combination m_1/m_2.

    (b) If the masses are both doubled, what happens to the accelerations?

    (c) How would you choose the mass ratio m_1/m_2 to make $\ddot{x}_1$ small and positive?

3. Each of the following answers is suggested for the tension in Example 1 in the text. Show in each case that the proposed answer is wrong by using one of the checks explained in the text.

    (a) $T = \dfrac{(m_1 m_2)^2}{m_1 + m_2} g$ (b) $T = \dfrac{m_1^2 + m_2^2}{m_1 + m_2} g$.

4. Suppose that in Example 1 the particles start from rest with the second particle at a height of 1 metre above ground level. If the second particle takes 1 second to reach the ground, what is the ratio of the two masses? Assume, of course, that the first particle does not run out of rope.
    [You are to solve this problem by solving a suitable differential equation with the relevant initial conditions.]

5. Two particles, of mass m_1 and m_2 respectively, are connected by a light rope passing over a smooth light pulley. A third particle, of mass m_3, hangs by a light rope from the second particle. The coordinates of the first two particles at time t are their respective distances x_1 and x_2 below the centre of the pulley.

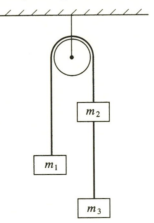

    (a) Copy the diagram above and show the coordinates x_1 and x_2 on it.

    (b) Use your physical intuition to find, in each case, a necessary and sufficient condition on m_1, m_2 and m_3 to ensure that throughout the motion

        (i) $\ddot{x}_1 > 0$ (ii) $\ddot{x}_1 = 0$.

    (c) Over which sections of the rope must the tension stay constant? How many different tensions are there?

    (d) Which two particles have the same acceleration?

6. For the mechanical system in Exercise 5, find the accelerations of the particles and the tension in the rope by carrying out Steps 1–6 (explained in Section 3.2).

To solve the linear equations you may assume that the matrix below on the left has the row echelon form on the right, where $\mu = (m_1 + m_2 + m_3)^{-1}g$.

$$
\left[
\begin{array}{cccc|c}
m_1 & 0 & 1 & 0 & m_1g \\
0 & m_2 & 1 & -1 & m_2g \\
0 & m_3 & 0 & 1 & m_3g \\
1 & 1 & 0 & 0 & 0
\end{array}
\right]
\qquad
\left[
\begin{array}{cccc|c}
1 & 0 & 0 & 0 & (m_1 - m_2 - m_3)\mu \\
0 & 1 & 0 & 0 & -(m_1 - m_2 - m_3)\mu \\
0 & 0 & 1 & 0 & 2m_1(m_2 + m_3)\mu \\
0 & 0 & 0 & 1 & 2m_1m_3\mu
\end{array}
\right]
$$

7. The contented worker shown below is using sound mechanical principles to help him raise a heavy load. Show the tension in the various sections of the rope when the man pulls with a force T. What is the net force then exerted by the ropes on the heavy load? Assume the ropes stay vertical and the load stays horizontal.

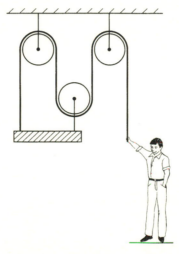

8. When a heavy load is raised by a system of pulleys, the ratio

$$\frac{\{\text{force exerted on load}\}}{\{\text{force exerted by worker}\}}$$

is called the *mechanical advantage* of the system.

  (a) What is the mechanical advantage of the system in Exercise 7?

  (b) Design a similar system with a mechanical advantage of 4.

9. Complete the solution of Example 2 in the text by carrying out Steps 1–6 from Section 3.2.

To solve the linear equations you may assume that the matrix below on the left has the row echelon form shown on the right, where $\mu = (m_1 + 4m_2)^{-1}g$.

$$
\left[
\begin{array}{ccc|c}
m_1 & 0 & 2 & m_1g \\
0 & m_2 & 1 & m_2g \\
2 & 1 & 0 & 0
\end{array}
\right]
\qquad
\left[
\begin{array}{ccc|c}
1 & 0 & 0 & (m_1 - 2m_2)\mu \\
0 & 1 & 0 & -2(m_1 - 2m_2)\mu \\
0 & 0 & 1 & 3m_1m_2\mu
\end{array}
\right]
$$

10. Each of the following answers is suggested for the acceleration of the first particle in Example 2. Show in each case that the proposed answer is wrong by using one of the checks explained in the text.

(a) $\ddot{x}_1 = \dfrac{m_1^2 - 2m_2^2}{m_1 + 4m_2}g$ (b) $\ddot{x}_1 = \dfrac{m_1 - 2m_2}{m_1 + m_2}g.$

11. A light rope passes over two smooth light pulleys suspended at a fixed height below the ceiling. Attached to the ends of the rope are two particles of mass m_1 and m_3 respectively. The central portion of the rope passes under a pulley which supports a particle of mass m_2 in such a way that the sections of the rope not touching the pulleys hang vertically, as in the diagram below.

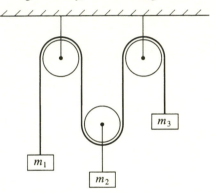

Choose coordinates for each particle so that their positive directions are downwards and then find the accelerations of the particles and the tension in the rope following Steps 1–6, explained in Section 3.2.

To solve the linear equations you may assume that the matrix below on the left has the row echelon form on the right, where $\mu = (m_1m_2 + m_2m_3 + 4m_3m_1)^{-1}$.

$$\begin{bmatrix} m_1 & 0 & 0 & 1 & | & m_1g \\ 0 & m_2 & 0 & 2 & | & m_2g \\ 0 & 0 & m_3 & 1 & | & m_3g \\ 1 & 2 & 1 & 0 & | & 0 \end{bmatrix} \qquad \begin{bmatrix} 1 & 0 & 0 & 0 & | & (1 - 4m_2m_3\mu)g \\ 0 & 1 & 0 & 0 & | & (1 - 8m_1m_3\mu)g \\ 0 & 0 & 1 & 0 & | & (1 - 4m_1m_2\mu)g \\ 0 & 0 & 0 & 1 & | & 4m_1m_2m_3\mu g \end{bmatrix}$$

12. Find a necessary and sufficient condition on the masses m_1, m_2, m_3 in Exercise 11 in order that the particles remain stationary if initially at rest.

13. Three particles of mass m_1, m_2 and m_3 respectively are firmly attached to a light rope — one at either end and the remaining one at an intermediate point along the rope, as shown below. The portion of the rope joining particles one and two passes over a smooth light pulley attached to the ceiling; that joining particles two and three passes over a similar pulley.

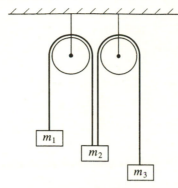

Find the tension in the rope and the accelerations of the particles by following Steps 1–6 of Section 3.2. [Note that this problem differs substantially from Exercise 11. It involves five equations in five unknowns.]

To solve the linear equations you may assume that the matrix

$$\begin{bmatrix} m_1 & 0 & 0 & 1 & 0 & m_1 g \\ 0 & m_2 & 0 & 1 & 1 & m_2 g \\ 0 & 0 & m_3 & 0 & 1 & m_3 g \\ 1 & 1 & 0 & 0 & 0 & 0 \\ 0 & 1 & 1 & 0 & 0 & 0 \end{bmatrix}$$

has the row echelon form

$$\begin{bmatrix} 1 & 0 & 0 & 0 & 0 & (m_1 - m_2 + m_3)\mu \\ 0 & 1 & 0 & 0 & 0 & -(m_1 - m_2 + m_3)\mu \\ 0 & 0 & 1 & 0 & 0 & (m_1 - m_2 + m_3)\mu \\ 0 & 0 & 0 & 1 & 0 & 2m_1 m_2 \mu \\ 0 & 0 & 0 & 0 & 1 & 2m_2 m_3 \mu \end{bmatrix}$$

where $\mu = (m_1 + m_2 + m_3)^{-1} g$.

14. (a) Refer to Example 1 in the text, and assume $m_1 < m_2$. Show that inequalities (2) and (3) hold. To show (3) observe first that T can be written in each of the forms

$$T = \frac{m_2 + m_2}{m_1 + m_2} m_1 g \qquad T = \frac{m_1 + m_1}{m_1 + m_2} m_2 g.$$

(b) Show that $\dfrac{2}{T} = \dfrac{1}{m_1 g} + \dfrac{1}{m_2 g}$. This equation may be expressed by saying that T is the *harmonic mean* of the two weights.

3.4 Symmetry

In everyday life, symmetry is seen in the patterns on wall paper, in the design of furniture and in the architecture of great buildings. In nature, symmetry is most evident in the crystalline structure of solids like common salt and snowflakes. Some of the most interesting problems in

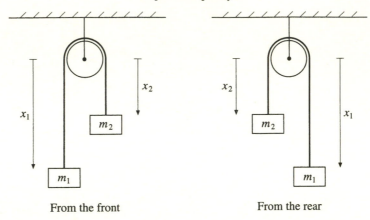

From the front　　　　　　　　　From the rear

Fig. 3.4.1. Example of a mechanical problem with symmetry.

mechanics also possess symmetry and this is reflected in special properties of the equations of motion for such systems.

An example of a mechanical problem with symmetry is provided by Example 1 of Section 3.3. If you look at this system from behind the page you find that you get exactly the same system except that the roles of the first and second particles have been interchanged. The second particle now appears on the *left* as shown in Figure 3.4.1.

The same system is observed from the front as from the rear. Thus the same process we applied to m_1 and m_2 (in that order) to get the tension T should give us the correct answer when applied to m_2 and m_1 (in this order). As a consequence, *our answer for T should not change when we interchange m_1 and m_2.*

Let us verify this. The answer actually obtained was

$$T = \frac{2m_1 m_2}{m_1 + m_2}.$$

If, on the right-hand side (RHS), the letters m_1 and m_2 are interchanged we get

$$\frac{2m_2 m_1}{m_2 + m_1}.$$

But this is obviously equal to what we had before the interchange. Thus, the symmetry of the original problem reflects itself in the answer for the tension.

Similar remarks apply to the accelerations. Because of symmetry, interchange of the subscripts '1' and '2' wherever they occur leaves the formula valid; the formula for $\ddot{x}_1$ changes into the formula for $\ddot{x}_2$ and vice versa.

Recognizing symmetry in a mechanical system thus leads to an additional check on the answers. Thus, for example, the formula

$$T = \frac{2m_1 m_2}{m_1 + 2m_2} g$$

cannot be the correct answer to the above problem since interchange of m_1 and m_2 transforms the RHS into

$$\frac{2m_1 m_2}{m_2 + 2m_1} g.$$

This is not equal to what we had before the interchange and so the required symmetry is lacking.

In more advanced courses, symmetry can occur in mechanical systems in quite complicated ways. The systematic study of symmetry was closely associated with the rise of modern algebra, particularly group theory.

Exercises 3.4

1. Is the mechanical system of Example 2 of Section 3.3 symmetric with respect to the first and second particles? Do you expect the formula for T to stay the same when m_1 and m_2 are interchanged? Verify your answer by carrying out this interchange in the formula for T.

2. Is the mechanical system of Exercise 3.3.11 symmetric with respect to the first and third particles? Do you expect the formula for T to stay the same when m_1 and m_3 are interchanged? Verify your answer by carrying out this interchange in the formula for T.

4

Friction

Together with gravity, friction forces are the ones which play the biggest rôle in shaping everyday life. Without friction, we would be unable to drive our cars, to walk, or even to hold our pens. The simple laws of friction on which our model is based seem to have been first stated by Leonardo da Vinci (1452–1519), who wrote prolifically about lots of things.

Although these laws for friction are very simple, they provide useful estimates and qualitative predictions for a wide range of behaviour associated with friction. More sophisticated models are sometimes used, however, in specialized areas (such as the design of bearings in engineering).

4.1 Coefficients of friction

Friction forces arise as a result of contact between two surfaces. A good way to get a feeling for these forces is to experiment with the two surfaces consisting of the top of the table and the palm of your hand.

An instructive experiment is to rest your hand on the table and then exert a gentle forwards pressure, but not enough to move your hand. You should then be able to feel the backwards reaction from the table opposing your forwards push. Now gradually increase the forwards pressure. The backwards reaction force builds up to a maximum. After this your hand slides forwards. The backwards reaction force which opposes your push in the forwards direction arises from the friction between the palm of your hand and the table.

In order to explain our model of friction, it is convenient to replace your hand in the above experiment by a block of some solid substance. It is free to move on a plane surface made of a second solid substance.

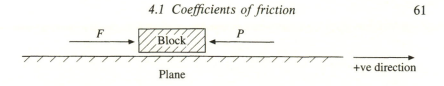

Fig. 4.1.1. Static friction F opposes a pushing force P.

The forces acting on the block which act parallel to the plane are shown in Figure 4.1.1, where *right* has been chosen as the positive direction.

The pushing force is denoted by P (where $P < 0$), while the friction force which opposes it is denoted by F (where $F > 0$). In the diagram it is assumed that the friction force acts in the positive direction while the pushing force acts in the opposite direction. While the block is at rest (or moving with uniform velocity), Newton's second law gives $P + F = 0$ and hence $F = -P$.

Thus, as $|P|$ increases from zero, $|F|$ increases by an equal amount until it reaches a maximum value, at which stage the block begins to slide in the direction of the push. The maximum magnitude of the friction force will be denoted by

$$F_{\max}.$$

The magnitude of the friction force depends on how hard one pushes the block and can assume any value between 0 and $F_{\max}$.

The value of $F_{\max}$ depends on how hard the two surfaces are pressed together. It is harder to make a heavy load slide than a lighter one. To measure the extent to which two surfaces are pressed together, we note that, by Newton's third law, the block and the plane exert equal and opposite forces on each other in the direction normal (perpendicular) to the plane. These forces are illustrated in Figure 4.1.2. Their common magnitude N is called the *normal reaction* between the plane and the block.

In the model we adopt for friction, it is assumed that the maximum friction is a linear function of the normal reaction.

Law of static friction. *For a given pair of substances (for the surfaces in contact) there is a constant $\mu_s > 0$ such that*

$$F_{\max} = \mu_s N.$$

Note that $F_{\max}$ and N are both positive quantities since they are defined as magnitudes.

The dimensionless constant μ_s is called the *coefficient of static friction*

Friction

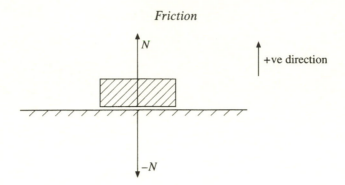

Fig. 4.1.2. Normal reaction.

between the two surfaces. It is assumed to be independent of the shape of the block and of the area of the portion which maintains contact. It depends only on the particular substances of which the block and plane are composed.

The law of static friction is essentially a pair of *inequalities*

$$-\mu_s N \le F \le \mu_s N$$

for the friction force F which acts when the block is stationary, relative to the surface. If no other force is applied in a direction along the plane, the friction force will be zero. When a pushing force is applied, however, the friction force acts in the opposite direction and assumes the extreme values only if the push is hard enough.

The reason for the adjective 'static' in the description of the above coefficient is as follows. As soon as the block begins to move, the magnitude of the friction force drops to a value F_{kin} which is smaller than F_{max}. In the model we adopt, F_{kin} is assumed to be a linear function of the normal reaction.

Law of kinetic friction. *For a given pair of substances there is a constant $\mu_k > 0$ such that*

$$F_{kin} = \mu_k N.$$

The dimensionless constant μ_k is called the *coefficient of kinetic friction* for the two substances. Sometimes it is called the *coefficient of dynamic friction* and sometimes the *coefficient of sliding friction*.

When the block is sliding over the plane, the friction force is given by one of the two *equalities*

$$F = \mu_k N \quad \text{or} \quad F = -\mu_k N,$$

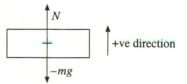

Fig. 4.1.3. Kinetic friction opposes the motion.

Fig. 4.1.4. On a horizontal surface the normal reaction force is opposite to gravity.

its direction being opposite to that of the velocity of the block, as illustrated in Figure 4.1.3.

For the same pair of substances, the following equality holds between the two coefficients of friction.

$$\mu_k < \mu_s.$$

Some typical values are shown in the following table.

Table 4.1.1. Values of coefficient of friction.

Pair of substances	μ_k	μ_s
Steel on ice	0.06	0.1
Rubber on dry concrete	0.7	1.0

This table makes it clear why you should never brake your car so hard that it begins to slide: the friction force is reduced to about 70% of its value prior to sliding.

Applying the laws of friction involves calculating the normal reaction between the block and the surface. In the special case in which the surface is horizontal, this is particularly easy since both the normal reaction and gravity act vertically, as in Figure 4.1.4. The net force on the block in the vertically upwards direction is $N - mg$ where m is the mass of the block. The surface being fixed, the vertical acceleration of the block is zero. Hence $N - mg = 0$ and so $N = mg$.

Exercises 4.1

1. A block of mass 3 kg lies on a horizontal table. The coefficients of friction between the block and the table are given by

$$\mu_s = 0.3 \quad \text{and} \quad \mu_k = 0.2.$$

State the direction of the friction force acting on the block in each of the following cases and state its magnitude.

(a) The block was given a push and is now moving to the right.

(b) The block is at rest but pressure is being exerted on it and it is on the point of moving to the left.

2. A car of mass 2600 kg is parked across a driveway with the brakes on and the owner has lost his keys. What is the magnitude of the force required to

(a) just start the car sliding?

(b) keep it sliding, once in motion?

Use the coefficients of friction given in the text.

3. Discuss the following statement (in which μ denotes the coefficient of static friction).

Values for μ depend on the materials in contact and the state of the surfaces, and range from about 0.04 for ski wax on dry snow, through 0.4 for brake lining on cast iron and 1 for rubber on a hard dry road, to values considerably greater than 1 for very wide drag-racing tyres.

4.2 Further applications

The following examples will illustrate how the coefficients of friction may be used to study the motion of a block sliding over a horizontal surface. The coefficient of static friction μ_s can be used to decide whether the force applied to the block is enough to set it in motion. Once the block is moving, however, the coefficient of kinetic friction μ_k is used to determine the magnitude of the friction force, which acts in the opposite direction to the velocity.

The steps used to get the equations of motion for the block are similar to those used in Section 3.2: *introduce a coordinate for the block (now regarded as a particle), draw a diagram showing the forces acting on the block and then apply Newton's second law of motion.* Only those problems leading to constant acceleration will be considered in this section, so that no new methods will be needed to solve the differential equations.

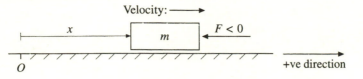

Fig. 4.2.1. Coordinates for Example 1.

Example 1. *A block of mass $m > 0$ is given a sufficiently hard push along a horizontal surface to set it in motion, with an initial speed of v. The coefficient of kinetic friction between the block and surface is $\mu > 0$. Show that the block comes to rest after a time $v/\mu g$, having covered a distance $\frac{1}{2}v^2/(\mu g)$.*

Solution. *Let x be the displacement of the block from the initial point O at time t, with the positive direction for x chosen to be the direction of the velocity of the block.*

The friction force F acts in the negative direction for the coordinate x, as shown in Figure 4.2.1. Hence

$$F = -\mu N$$

where N is the normal reaction between the plane and the block. Since the plane is horizontal, $N = mg$. Hence

$$F = -\mu mg.$$

By Newton's second law,

$$m\ddot{x} = F = -\mu mg$$

hence the equation of motion is

$$\ddot{x} = -\mu g. \tag{1}$$

The initial conditions are $x = 0$ and $\dot{x} = v$ when $t = 0$; hence by antidifferentiation (1) is equivalent to

$$\dot{x} = -\mu g t + c \qquad \text{where } c = v \tag{2}$$

$$x = -\frac{1}{2}\mu g t^2 + vt + d \qquad \text{where } d = 0. \tag{3}$$

The block comes to rest when $\dot{x} = 0$; hence by (2) when

$$t = v/\mu g.$$

Its distance from the initial point O at this time is now obtained from (3) as

$$x = -\frac{1}{2}\mu g \left(\frac{v}{\mu g}\right)^2 + v\left(\frac{v}{\mu g}\right)$$

$$= \frac{1}{2}\frac{v^2}{\mu g}$$

which is the required distance.

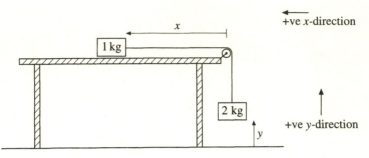

Fig. 4.2.2. Coordinates for Example 2.

As a check on these answers note that both the time and distance taken for the block to stop increase with the initial speed v, but decrease as the friction μ increases. The dimensions of the answers are correct also.

Example 2. *A block of mass 1 kg lies on a table and is attached to one end of a light string. The string passes over a smooth light pulley at the edge of the table and supports a block of mass 2 kg hanging down over the edge of the table. The coefficient of kinetic friction between the block and the table is 0.1.*
  Find the acceleration of the block on the table when it is projected (a) away from the pulley, and (b) towards the pulley, assuming the string remains taut.

Solution. *Let x metres be the distance of the block on the table from the pulley, at time t seconds, and let y metres be the height above ground level of the suspended block — as in Figure 4.2.2. To any change in x will correspond an equal change in y, hence $\dot{x} = \dot{y}$ always and so*

$$\ddot{x} = \ddot{y}. \tag{1}$$

  The tension in the string will be denoted by T newton.
Case (a): the block on the table is projected away from the pulley.
  The normal reaction N newton between the block and the table exactly balances the weight of the block so

$$N = 1g.$$

The remaining forces on the blocks are shown in Figure 4.2.3, where F newton is the friction force opposing the motion of the block. Since the friction acts in the opposite direction to the velocity, F acts in the negative x-direction and so F is negative. Hence

$$F = -0.1N$$
$$= -0.1g.$$

An application of Newton's second law to each block in turn now gives

$$\ddot{x} = F - T = -0.1g - T \tag{2}$$
$$2\ddot{y} = T - 2g \tag{3}$$

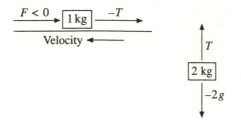

Fig. 4.2.3. Forces on each mass.

Solution of the simultaneous linear equations (1), (2) and (3) for $\ddot{x}, \ddot{y}$ and T gives

$$\ddot{x} = \ddot{y} = -0.7g, \qquad T = 0.6g.$$

Thus the block on the table has an acceleration of $0.7g$ m/s^2 towards the pulley and the tension in the string is $0.6g$ newton.

Case (b): the block on the table is projected towards the pulley. *The direction of the friction force F is now reversed. Thus F acts in the positive x-direction and so $F = 0.1g$. This leads to the answers*

$$\ddot{x} = \ddot{y} = -0.63g, \qquad T = 0.73g.$$

Thus the block on the table has an acceleration of $0.63g$ m/s^2 towards the pulley and the tension in the string is $0.73g$ newton.

The above answers are valid while the blocks continue to move in the original directions of projection. They will cease to be valid when the velocity of the block on the table becomes zero or when the suspended mass hits the ground.

Exercises 4.2

1. Repeat the solution to Example 1 in the text, but this time choose the positive direction for the x-coordinate to be opposite to the direction of the velocity.

2. Verify that the answers obtained in Example 1 in the text, for the time and distance needed by the block to come to rest, have the correct dimensions.

3. Suppose that in Example 2 in the text, the suspended block is given an initial downwards velocity of 1 m/s. By solving a suitable differential equation find how long it takes for the block to descend $\frac{1}{2}$ metre.

4. As a generalization of Example 2 in the text, suppose that the block on the table has mass $m_1 > 0$, the suspended block has mass $m_2 > 0$
    and the coefficient of friction is $\mu > 0$. Show that

    (a) when the block on the table is projected away from the pulley its acceleration is

$$\ddot{x} = \frac{-\mu m_1 - m_2}{m_1 + m_2} g,$$

(b) when it is projected towards the pulley its acceleration is

$$\ddot{x} = \frac{\mu m_1 - m_2}{m_1 + m_2} g.$$

(c) In which case does the acceleration have the larger magnitude? Is this reasonable?

5. Apply appropriate checks to the answers given in Exercise 4.

6. Suppose that in Example 1 in the text, the answer for the time the block takes to come to rest had been given as $v\mu/g$. How could you tell this was wrong (short of working out the correct answer)?

4.3 Why does the wheel work?

In particle mechanics, the modelling process reduces complicated bodies such as cars, space ships and planets to single points. The study of the wheel reveals the limitations of this procedure, and to explain why the wheel works it will be necessary to go outside particle mechanics, at least temporarily. Instead of modelling both a block and a wheel as single points, it will be necessary to take into account the basic differences in their geometry.

The wheel was discovered at some unknown time very early in the history of civilization. Stylish wheels complete with spokes appear in pictorial records of horse-drawn chariots from ancient Egypt and Mesopotamia. More primitive looking wheels, solid without spokes, appear on chariots belonging to the Philistines. At the other extreme, the American Indians were apparently unaware of the wheel before the arrival of the Europeans.

The discovery of the wheel is widely regarded as one of mankind's great technological advances. The reason why the wheel works, however, does not appear to be so widely understood.

First, one must understand that the wheel is basically a device for by-passing friction. In a frictionless world, wheels would be unnecessary because the heaviest load could be set in motion by the slightest push. Once started, it could maintain its velocity on a horizontal plane without further pushing. The smallness of the coefficient of friction between wood and ice provides an approximation to this ideal situation and has enabled Inuit (Eskimos) to transport things efficiently by sledge, without the use of wheels.

Second, the geometry of the circle makes it possible for a wheel to

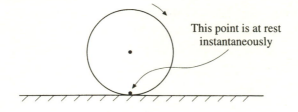

Fig. 4.3.1. The instantaneous point of contact on a rolling wheel.

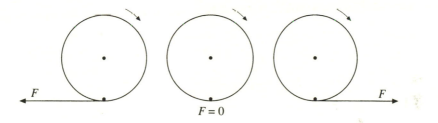

Fig. 4.3.2. Possible stationary frictional forces for a wheel.

move along a road by *rolling*. When this happens, at each instant

  the points of the wheel in contact with the road are stationary,

as illustrated in Figure 4.3.1. Because the points of the wheel in contact
with the road are instantaneously at rest, no slipping occurs. Hence it
seems reasonable to try to explain the friction force acting on the wheel
by using our model for *stationary* friction, rather than that for *kinetic*
friction.

This model of friction permits the friction force F between the wheel
and the road to have any value consistent with the inequalities

$$-\mu_s N \leq F \leq \mu_s N$$

(where μ_s is the coefficient of static friction between the wheel and the
road, and N is the normal reaction). Thus, the model for stationary
friction permits the friction force to act in either direction and *does not
exclude the possibility of its being zero.* Some possibilities are illustrated
in Figure 4.3.2.

By way of contrast, note that to explain the friction forces on a sledge
being dragged along a road we would need to use the model for kinetic
friction, because slipping occurs between the road and the sledge. This

model permits only one value for the magnitude of the friction force, namely $\mu_k N$, and so precludes it from having the value zero.

Exercises 4.3

1. Discuss the following statement.

The frictional force that opposes one body rolling over another is much less than that for a sliding motion and this, indeed, is the advantage of the wheel over the sledge. This reduced friction is due in large part to the fact that, in rolling, the microscopic welds are 'peeled' apart rather than 'sheared' apart as in sliding friction. This may reduce the frictional force by as much as 1000-fold.

[This quote is taken from Halliday and Resnick (1974), pages 80 and 81.]

5

Differential equations: linearity and SHM

The differential equations which arise in Chapter 6, where the motion produced by springs is studied, are of a different type from those encountered so far. In particular, they can no longer be solved by the simple process of antidifferentiation. They are called the equations of simple harmonic motion (SHM) equations. The goal of this chapter is to provide practice at recognizing SHM equations and writing down their solutions.

SHM equations are autonomous and linear. These important properties make it very easy to find all the solutions of the SHM equations.

This chapter assumes familiarity with the sine and cosine functions, including their graphs and their derivatives. It does not depend on Chapter 3 or 4, however, and may be started immediately after Chapter 2.

5.1 Guessing solutions

The differential equations in previous chapters have been of the form

$$\ddot{x} = f(t) \tag{1}$$

where $f(t)$ is a known function of t. Such differential equations may be solved by antidifferentiating both sides with respect to t.

To extend the idea of a differential equation, we now permit the right-hand side to be a known function of x to get, say,

$$\ddot{x} = g(x) \tag{2}$$

where $g(x)$ is a known function of x. It is no longer possible to solve such differential equations by antidifferentiation. The trouble is that the right-hand side of (2), unlike that of (1), involves the unknown function

71

$x = \phi(t)$. You can't antidifferentiate a function if you don't even know which function it is!

Thus we are forced to look for a different way of solving differential equations of the form (2). A crude, but none the less effective, way is to *guess a solution, then substitute it back into the differential equation to see if it works.* Even though your first guess may be not completely correct, it will often be easy to see how to modify it to give a correct solution. Constant solutions are particularly easy to guess.

Example 1. *Find two solutions of the differential equation*

$$\ddot{x} = x(x - 1). \tag{3}$$

Solution. *If x is a constant function of t then*

$$LHS = \ddot{x} = 0$$

while

$$RHS = x(x - 1).$$

The two sides will thus be equal if the constant x satisfied the algebraic equation $x(x - 1) = 0$. Thus the differential equation has the two constant solutions

$$x = 0 \qquad and \qquad x = 1.$$

A useful aid to guessing solutions of differential equations is familiarity with the derivatives of the standard elementary functions. For example, the successive derivatives of sin and cos are shown in the following table:

$x =$	$\sin(t)$	$x =$	$\cos(t)$
$\dot{x} =$	$\cos(t)$	$\dot{x} = -\sin(t)$
$\ddot{x} = -\sin(t)$	$\ddot{x} = -\cos(t)$

$$\vdots \qquad\qquad \vdots$$

Repeated differentiation of either sin or cos produces the negative of the original function and then, later, the original function itself.

The differential equation in the following example has the constant solution in which $x = 0$ for all values of t. The above table helps us guess two other solutions.

Example 2. *Find two solutions, other than the constant solution, of the differential equation.*

$$\ddot{x} = -x.$$

Solution. *The second derivative is to be the negative of the function, suggesting* sin *and* cos. *Choose* $x = \sin(t)$, *to get*

$$LHS = \ddot{x} = -\sin(t)$$
$$RHS = -x = -\sin(t).$$

Thus $x = \sin(t)$ *is a solution. That* $x = \cos(t)$ *is a solution may be checked similarly.*

Along with sin and cos, the exponential function figures prominently in the solution of a lot of commonly occurring differential equations. Successive differentiations give the following table

$$x = e^t \qquad\qquad x = e^{-t}$$
$$\dot{x} = e^t \qquad\qquad \dot{x} = -e^{-t}$$
$$\ddot{x} = e^t \qquad\qquad \ddot{x} = e^{-t}$$
$$\vdots \qquad\qquad\qquad \vdots$$

In each of the above cases, two differentiations give back the original function. This enables us to guess the answer to the following problem.

Example 3. *Find two non-constant solutions of the differential equation*

$$\ddot{x} = x.$$

Solution. *The second derivative is to equal the original function. Choose* $x = e^t$ *and substitute into the differential equation to get*

$$LHS = \ddot{x} = e^t$$
$$RHS = x = e^t.$$

Thus $x = e^t$ *is a solution. That* $x = e^{-t}$ *is also a solution may be checked similarly.*

Further useful tables of derivatives may be obtained by first scaling the time t by a constant factor. For example, the chain rule for differentiating composite functions gives the following table:

$$x = \cos(2t)$$
$$\dot{x} = -2\sin(2t)$$
$$\ddot{x} = -4\cos(2t)$$
$$\vdots$$

Such results greatly extend the number of differential equations for which we can guess non-constant solutions.

Exercises 5.1

In each case guess the required solutions and then substitute back into the differential equation. Work out both LHS and RHS and check whether they are equal.

1. In each case find all the constant solutions of the differential equation.

   (a) $\dot{x} = x(x + 1)$ (b) $\ddot{x} + x^2 = 1$

   (c) $\ddot{x} = x$ (d) $\ddot{x} + 3\dot{x} + 2x = 4$

2. Find a non-constant solution in each case.

   (a) $\dot{x} = x$ (b) $\dot{x} + x = 0$

   (c) $\dot{x} = 2x$ (d) $\dot{x} = -3x$

3. Find two non-constant solutions in each case.

   (a) $\ddot{x} + x = 0$ (b) $\ddot{x} - x = 0$

   (c) $\ddot{x} = -4x$ (d) $\ddot{x} = 4x$

   (e) $\ddot{x} = -9x$ (f) $\ddot{x} + 16x = 0$

4. Let ω be a fixed real number. Find two non-constant solutions for the differential equation

$$\ddot{x} = -\omega^2 x$$

in the cases

   (a) $\omega = 0$ (b) $\omega \neq 0$

5. Let $x^{(iv)}$ denote the fourth derivative of x with respect to t. In each case give five solutions of the differential equation.

   (a) $x^{(iv)} = x$ (b) $x^{(iv)} = 16x$

5.2 How many solutions?

In the previous section a number of solutions of various differential equations were found by guessing. There may clearly be other solutions besides those that were guessed. How can we tell when we have found all the solutions? To answer this question it is necessary to introduce the idea of the *order* of a differential equation.

A *first-order* differential equation is one which expresses the first derivative $\dot{x}$ in terms of t *and* x. It may thus be written in the form

$$\dot{x} = F(t, x) \tag{1}$$

where $F(t, x)$ denotes a known function of t and x (for example, $F(t, x) = tx^2 + x$).

A *second-order* differential equation is one which expresses the second derivative $\ddot{x}$ in terms of t, x *and* $\dot{x}$. It may thus be written in the form

$$\ddot{x} = G(t, x, \dot{x}) \tag{2}$$

where $G(t, x, \dot{x})$ denotes a known function of t, x and $\dot{x}$.

Third- and *higher-order* differential equations are defined in a similar way.

Example 1. *Each of the differential equations*

$$\dot{x} = t^3, \quad \dot{x} = x^3, \quad \dot{x} = x + 5t$$

is of first order, while each of the differential equations

$$\ddot{x} = t^3, \quad \ddot{x} = x^3, \quad \ddot{x} = \dot{x} + \cos(x)t^4$$

is of second order.

Most of the differential equations which arise in practice (or rather, the functions which define their right-hand sides such as F and G above) have a property known as *smoothness*. In first-year calculus, examples are given of functions which fail to have a derivative at some point of their domains, or which fail to be continuous. Roughly speaking, a *smooth* function is one which does not behave in such nasty ways. In courses on advanced calculus the concept of smoothness is given a precise definition, applicable to functions of several variables. For the purpose of this book, however, it is sufficient for you to know that the differential equations we shall use have all been prechecked for smoothness. Each solution of a differential equation will be assumed to have an interval for its domain, which is chosen to be as large as possible.

The question as to how many solutions a differential equation has is given a precise answer by the following theorem.

Theorem (Existence–Uniqueness). *Let t_0 and x_0 be any real numbers. If the function F is sufficiently smooth then the first-order differential equation (1) has a unique solution $x = \phi(t)$ satisfying the initial condition*

$$x = x_0 \quad \text{when} \quad t = t_0.$$

If, furthermore, $\dot{x}_0$ is any real number and the function G is sufficiently smooth then the second-order differential equation (2) has a unique solution satisfying the initial conditions

$$x = x_0 \quad \text{and} \quad \dot{x} = \dot{x}_0 \quad \text{when} \quad t = t_0.$$

Similar results hold for differential equations of third and higher orders, the number of initial conditions being equal to the order of the differential equation. Although the *proof* of this theorem lies outside the scope of this book, this should not prove an impediment to those who merely wish to *use* the theorem.

Example 2.   *The existence–uniqueness theorem implies that the second-order differential equation*

$$\ddot{x} = -x$$

has a solution $x = \phi(t)$ *satisfying the initial conditions*

$$x = 1 \quad and \quad \dot{x} = 1 \quad when \quad t = 0.$$

Since none of the solutions of $\ddot{x} = -x$ found in Section 5.1 satisfies these initial conditions, there must be more solutions than those already guessed. One way of getting new solutions will now be discussed.

A differential equation is said to be *autonomous* if it does not involve the time t explicitly on the RHS. For example, each of the differential equations

$$\dot{x} = x, \quad \ddot{x} = x^2 + \dot{x}, \quad \ddot{x} = 0$$

is autonomous, whereas none of the differential equations

$$\dot{x} = t, \quad \dot{x} = tx, \quad \ddot{x} = x + t^2$$

is autonomous.

Theorem (Phase-Shift). *If* $x = \phi(t)$ *is a solution of an autonomous differential equation and* ε *is any real number, then* $x = \phi(t + \varepsilon)$ *is also a solution.*

Although the proof of this theorem is omitted, it involves little more than an application of the chain rule for differentiating composite functions. To see a typical application of the theorem, recall from Example 2 in Section 5.1 that the autonomous differential equation

$$\ddot{x} = -x$$

has the solution $x = \sin(t)$. From the theorem it therefore follows that, for each ε,

$$x = \sin(t + \varepsilon)$$

is also a solution. (You may — and should — also verify this by direct substitution in the differential equation.) Thus the theorem has enabled us to get *infinitely many* solutions from *one* solution — not a bad trick! As yet, however, we have not obtained *all* the solutions.

Exercises 5.2

1. In each case give the order of the differential equation and say whether the differential equation is autonomous.

(a) $\dot{x} = tx^3$ (b) $\ddot{x} = x^4$

(c) $\dot{x} = \cos(x)$ (d) $\ddot{x} + 2\dot{x} + x = t^3$

2. Recall from Example 3 of Section 5.1 that $x = e^t$ is a solution of the differential equation

$$\ddot{x} = x.$$

Now use the phase-shift theorem to show that, for each $c > 0$, $x = ce^t$ is also a solution. Is this also a solution for $c \leq 0$?

3. (a) Verify, by direct substitution, that the differential equation

$$\dot{x} = -x^2$$

has the solution $x = 1/t$, defined on the interval for which $t > 0$. Sketch the graph of the solution.

(b) State why the phase-shift theorem is applicable and hence give infinitely many solutions of the differential equation.

(c) Hence find the solution satisfying the initial conditions $x = 1$ when $t = 0$.

5.3 Linearity

A differential equation of the second order is said to be *linear* if it can be written in the form

$$\ddot{x} = f(t)\dot{x} + g(t)x + h(t)$$

where $f(t), g(t)$ and $h(t)$ are known functions of t, which may be constants and may be zero. Thus the *unknown* function and its derivatives occur in a linear way. Linear differential equations are usually written with all the terms involving the unknown function transferred to the *left-hand* side to give

$$\ddot{x} - f(t)\dot{x} - g(t)x = h(t). \tag{1}$$

The function $h(t)$ is then referred to as the *right-hand* side of the differential equation. A linear differential equation is said to be *homogeneous* if its right-hand side is the zero function so that it can be written

$$\ddot{x} - f(t)\dot{x} - g(t)x = 0. \tag{2}$$

The homogeneous linear differential equation (2) is called the *homogenized version* of (1).

Table 5.3.1. *Examples of differential equations.*

Differential equation	Order	Linear	Homogeneous	Homogenized version
$\ddot{x} + 2\dot{x} + x = 1$	2	Yes	No	$\ddot{x} + 2\dot{x} + x = 0$
$\ddot{x} = -x$	2	Yes	Yes	
$\ddot{x} = -t$	2	Yes	No	$\ddot{x} = 0$
$\ddot{x} = x^2 + t$	2	No		
$\ddot{x} = t^2 + x$	2	Yes	No	$\ddot{x} = x$
$\dot{x} + \cos(t)x = 0$	1	Yes	Yes	
$\dot{x} + \cos(x)t = 0$	1	No		

Similar definitions apply to differential equations of other orders. Thus a first-order differential equation is *linear* if it can be written as

$$\dot{x} - g(t)x = h(t)$$

and the *homogenized version* has $h(t)$ replaced by zero. Table 5.3.1 illustrates the above definitions.

Since the terms 'homogeneous' and 'homogenized version' refer only to linear equations, the above spaces have been left blank for the non-linear ones. The following theorem shows how we can write down lots of solutions of a linear homogeneous differential equation. The proof is omitted.

Theorem (Homogeneous Superposition). *If $x = \phi_1(t)$ and $x = \phi_2(t)$ are solutions of a linear homogeneous differential equation then*

$$x = c_1\phi_1(t) + c_2\phi_2(t)$$

is also a solution for each $c_1 \in \mathbb{R}$ and $c_2 \in \mathbb{R}$.

For a *second-order* differential equation which is homogeneous linear, it can be shown that the above formula gives *all* the solutions provided that neither of the pair of ϕ_1 and ϕ_2 is a constant multiple of the other. A pair of functions satisfying this condition is said to be *linearly independent*.

For a *first-order* differential equation which is homogeneous linear, things are a bit simpler. If $x = \phi_1(t)$ is any solution of the differential equation, not identically zero, then *all* solutions are given by

$$x = c_1\phi_1(t)$$

for some arbitrary constant $c_1 \in \mathbb{R}$.

Example 1. *Find all solutions of the differential equation*

$$\ddot{x} + x = 0.$$

Solution. *This differential equation is linear and homogeneous. From Section 5.1, it has the solutions $x = \cos(t)$ and $x = \sin(t)$. Hence, by the homogeneous superposition theorem for each choice of the constants c_1 and c_2,*

$$x = c_1 \cos(t) + c_2 \sin(t)$$

is a solution. Since the differential equation is also second order, and cos and sin are linearly independent, it follows that this formula gives all the solutions.

The final theorem gives a way to solve linear differential equations which are not necessarily homogeneous. The proof is omitted.

Theorem (Inhomogeneous Superposition). *Let $x = \phi_P(t)$ be a particular solution of a linear differential equation. If $x = \phi_H(t)$ is any solution of the homogenized differential equation, then*

$$x = \phi_P(t) + \phi_H(t)$$

is a solution of the original linear differential equation. Each solution of the differential equation can be obtained in this way, moreover, by suitable choice of ϕ_H.

For a *second-order* linear differential equation, it follows that, if ϕ_1 and ϕ_2 is a linearly independent pair of solutions of the homogenized equation, then every solution of the original equation can be written as

$$x = \phi_P(t) + c_1 \phi_1(t) + c_2 \phi_2(t)$$

for some arbitrary constants c_1 and $c_2 \in \mathbb{R}$.

Similarly, for a *first-order* differential equation, if ϕ_1 is a non-zero solution of the homogenized equation, then every solution of the equation can be written as

$$x = \phi_P(t) + c_1 \phi(t)$$

for some arbitrary constant $c_1 \in \mathbb{R}$.

The following example illustrates how this theorem can be used to solve linear differential equations.

Example 2. *Find all solutions of the differential equation*

$$\ddot{x} + x = 1.$$

Solution. *This differential equation is linear, but not homogeneous. Hence the inhomogeneous superposition theorem is the one to use.*

To guess a particular solution, try a constant so the second derivative is zero. This suggests the choice

$$x = 1 \tag{3}$$

which gives

$$LHS = \ddot{x} + x = 0 + 1 = 1$$
$$RHS = 1.$$

Thus the guess is correct and (3) gives a particular solution.

Next, the homogenized linear differential is $\ddot{x} + x = 0$, which was solved in Example 1 of Section 5.3, each solution being given by

$$x = c_1 \cos(t) + c_2 \sin(t) \tag{4}$$

for suitable c_1 and $c_2 \in \mathbb{R}$.

Finally, the sum of the particular solution (3) and the solution (4) of the homogenized equation gives the solution

$$x = 1 + c_1 \cos(t) + c_2 \sin(t).$$

In view of the inhomogeneous superposition theorem, each solution is given by this formula for a suitable choice of c_1 and c_2.

Exercises 5.3

1. Copy and complete the following table giving the classification of the differential equations for x as a function of t.

Differential equation	Order	Linear	Homogeneous	Homogenized version
$\ddot{x} + \dot{x} + x = e^t$				
$\ddot{x} + \dot{x} = e^t$				
$\dot{x} + x = 2$				
$\dot{x} + x^2 = t$				
$\ddot{x} + 7x + tx = 0$				
$\ddot{x} = \sin(2t)$				
$\ddot{x} = \sin(2x)$				

2. Find all the solutions of the differential equation $\ddot{x} + x = 2$. Hence find the solution which satisfies the initial conditions $x = 1$ and $\dot{x} = 1$ when $t = 0$.

3. For each of the following differential equations (i) guess a linearly independent pair of solutions, as in Section 5.1 and then (ii) find all solutions:

    (a) $\ddot{x} + x = 0$ (b) $\ddot{x} - x = 0$

    (c) $\ddot{x} = -4x$ (d) $\ddot{x} = 4x$

    (e) $\ddot{x} = -9x$ (f) $\ddot{x} + 16x = 0$.

4. Find all the solutions of each of the following differential equations:

    (a) $\ddot{x} + x = 5$ (b) $\ddot{x} - x = 1$

    (c) $\ddot{x} = -4x + 2$ (d) $\ddot{x} = 4x + 1$

    (e) $\ddot{x} = -9x + 3$ (f) $\ddot{x} + 16x = 32$.

5.4 The SHM equation

This section makes a special study of differential equations of the form

$$\ddot{x} = -\omega^2 x \tag{1}$$

where $\omega > 0$. Although (1) is called the *SHM (simple harmonic motion)* equation, it is more accurately an equation involving a parameter and represents infinitely many equations, one for each choice of the parameter ω. Examples which have arisen earlier in this chapter are the differential equations

$$\ddot{x} = -x, \quad \ddot{x} = -4x \quad \text{and} \quad \ddot{x} = -9x,$$

which correspond to the cases $\omega = 1$, $\omega = 2$ and $\omega = 3$, respectively.

    With the aid of the results given in Section 5.1, it is easy to guess the following pair of solutions to (1):

$$x = \phi_1(t) = \sin(\omega t) \quad \text{and} \quad x = \phi_2(t) = \cos(\omega t). \tag{2}$$

There are now two ways in which to obtain the remaining solutions using properties discussed previously.

Two forms of solution

The first way uses the fact that the SHM equation is *homogeneous linear* and so, by the superposition theorem, it has the solution

$$\boxed{x = c_1 \sin(\omega t) + c_2 \cos(\omega t)} \tag{3}$$

for each choice of $c_1, c_2 \in \mathbb{R}$. As the pair of solutions (2) is linearly independent and the SHM equation is second order, all its solutions are given by (3).

The second way uses the fact that the SHM equation is also *autonomous*. Given an arbitrary constant $A \geq 0$, we choose $c_1 = A$ and $c_2 = 0$ in (3) to get the solution

$$x = A\sin(\omega t).$$

We may add an arbitrary constant $\varepsilon \in \mathbb{R}$ to the time t to get the solution

$$x = A\sin(\omega(t + \varepsilon))$$

or, alternatively,

$$\boxed{x = A\sin(\omega t + \delta)} \tag{4}$$

where $\delta = \omega\varepsilon$ can be any real number.

Note that (3) involves two arbitrary constants c_1 and c_2, while (4) also involves two arbitrary constants $A \geq 0$ and δ. It can be shown, moreover, that, by suitable choice of A and δ, it is possible to satisfy any initial conditions on x and $\dot{x}$. Hence, by the existence–uniqueness theorem, (4) also gives *all* the solutions of the SMH equation. An alternative way to show the equivalence of (3) and (4) is to use the well-known trigonometric identities (see Exercise 3). One then finds the constants are related by

$$A = \sqrt{c_1^2 + c_2^2} \quad \text{and} \quad \cos(\delta) = \frac{c_2}{A}, \quad \sin(\delta) = \frac{c_1}{A}$$

The form (4) makes it easier to sketch the graphs of the solutions. The reason (4) is easier to graph is that it involves only one trigonometric function whereas to graph (3) you must add the graphs of two trigonometric functions.

Period and amplitude

In the case $\delta = 0$, the graph of (3) has the general shape shown in Figure 5.4.1 when $A > 0$. As t increases, x oscillates between $-A$ and A and the graph repeats itself after each time interval of length $2\pi/\omega$. For these reasons, A is called the *amplitude* of the solution and $2\pi/\omega$ is called its *period*. The reciprocal of the period ($\omega/2\pi$) is the number of complete oscillations per unit time and is called the *frequency* of the solution.

If $\delta > 0$, the graph of (4) is obtained by shifting the graph in Figure 5.4.1 a distance δ/ω to the right. The number δ is called the *phase* of the solution.

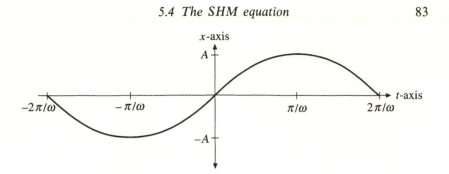

Fig. 5.4.1. Graph of a solution of the SHM equation of amplitude A and period $2\pi/\omega$.

Example 1. *Find the amplitude and the phase of the solution of the differential equation*

$$\ddot{x} = -4x$$

which satisfies the initial conditions $x = 1$ and $\dot{x} = 2$ when $t = 0$.

Solution. *The differential equation is the SHM equation with parameter $\omega = 2$, so by (4) we look for a solution of the form*

$$x = A\sin(2t + \delta)$$

where $\delta \in \mathbb{R}$ and $A \geq 0$. Since

$$\dot{x} = 2A\cos(2t + \delta)$$

the initial conditions are equivalent to the equations

$$1 = A\sin(\delta)$$
$$1 = A\cos(\delta).$$

Squaring these equations and adding them and then using $\cos^2(\delta) + \sin^2(\delta) = 1$ shows that A can only be $\sqrt{2}$. Hence

$$\sin(\delta) = \cos(\delta) = 1/\sqrt{2}$$

which have a solution $\delta = \pi/4$. This gives

$$x = \sqrt{2}\sin(2t + \pi/4)$$

which clearly satisfies the required initial conditions. Hence the amplitude of the desired solution is $\sqrt{2}$ and its phase is $\pi/4$.

Exercises 5.4

1. State which of the following differential equations are of the SHM type and, if they are, give the period and frequency of their oscillatory solutions:

(a) $\ddot{x} + 4x = 0$ (b) $\ddot{x} + 4\dot{x} = 0$

(c) $\ddot{x} = -9t$ (d) $\ddot{x} = -9x$

(e) $\ddot{x} = x$ (f) $\ddot{x} - 9x = 0$

(g) $\ddot{x} + \frac{k}{m}x = 0$ where k and m are positive constants.

2. In each case, find the solution of the differential equation

$$\ddot{x} = -9x$$

which satisfies the initial conditions:

(a) $x = 0$ and $\dot{x} = 3$ when $t = 0$

(b) $x = 1$ and $\dot{x} = 0$ when $t = 0$

(c) $x = 1$ and $\dot{x} = 3$ when $t = 0$

(d) $x = 1$ and $\dot{x} = -3$ when $t = 0$.

3. By first equating the two forms of solution

$$x = c_1 \cos(\omega t) + c_2 \sin(\omega t) \qquad \text{and} \qquad x = A \sin(\omega t + \delta),$$

use standard trigonometric identities to show that

$$A = \sqrt{c_1^2 + c_2^2} \qquad \text{and} \qquad \sin(\delta) = \frac{c_1}{A}, \quad \cos(\delta) = \frac{c_2}{A}.$$

4. What is the general solution of the equation

$$\ddot{x} + \omega^2 x = 0?$$

Include *all* cases for the value of ω.

5. (a) Find the solution of the differential equation

$$\ddot{x} + \omega^2 x = 0$$

which satisfies the initial conditions $x = 1$ and $\dot{x} = 0$ when $t = 0$ in each of the cases $\omega = 1$, $\omega = \frac{1}{2}$, $\omega = \frac{1}{4}$. Sketch the graphs of these solutions on the same diagram. Guess what happens to the solution as ω approaches 0.

(b) For each fixed t, calculate

$$\lim_{\omega \to 0} \sin(\omega t + \pi/2).$$

Does the answer substantiate your guess in part (a)?

6

Springs and oscillations

This chapter is based on a simple model for the force in a spring which was first proposed by Robert Hooke, a contemporary of Newton. In this model, the force exerted by a spring is assumed to be directly proportional to the distance by which the spring is extended. It is quite a useful model when the extension of the spring is not too large.

Some interesting mechanical systems arise when particles are attached to the ends of springs. A consequence of Hooke's law is that the equations of motion for such particles are linear differential equations, usually the SHM equation or some simple variant of it. The solutions of these differential equations can be expressed in terms of trigonometric functions and hence the model predicts oscillatory motion for the particles.

Later in the book, when you have learned more about differential equations, you will be able to include the effect of damping forces in the model.

Oscillatory phenomena occur widely in nature: the alternate rising and setting of the sun, the waxing and waning of the moon, the ebb and flow of the tides are examples from physics, while the regular beating of your heart is an equally familiar example from biology. Although these phenomena may all be described by differential equations, these differential equations turn out to be non-linear and hence much harder to solve than the simple linear ones used in this chapter.

6.1 Force in a spring

A spring which is neither extended nor compressed, but is just lying on a table for example, has a certain length. This length is called the *natural length* of the spring. Some springs are so tightly coiled that they cannot be compressed to anything shorter than their natural lengths. The springs

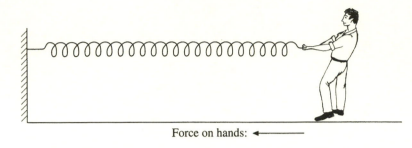

Force on hands: ←————

Fig. 6.1.1. Pulling a spring.

used to pull wire doors shut are of this type. On the other hand, springs
used in the suspension of a car permit both expansion to longer than,
and compression to shorter than, their natural lengths. It is with springs
of this latter type that we shall be mainly concerned.

Hooke's law

If you attach one end of a spring to a fixed object, say the wall, and pull
on the other end so as to stretch the spring beyond its natural length,
the spring will exert a force on your hand which tends to pull it back
towards the wall — as in Figure 6.1.1.

The further you stretch the spring, the larger this force becomes. If
you pull too far, however, the spring may become permanently stretched
and thereby be reduced to just a twisted piece of wire. In the problems
in this book, it will be assumed that the spring is never stretched to this
extent.

If, on the other hand, you push the spring back towards the wall so as
to compress it to less than its natural length, the spring will exert a force
on your hand which tends to push it back from the wall — as in Figure
6.1.2.

The further you compress the spring, the larger this force becomes.
Eventually the spring becomes so compressed that each coil of the spring
touches the next one; after this, no further compression is possible. In
the problems in this book it will be assumed that springs are never
compressed as far as this.

Thus, whether extended or compressed, the force exerted by the spring
on your hand acts in a direction which tends to restore the spring to its
natural length. For this reason, the force exerted by the spring is often

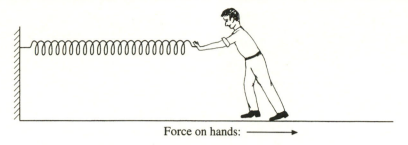

Force on hands: ──────▶

Fig. 6.1.2. Pushing a spring.

called a *restoring force*. For extensions and compressions which are not too large, the magnitude of the force is given (to a reasonable degree of accuracy) by the law first stated by Robert Hooke (1638–1703).

Hooke's law. *The magnitude of the restoring force in a spring is directly proportional to the length by which the spring is extended or compressed.*

By introducing a constant of proportionality $k > 0$ we may write Hooke's law as an equality:

$$|\{\text{restoring force}\}| = k \times \left\{\begin{array}{l}\text{length by which spring is}\\\text{extended or compressed}\end{array}\right\}.$$

The constant k is called the *stiffness* of the spring. Its dimensions are those of force per unit length so that

$$[k] = \frac{\text{MLT}^{-2}}{\text{L}} = \text{MT}^{-2}.$$

In the SI system the unit of k are units are newtons/metre or kg/s^2. The stiffness of a spring depends on the composition of the steel of which it is made, the processes used in its manufacture, the thickness of the wire, the number of coils in the spring, and so on.

Hooke established his claim to the discovery of his law by publishing a famous anagram in 1676 consisting of the letters

$$c\, e\, i\, i\, i\, n\, o\, s\, s\, s\, t\, t\, u\, v,$$

which are a rearrangement of the letters in the Latin phrase

ut tensio, sic vis

which means: *as the extension, so the force.* It was not till two years later that he revealed the meaning of the anagram to his colleagues.

Hooke's law provides the theoretical basis for the spring balance, which is commonly used to measure weights and other forces. The fact that

Springs and oscillations

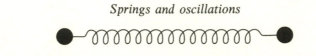

Fig. 6.1.3. A spring with both ends free to move.

equal increments in the force produce equal increments in the length of
the spring makes the spring balance easy to calibrate.

In the discussion so far, it has been tacitly assumed that one end of
the spring stays fixed. A possible situation in which both ends are free to
move is shown in Figure 6.1.3, where particles are attached to each end
of the spring.

To simplify the modelling for this situation, we assume henceforth that
all springs are *light* — that is, have zero mass. An argument similar
to that given in Section 3.1 for ropes then shows that *the forces exerted
by the spring at either end have equal magnitude*, which we assume to be
given by Hooke's law. As to the directions of the forces acting on the
two particles: both will be inwards towards the centre of the spring if the
spring is extended beyond its natural length, but both will be outwards
away from the centre if the spring is compressed.

Exercises 6.1

1. How would you test the validity of Hooke's law for a particular spring and
how would you determine its stiffness?

6.2 A basic example

This section will investigate the motion of a particle attached to one
end of a spring when the other end is fixed. This will introduce various
ideas and techniques which can later be used to solve more complicated
problems involving springs.

Many different choices of coordinate system are possible in spring
problems. Some of these choices will lead to simpler equations of motion
than others. In Example 1 below we choose the coordinate of the particle
to be the extension of the spring. This choice fits in naturally with
Hooke's law.

A frequent source of difficulty in solving spring problems lies in attaching the correct sign to the force exerted by the spring on the particle.
A safe way to achieve this is to consider separately the two cases

(a) the spring is extended,

(b) the spring is compressed.

Let x metres denote the extension of the spring. A negative extension is a positive compression; hence, in the case $x > 0$ the spring is *extended*, while in the case $x < 0$ the spring is *compressed*.

    The direction of the force exerted by the spring will thus depend on the sign of the extension x. It is important to realize that

$$\textit{if } x < 0 \textit{ then } -x \textit{ is positive.}$$

This is so because if $x < 0$ then x contains a 'built-in' negative sign. For example: if $x = -2$ then $-x = -(-2) = 2$, which is *positive*. Thus the *absolute value* or *magnitude* of the extension is given by

$$|x| = \begin{cases} x & \text{if } x > 0 \\ 0 & \text{if } x = 0 \\ -x & \text{if } x < 0. \end{cases}$$

These basic ideas about signed numbers will show us how to attach the correct sign to the forces arising from springs, whether under extension or compression.

Example 1. *A light spring has stiffness $k > 0$. One end of the spring is attached to a wall and the other end is attached to a particle of mass $m > 0$. The particle and the spring lie on the floor, assumed smooth, and are free to move in a line perpendicular to the wall. Find the equation of motion of the particle when the coordinate for the particle is the extension of the spring.*

Solution. *At any instant the spring may be extended or compressed. To define a coordinate x for the particle, put*

$$\ell + x = \{\text{length of the spring}\}$$

where ℓ denotes the natural length of the spring. Since the spring assumes its natural length when $x = 0$, the origin for the x-coordinate is at a distance ℓ from the wall. If x increases, moreover, the particle will move away from the wall and so this direction is the positive direction for the x-coordinate.

    Our aim now is to express the force exerted on the particle by the spring as a function of the coordinate x, distinguishing between the cases $x > 0$ and $x < 0$. Put

$$F = \{\text{force acting on the particle}\}, \qquad \text{measured in the positive } x\text{-direction.}$$

Case (a): $x > 0$. This case is illustrated in Figure 6.2.1. In this case the spring is extended beyond its natural length, its extension being x metres. The spring pulls the particle back towards the wall and so the force on the particle acts in the negative x-direction, hence $F < 0$. But, by Hooke's law, the magnitude of F is

$$\{\text{stiffness}\} \times \{\text{extension}\} = kx,$$

Springs and oscillations

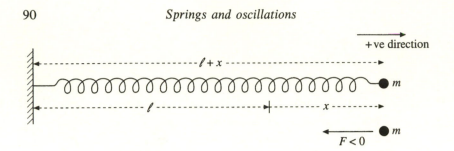

Fig. 6.2.1. Coordinates for Example 1, Case (a): spring extended.

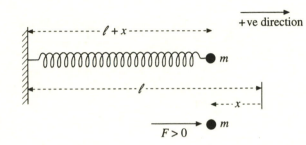

Fig. 6.2.2. Coordinates for Example 1, Case (b): spring compressed.

which is clearly positive. Since F is negative, however, this implies that

$$F = -kx \qquad when \qquad x > 0. \tag{1}$$

Case (b): $x < 0$. *This case is illustrated in Figure 6.2.2.*

In this case the spring is compressed shorter than its natural length by a distance of $|x| = -x$ metres. The spring pushes the particle away from the wall and so the force on the particle acts in the positive x-direction, hence $F > 0$. But, by Hooke's law, the magnitude of F is

$$\{stiffness\} \times |\{extension\}| = k|x| = k(-x)$$

which is positive. Since F is positive, this implies that

$$F = k(-x) = -kx \text{ when } x < 0. \tag{2}$$

From (1) and (2) we have in both cases

$$F = -kx \tag{3}$$

for the force F acting on the particle (provided the magnitude of x stays sufficiently small for Hooke's law to be applicable). This is also true when $x = 0$.

Now regard the coordinate x of the particle as a function of the time t and apply Newton's second law to get

$$m\ddot{x} = F$$
$$= -kx$$

provided |x| remains sufficiently small. Hence

$$\ddot{x} = -\frac{k}{m}x \qquad (4)$$

which is the desired equation of motion for the particle.

As a simple check on the equation of motion, note that it admits the constant solution in which x stays zero. At this point the spring exerts no force on the particle. If the particle is initially placed at rest in this position, it will not move. We call this point an *equilibrium point* for the particle.

As a further check, note that in the limit as $k/m \to 0$ the equation of motion becomes simply

$$\ddot{x} = 0$$

so that the particle moves with constant velocity. This seems physically plausible since, if the stiffness of the spring is small compared with the mass of the particle, we would expect the spring to exert a relatively small force on the particle.

The above solution, of the problem in Example 1, may seem excessively laborious in as much as the formula (4) for the force F has been derived by considering each case separately. It should, none the less, be instructive for beginners in mechanics to work through the various cases since one of their main difficulties lies in attaching the correct signs to the various forces. With the benefit of hindsight, however, it is now possible to see why each case leads to the same formula (3),

$$F = -kx.$$

The 'minus' sign is valid here in every case because *F is a restoring force and so must have the opposite sign to the extension x.*

Our discussion of Hooke's law began with its statement in English and Latin and now includes the above translation into an algebraic formula. To round off the discussion it therefore seems appropriate to represent the law as a graph.

When restricted to the domain for which Hooke's law is valid, F is a linear function of x with negative slope. Hence the graph of F against x has the form shown in Figure 6.2.3.

The types of motion which may be predicted on the basis of Hooke's law will now be discussed.

Example 2. *Describe the possible types of motion for the particle in Example 1.*

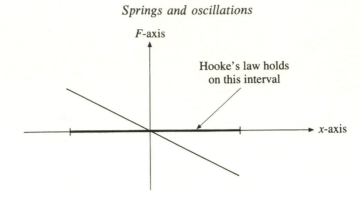

Fig. 6.2.3. Hooke's law holds on some interval.

Solution. *Let the coordinate x for the particle be the extension of the spring beyond its natural length ℓ, at time t. Hence, by the solution to Example 1, the possible motions are obtained by solving the differential equation*

$$\ddot{x} + \frac{k}{m}x = 0$$

for x as a function of t. This is the SHM equation, discussed in Section 5.4, with parameter $\omega = \sqrt{k/m}$. The solutions are given by

$$x = A \sin\left(\sqrt{\frac{k}{m}}\, t + \delta\right)$$

where $A, \delta \in \mathbb{R}$ are arbitrary constants with $A \geq 0$. The values of these constants are determined from initial conditions.

It follows that the particle either remains at the equilibrium point where $x = 0$ (when $A = 0$) or oscillates about the equilibrium point with SHM of amplitude A (when $A > 0$) as shown in the tracking diagram in Figure 6.2.4. It is assumed that A is small enough to lie in the interval for which Hooke's law is valid. Changing ε corresponds to changing the origin of time and in the diagram we have put $\varepsilon = 0$. The period of the oscillations is $2\pi\sqrt{m/k}$. It is interesting to note that this period is independent of the amplitude A of the oscillations.

As a check on our answer for the period, note that its dimensions are correct, being given by

$$[2\pi\sqrt{m/k}] = [m/k]^{\frac{1}{2}} = (\mathrm{M}/\mathrm{M}\mathrm{T}^{-2})^{\frac{1}{2}} = \mathrm{T}.$$

Our answer also shows that the period increases with the ratio m/k, which seems physically plausible: particles with larger mass would take longer to complete an oscillation, for a given stiffness.

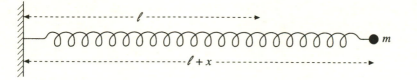

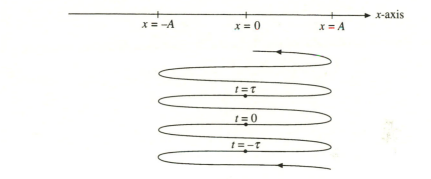

Fig. 6.2.4. Tracking diagram where the period $\tau = 2\pi\sqrt{m/k}$.

Exercises 6.2

1. A particle moves on a line and the force F acting on the particle is given as a function of its displacement x to the right of an origin O by the formula

$$F = kx, \qquad k > 0.$$

(a) Which is the equilibrium point for the particle?

(b) In each of the cases $x > 0$ and $x < 0$ sketch a diagram showing the particle in a typical position, the direction of the force and the direction in which the particle would move if started from rest.

(c) Is the force *restoring* (in the sense that it always pushes the particle back towards the equilibrium point)?

2. In each of the following cases, repeat Exercise 1 but use the new law of force:

(a) $F = -kx^2$

(b) $F = -kx^3$

(c) $F = k(1 - x)$ (distinguish now the cases $x > 1$ and $x < 1$).

3. A particle moving along a line is a distance y to the right of an origin 0 at time t.

(a) In each of the following cases state which point is the equilibrium point:

(i) the equation of motion is $\ddot{y} + 3y = 6$,

(ii) the equation of motion is $\ddot{y} + 3y = 0$.

(b) If the equation of motion is linear *and homogeneous*, which point must
be an equilibrium point?

4. Repeat Example 1, but this time choose as coordinate for the particle its
distance y from the wall and so obtain the equation of motion

$$\ddot{y} + \frac{k}{m}y = \frac{k}{m}\ell.$$

[*Hints:* Follow the steps given at the start of the section, making whatever
changes are necessary in the notation. Be sure to get the correct formula for the
extension of the spring in terms of y and the natural length.]

5. How does the equation of motion obtained in Exercise 4 differ from that
obtained in Example 1? Solve the equation of motion in Exercise 4 and show
that the resulting description of the possible types of motion for the particle
agrees with that found in Example 2.

6.3 Further spring problems

The procedure used in Section 6.2 to obtain the equation of motion
for the particle attached to the spring will now be analysed to provide a
number of simple steps. These steps will provide a useful guide for solving
other spring problems, even when the problems are more complicated.

In reading through these steps, you may find it helpful to refer back to
Example 1 in Section 6.2. The effect of each step, as it occurred in that
example, is shown below in parenthesis.

STEP 1: *Introduce notation* for the natural length of the spring (ℓ),
a coordinate for the particle (x), the spring force (F) on the particle
(measured in the positive coordinate direction).

STEP 2: In case (a) *(extended spring)*, *draw a diagram* to show the spring,
the coordinate of the particle, an arrow giving the direction of the spring
force, the extension of the spring, and *apply Hooke's law* to get a formula
for the spring force.

In case (b) *(compressed spring)*, draw a similar diagram showing the
above quantities and get a formula for the spring force.

Finally give a formula for the spring force covering all cases. (These
steps resulted in Figures 6.2.1 and 6.2.2 and the formula $F = -kx$.)

STEP 3: Write down the net force on the particle and *apply Newton's
second law* to get the equation of motion $(m\ddot{x} = -kx)$.

STEP 4: *Checks:* is the equation of motion dimensionally correct? Does
it give the correct equilibrium point?

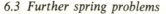

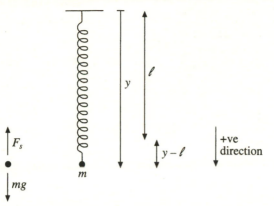

Fig. 6.3.1. Coordinates for Example 1, Case (a): spring extended.

These steps should be regarded as a guide, to help ensure nothing essential is missing from your solution, rather than as inflexible instructions which must always be carried out. In subsequent exercises some of the steps may already be done for you. The following example provides a further illustration of their use.

Example 1. *A light spring has stiffness $k > 0$. One end of the spring is attached to the ceiling, and to the other end, hanging vertically below it, there is attached a particle of mass m. Find the equation of motion of the particle when its distance below the ceiling is taken as coordinate.*

Solution.

STEP 1: *Let ℓ be the natural length of the spring, let y be the distance of the particle below the ceiling at time t. Since the particle moves downwards if y increases, downwards is the positive direction for the y-coordinate. Let F_s be the force on the particle due to the spring, measured in the downwards direction.*

STEP 2: *Here we get a formula for the spring force F_s by considering the possible cases for y.*
Case (a): $y > \ell$, spring extended. *This case is illustrated in Figure 6.3.1.*
The spring force acts upwards, opposite to the positive direction for the y-coordinate. Hence $F_s < 0$. The extension of the spring is $y - \ell > 0$. Hence, by Hooke's law, the magnitude of F_s is

$$\{\text{stiffness}\} \times |\{\text{extension}\}| = k(y - \ell)$$

which is positive. Since F_s is negative, this implies

$$F_s = -k(y - \ell) \qquad when \quad y > \ell. \tag{1}$$

Case (b): $y < \ell$, spring compressed. *This case is illustrated in Figure 6.3.2.*
The spring force now acts downwards, in the positive direction for the y-coordinate.

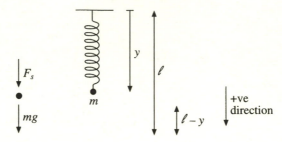

Fig. 6.3.2. Coordinates for Example 1, Case (b): spring compressed.

Hence $F_s > 0$. *The spring is compressed a distance* $\ell - y > 0$. *Hence, by Hooke's law, the magnitude of* F_s *is*

$$\{\text{stiffness}\} \times \{\text{extension}\} = k(\ell - y)$$

which is positive. Since F_s *is positive, this implies*

$$F_s = k(\ell - y) = -k(y - \ell) \quad when \quad y < \ell. \tag{2}$$

Thus by (1) and (2) the spring force is given in all cases by

$$F_s = -k(y - \ell)$$

provided that $|y - \ell|$ *is sufficiently small.*

STEP 3: *The net force* F *on the particle is the sum of the spring force and gravity. Hence*

$$F = -k(y - \ell) + mg$$

By Newton's second law,

$$m\ddot{y} = F = -k(y - \ell) + mg$$

so the equation of motion is

$$\ddot{y} = -\frac{k}{m}(y - \ell) + g. \tag{4}$$

STEP 4: *As a check, note that* $[k/m] = \mathrm{MT^{-2}M^{-1}} = \mathrm{T^{-2}}$ *so that each term in the equation of motion has the dimensions of acceleration.*

As another check, consider the equilibrium point of the system. This will be at some distance, say d, *below the natural length of the spring where the spring force is balanced by gravity. At this point,*

$$y = \ell + d \qquad and \qquad mg = kd.$$

Thus $d = mg/k$ *and*

$$y = \ell + mg/k.$$

Thus it lies a distance $\frac{m}{k}g$ *below the natural length position as we found earlier. This physical argument thus gives the same value* $y = \ell + mg/k$ *as that obtained by putting* $\ddot{y} = 0$ *in (4).*

Fig. 6.3.3. Spring with two particles.

Solving the differential equation

The equation of motion (4), although linear, is not homogeneous. The reason for the lack of homogeneity is that we did not choose the y-coordinate to have its origin at the equilibrium point where, in fact, $y = \ell + mg/k$. This value of y, however, provides a constant solution of the non-homogeneous equation, thereby facilitating its complete solution.

Example 2. *Describe the possible types of motion for the particle in Example 1.*

Solution. *The equation of motion (4) obtained in Example 1 for the y-coordinate of the particle may be written as*

$$\ddot{y} + \frac{k}{m}y = \frac{k}{m}\ell + g.$$

A particular solution of this second-order linear differential equation is the constant solution $y = \ell + mg/k$. The homogenized differential equation is the SHM equation

$$\ddot{y} + \frac{k}{m}y = 0$$

with the solutions $y = A\sin\left(\sqrt{\frac{k}{m}}t + \delta\right)$ where $A, \delta \in \mathbb{R}$ with $A \geq 0$. Hence the solutions of the equation of motion are given by

$$y = \ell + \frac{mg}{k} + A\sin\left(\sqrt{\frac{k}{m}}t + \delta\right).$$

Putting $A = 0$ gives the equilibrium solution. For $A > 0$ and sufficiently small we get a solution in which the particle oscillates up and down with SHM about the equilibrium point with amplitude A and period $2\pi\sqrt{m/k}$.

The steps used to solve Example 1 may be adapted to help you with the solution of more complicated spring problems. Such problems may involve two particles attached to opposite ends of the same spring, as in Figure 6.3.3, or several springs attached to the same particle, as in Figure 6.3.4. We assume the springs slide on a smooth horizontal table. The modifications needed to solve these problems are as follows.

For the system consisting of *two particles attached to the same spring*, two coordinates are needed — one for each particle. The extension of the spring can then be expressed in terms of these coordinates and the natural length of the spring.

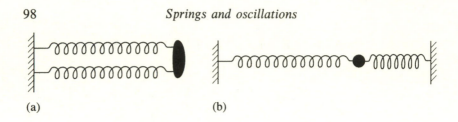

(a)   (b)

Fig. 6.3.4. One particle and two springs.

If, as we assume, the spring is light, the forces exerted on the particles at opposite ends of the spring will act in opposite directions. Since two coordinates are involved, the motion of the particles will be described by a pair of simultaneous differential equations.

For systems with *several springs attached to the same particle*, there may be an increase in the number of cases needed to obtain all possible combinations of extensions and compressions for the two springs. To reduce the number of cases, we shall assume certain relationships between the natural lengths of the springs.

In Figure 6.3.4, if for system (a) the two springs have the same natural length, then they will both be extended or both be compressed (or both be in the equilibrium position at any given time). Thus there are still only two cases to consider. To simplify system (b), however, we assume the sum of the natural lengths is equal to the distance between the walls. If one spring is extended, the other is then compressed. Hence, once again, there are only two cases to consider.

Exercises 6.3

1. Repeat Example 1, but this time choose as coordinate for the particle its distance x below the equilibrium position and so obtain the equation of motion

$$\ddot{x} + \frac{k}{m}x = 0.$$

[*Hints:* Again follow the steps given. When sketching diagrams showing the spring and the forces on the particle in the various cases, recall that the equilibrium point is a distance d below the natural length where $d = mg/k$. When the coordinate of the particle is x, the particle is a distance x below this again. Hence obtain the extension of the spring in terms of x and d.]

2. How does the equation of motion obtained in Exercise 1 differ from that obtained in Example 1? Solve the equation of motion in Exercise 1 and show that the resulting description of the possible types of motion for the particle agrees with that found in Example 2.

3. Two springs are attached to a wall at one end and to a single particle at the other, as shown below. Their natural lengths are ℓ_1 and ℓ_2 respectively, where $\ell_1 < \ell_2$. The springs lie in a line perpendicular to the wall and the distance of the particles from the wall is y.

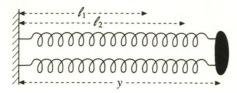

(a) Find the extension (or compression) of the springs in each of the cases

   (i) $\ell_2 < y$

  (ii) $\ell_1 < y < \ell_2$

 (iii) $y < \ell_1$.

(b) Indicate the directions of the forces F_1 and F_2 exerted by the springs on the particle in each of the above cases. What further information would you need, to be able to write down their magnitudes?

4. A particle of mass m is placed on a small table and is attached to a wall at the left by a pair of light springs with the same natural length ℓ but possibly different stiffnesses k_1 and k_2. The springs are free to move in a line perpendicular to the wall. Choose as coordinate for the particle the extension x of the spring, as shown below.

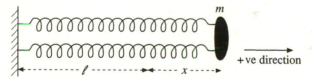

(a) Show, giving all relevant steps, that the equation of motion of the particle is

$$\ddot{x} + \frac{k_1 + k_2}{m}x = 0.$$

(b) Solve the equation of motion and hence find the possible types of motion for the particle.

(c) If the pair of springs were to be replaced by a single spring having the same net effect, what should be its natural length? Its stiffness?

5. A pair of springs have natural lengths ℓ_1 and ℓ_2 and stiffnesses k_1 and k_2 respectively. One end of either spring is attached to a particle of mass m while the remaining ends of the springs are attached to opposite walls, the springs remaining in a line perpendicular to the walls. The distance between the walls is the sum of the natural lengths of the springs. Choose as coordinate for the particle its distance x to the right of the point where each spring has its natural length, as shown below.

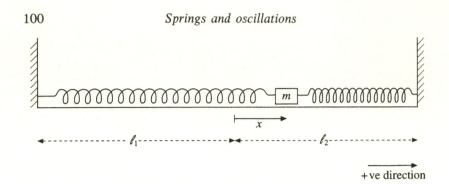

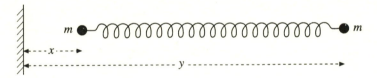

+ve direction

(a) Find the equation of motion of the particle, giving all relevant steps.

(b) A student gets the answer for (a) as

$$\ddot{x} + \frac{k_1 - k_2}{m} x = 0.$$

What is obviously wrong with this answer? Where did he go wrong?

6. As a model of a vibrating molecule consider two particles of equal mass m connected by a light spring which stays in a fixed horizontal line. The spring has natural length ℓ and stiffness k. Choose as coordinates for the particles their distances x and y from some fixed origin 0, as shown below.

(a) Show, giving all relevant steps, that in these coordinates the equations of motion of the particles are

$$\ddot{x} = \frac{k}{m}(y - x - \ell)$$

$$\ddot{y} = -\frac{k}{m}(y - x - \ell)$$

(b) To 'uncouple' these simultaneous differential equations introduce new coordinates u and v by putting

$$u = y + x$$
$$v = y - x.$$

Since x and y are functions of t, so are u and v so you may differentiate with respect to t to find $\ddot{u}$ and $\ddot{v}$ in terms of $\ddot{x}$ and $\ddot{y}$. Hence show from (a) that u and v satisfy the 'uncoupled' differential equations

$$\ddot{u} = 0$$

$$\ddot{v} + \frac{2k}{m}v = 2\frac{k}{m}\ell.$$

(c) Solve these differential equations for u and v as functions of t.

(d) What do the solutions tell you about the motion of the point mid-way between the two particles? Describe the possible motions of the particles.

Part two
Models with Difference Equations

7
Difference equations

The quantities involved in mechanics — such as displacement, velocity and acceleration — are typically related to time by smooth functions defined on an entire interval. Problems in mechanics lead, via Newton's second law, to differential equations. By way of contrast, the mathematical models to be studied in this part of the course involve quantities whose values are known only at certain specified times, equally spaced. Such quantities are expressed as functions of the time via sequences. The assumptions in the models can then be expressed as difference equations for these sequences. The difference between models leading to differential equations and those leading to difference equations is often expressed by saying that the former are continuous whereas the latter are discrete.

This chapter introduces the idea of a difference equation via a problem involving rabbit populations. Basic ideas regarding the solutions of these equations are then explained.

7.1 Introductory example

Leonardo of Pisa, or Fibonacci as he was better known, is often claimed to be the greatest mathematician of the Middle Ages. His book, *Liber abaci*, completed in 1202, took advantage of the Hindu–Arabic numerals. Among the problems which the book contains, the one of greatest interest to later mathematicians is as follows:

How many pairs of rabbits will be produced in a year, beginning with a single pair, if every month each pair produces a new pair, which becomes productive two months after birth?

Our interest in this problem stems from the fact that it leads to the idea of a difference equation.

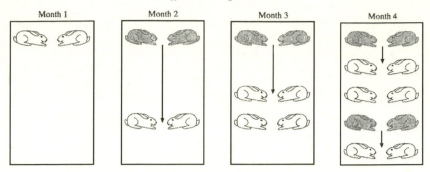

Fig. 7.1.1. A mature pair of rabbits (shaded grey) each month produce a new pair of rabbits (shaded white). The new rabbits mature after two months.

The way in which the rabbits breed for the first few months is illustrated in Figure 7.1.1. The pairs of rabbits are shown month by month within their enclosure. It is assumed that none of the rabbits die and, for ease of identification, each pair of rabbits is shown in the same position within the enclosure from month to month. The newly born rabbits are shown directly under their parents with an arrow pointing to them.

At the top of the enclosure each month is the original pair of rabbits — the only pair during the first month. This pair is assumed to be new born when placed in the enclosure and hence produces a new pair of baby rabbits in the second and in each subsequent month. The first pair of baby rabbits, born in the second month, do not produce any offspring till the fourth month.

Example 1. *What is the total number of pairs of rabbits, each month, up to and including the fourth month?*

Solution. *It is convenient to introduce some notation. Put*

$$y_k = \{\text{number of pairs of rabbits in the month } k\}$$

for each integer $k \geq 1$. Inspection of the enclosures in Figure 7.1.1 shows that

$$y_1 = 1, \quad y_2 = 2, \quad y_3 = 3, \quad y_4 = 5. \tag{1}$$

By continuing to sketch the rabbits in the enclosure month by month up till the twelfth, we could find successively the numbers

$$y_5, \quad y_6, \quad y_7, \quad y_8, \quad y_9, \quad y_{10}, \quad y_{11}, \quad y_{12}$$

and thereby solve the original problem. This would be quite laborious, however, and a more efficient way is to first derive a mathematical formula showing how the numbers in any month depend on those in previous months.

Example 2. *Develop a formula relating the number of rabbits present this month to the number of rabbits present in the previous month.*

Solution. *Note first that, since we neglect rabbits' deaths, their total number can only be affected by births. Hence, for the pairs of rabbits,*

$$\left\{ \begin{matrix} \text{number present} \\ \text{this month} \end{matrix} \right\} = \left\{ \begin{matrix} \text{number present} \\ \text{last month} \end{matrix} \right\} + \left\{ \begin{matrix} \text{number born} \\ \text{this month} \end{matrix} \right\} \qquad (2)$$

Since the rabbits take two months to become productive and then produce only one pair per month, the last term on the RHS of (2) is given by

$$\left\{ \begin{matrix} \text{number born} \\ \text{this month} \end{matrix} \right\} = \left\{ \begin{matrix} \text{number present} \\ \text{two months ago} \end{matrix} \right\} \qquad (3)$$

provided the current month is at least the third month. Hence, substituting (3) back into the RHS of (2) gives

$$\left\{ \begin{matrix} \text{number present} \\ \text{this month} \end{matrix} \right\} = \left\{ \begin{matrix} \text{number present} \\ \text{last month} \end{matrix} \right\} + \left\{ \begin{matrix} \text{number present} \\ \text{two months ago} \end{matrix} \right\}. \qquad (4)$$

This is the desired relationship between the numbers this month and those for the previous two months.

To express (4) in terms of the notation introduced previously, let the current month be the kth month ($k \geq 3$). Hence the last month was the $(k-1)$th and the one before that was the $(k-2)$th. Thus (4) may be abbreviated to

$$y_k = y_{k-1} + y_{k-2} \qquad (5)$$

where $k = 3, 4, 5, \ldots$ and we shall refer to (5) as the Fibonacci equation *(despite the fact that it was first written down explicitly by Kepler).*

Equation (5), which relates the numbers this month to those in the two preceding months, is an example of a *difference equation*.

To see how the difference equation helps in the calculation of the remaining numbers, take $k = 5$ in (5) to get

$$y_5 = y_4 + y_3.$$

Since y_3 and y_4 are already known from (1), this gives $y_5 = 5 + 3 = 8$. The next step is to take $k = 6$ in (5) to get

$$y_6 = y_5 + y_4.$$

But now both y_5 and y_4 are known so this gives $y_6 = 8 + 5 = 13$. By continuing in this way we can calculate in turn $y_7, y_8, \ldots$ stopping where we please. In particular we can find y_{12} in this way and hence solve the original problem. The answer turns out to be 233 pairs of rabbits.

A sequence may be regarded as a function whose domain is the

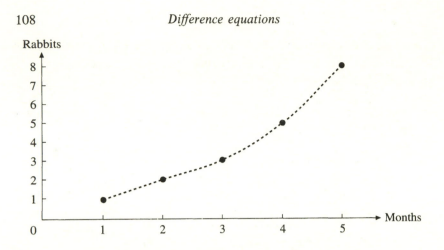

Fig. 7.1.2. Graph shows the total number of pairs of rabbits present each month. Because the time is *discrete* the graph consists of discrete points.

positive integers. A sequence thus has a graph, but it consists of isolated points rather than a continuous curve. As an illustration, the solution to Fibonacci's equation is graphed in Figure 7.1.2.

Despite its mathematical interest, Fibonacci's problem gives a greatly oversimplified view of the breeding patterns of rabbits. In practice, there is a wide variation in the size of their litters, with seven or eight per litter being quite common. Rabbits, moreover, normally take considerably longer to reach sexual maturity than the two months stated in the problem. One aspect of the model which does fit the facts, however, is the gestation period of a month: in practice it is normally 31 days.

An interesting article on the solutions of the Fibonacci equation and their relevance to areas quite remote from rabbit populations is given in Gardner (1981), Chapter 13. Further information about Fibonacci's mathematical achievements may be found in Boyer (1968), Chapter 14.

Exercises 7.1

In each problem, y_k denotes the number of pairs of rabbits present in the enclosure in the kth month.

1. For the original Fibonacci rabbit problem, it was shown in the text that $y_5 = 8$ and $y_6 = 13$.

   (a) Use a suitable choice of k in Fibonacci's equation, (5) in the text, to find y_7.

   (b) In a similar way find in succession the numbers $y_8, y_9, y_{10}, y_{11}, y_{12}$ thereby completing the solution of Fibonacci's problem.

2. Suppose that we modify Fibonacci's problem by increasing the number of rabbits born each month from *one* pair to *two* pairs.

 (a) What change (if any) is required in equations (2), (3), (4) and (5) in the text?

 (b) Starting from $y_1 = 1$ and $y_2 = 3$, find in succession the number of pairs of rabbits each month from the third to the sixth by using the difference equation found in part (a).

3. Repeat Exercise 2, but now modify the original Fibonacci's problem by increasing *one* pair to *three* pairs.

4. Suppose that now we modify the original Fibonacci problem by increasing the time taken for the rabbits to become productive from *two* months to *three* months.

 (a) What change (if any) is required in

 (b) equations (2), (3), (4) and (5) in the text?

 (c) Starting from $y_1 = 1$, $y_2 = 1$ and $y_3 = 2$, find in succession the number of pairs of rabbits each month up to the sixth month by using the difference equation found in part (a).

5. Repeat Exercise 4, but this time suppose the time taken for the rabbits to become productive is *four* months. In part (b) you will need t choose for y_1, y_2, y_3, y_4 appropriately.

7.2 Difference equations — basic ideas

The idea of a difference equation will now be formulated in a general way, applicable to a wide variety of problems. Difference equations arise in problems like that of Fibonacci concerning the rabbits, where the solution leads to a sequence of numbers

$$y_1, \quad y_2, \quad y_3, \quad y_4, \quad \dots$$

which we can imagine as never ending. Typically the sequence of numbers will represent the measurements of some quantity, made at equal intervals of time; hence the phrase 'discrete-time problems' is often used to describe the practical problems from which difference equations arise.

A *difference equation* may be defined as a rule which expresses each member of the sequence, from some point on, in terms of the previous members of the sequence. If the rule defines the kth member of the sequence in terms of the $(k-1)$th member (and possibly also the number k itself), then it is said to be a *first-order* difference equation. Once a value is specified for y_1, the difference equation then determines the rest of the sequence uniquely. The value given for y_1 is called an *initial condition* and the sequence obtained is called a *solution* of the difference equation.

Example 1. *For the first-order difference equation*

$$y_k = k(y_{k-1})^2 \qquad (k = 2, 3, 4, \ldots)$$

and the initial condition $y_1 = 1$, determine the solution as far as y_5.

Solution. *In the difference equation take successively $k = 2, 3, 4, 5$ to get*

$$y_2 = 2y_1^2 = 2 \times 1 = 2$$
$$y_3 = 3y_2^2 = 3 \times 2^2 = 12$$
$$y_4 = 4y_3^2 = 4 \times 12^2 = 576$$
$$y_5 = 5y_4^2 = 5 \times 576^2 = 1\,658\,880.$$

A rule which defines the kth member of the sequence in terms of the $(k-2)$th member (and possibly also the $(k-1)$th member or the number k itself) is called a *second-order* difference equation. A unique solution of such a difference equation is determined once the values of both y_1 and y_2 are specified. Difference equations of the third and higher orders may be defined in a similar way.

Example 2. *Note that the Fibonacci difference equation*

$$y_k = y_{k-1} + y_{k-2} \qquad (k = 3, 4, 5, \ldots)$$

is of second order. Given the initial conditions $y_1 = 1$ and $y_2 = 1$, determine the solution as far as y_5. (The solution is called the Fibonacci sequence.)

Solution. *Taking successively $k = 3, 4, 5, 6$ in the difference equation gives the following equations*

$$y_3 = y_2 + y_1$$
$$y_4 = y_3 + y_2$$
$$y_5 = y_4 + y_3$$
$$y_6 = y_5 + y_4.$$

The initial conditions, together with these equations used successively, give

$$y_1 = 1, \quad y_2 = 1, \quad y_3 = 2, \quad y_4 = 3, \quad y_5 = 5, \quad y_6 = 8.$$

The above process of repeatedly substituting old values back into the difference equation to produce new ones is known as *iteration*. It is clear that this process will eventually produce y_k for any prescribed value of k.

While iteration has the advantage of being repetitive, and therefore easy to apply, it has some drawbacks. For example, if you wished to calculate y_{100}, say, by iteration then you would also have to write down all the preceding members of the sequence $y_1, y_2, \ldots, y_{98}, y_{99}$ regardless of whether you wanted them or not.

For some difference equations it is possible to find a simple formula giving the solution y_k as a function of k. Such a formula is said to provide a 'closed-form' solution of the difference equation and enables y_{100}, say, to be calculated directly, without the need to calculate all the preceding members of the sequence.

The following examples illustrate how a closed-form solution may be guessed and then checked.

Example 3. *Guess a formula for the solution of the first-order difference equation*

$$y_k = 1 + y_{k-1} + 2\sqrt{1 + y_{k-1}} \qquad (k = 2, 3, 4, \ldots) \qquad (1)$$

which satisfies the initial condition $y_1 = 0$.

Solution. *To ensure our guess will be an informed one, we first calculate the numbers y_1, y_2, y_3, y_4. From the difference equation and the initial condition it follows by iteration that these four numbers are respectively 0, 3, 8, 15. These numbers look very close to the perfect squares 1, 4, 9, 16. More precisely*

$$y_1 = 1^2 - 1$$
$$y_2 = 2^2 - 1$$
$$y_3 = 3^2 - 1$$
$$y_4 = 4^2 - 1.$$

This leads to the guess

$$y_k = k^2 - 1 \qquad (2)$$

for all $k \geq 1$.

The formula (2) remains a guess at this stage since it has only been verified to hold for four of the infinitely many possible values of k. The following example shows how to verify it is valid for *all $k \geq 1$*.

Example 4. *In the previous example, verify that the formula (2) for y_k gives the correct solution to the difference equation (1).*

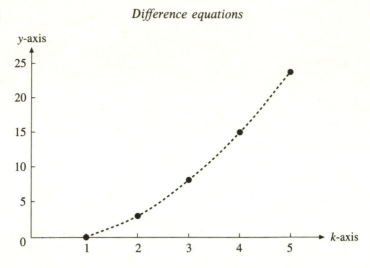

Fig. 7.2.1. Graph of the solution (2) of the difference equation (1).

Solution. *It will be shown that the formula for y_k satisfies both the initial condition and the difference equation.*

Putting $k = 1$ in (2) gives $y_1 = 0$; hence the initial condition is satisfied. Next, to check that the difference equation is satisfied, first replace k in (2) by $k - 1$ to get

$$y_{k-1} = (k - 1)^2 - 1 \qquad (k = 2, 3, 4, \ldots) \tag{3}$$

(this replacement being valid since (2) was assumed to hold for all $k \geq 1$). Substitution of these formulae in the difference equation (1) now gives, for $k \geq 1$,

$$
\begin{aligned}
RHS &= 1 + y_{k-1} + 2\sqrt{1 + y_{k-1}} \\
&= (k - 1)^2 + 2\sqrt{(k - 1)^2}, && \text{by (3)}, \\
&= (k - 1)^2 + 2(k - 1), && \text{as } k \geq 1, \\
&= k^2 - 1, \\
&= y_k, && \text{by (2)}. \\
&= LHS.
\end{aligned}
$$

Thus the difference equation is satisfied, both sides being equal.

The graph of the solution (2) is sketched in Figure 7.2.1. Since the solution is a sequence, its graph is a discrete set of points rather than a continuous curve. Note the reduced scale on the vertical axis to accommodate the points.

Exercises 7.2

1. In each case state the order of the difference equation and the number of initial conditions needed to determine a solution uniquely:

    (a) $y_k = 2y_{k-1} + (y_{k-1})^3$ $(k \geq 2)$,

    (b) $y_k = 2y_{k-1} + (y_{k-2})^3$ $(k \geq 3)$,

    (c) $y_k = 2y_{k-2}$ $(k \geq 3)$.

2. Suppose that $y_k = y_{k-1}$ for all $k \geq 2$. What can be said about the sequence $y_1, y_2, y_3, \ldots$?

3. Suppose that $y_k = (y_{k-1})^2$ for $k \geq 2$. For which values of k does it follow that $y_{k+1} = (y_k)^2$?

4. For the first-order difference equation

$$y_k = y_{k-1} + 2 \qquad (k \geq 2)$$

and the initial condition $y_1 = 0$:

    (a) calculate y_2, y_3, y_4, y_5 and then guess a formula for y_k in general,

    (b) verify that your formula satisfies the difference equation and the initial condition.

5. Repeat Exercise 4, but with the difference equation

$$y_k = 2y_{k-1} \qquad (k \geq 2)$$

and the initial condition $y_1 = 1$.

6. Repeat Exercise 4, but with the difference equation

$$y_{k+1} = \frac{y_k}{1 + y_k} \qquad (k \geq 1)$$

and the initial condition $y_1 = 1$.

7. Write down a first-order difference equation and one initial condition having the sequence $1, -1, 1, -1, 1, -1, \ldots$ as the solution.

8. Write down a second-order difference equation and two initial conditions having the sequence $1, -1, 1, -1, 1, -1, \ldots$ as the solution.

9. Suppose that a sequence $y_1, y_2, y_3, \ldots$ is defined by putting

$$y_k = \sqrt{k} \qquad (k \geq 1).$$

Write down a similar formula which must be satisfied by y_{k-1}. Hence obtain k in terms of y_{k-1} and then obtain a first-order difference equation which the sequence satisfies.

7.3 Constant solutions and fixed points

In the applications of difference equations, a solution represents some quantity measured at equal intervals of time. A solution in which the measured values do not change with time is called a *constant* or *steady-state* solution. Although a solution chosen at random is unlikely to be of this type, it may approach a steady-state solution over a long period of time. For this reason, steady-state solutions play a major rôle in the study of difference equations. To help place them in a geometrical context, we shall show how they correspond to fixed points of functions.

To encourage notational flexibility, we shall change our notation and denote a typical sequence of real numbers by

$$x_0, \quad x_1, \quad x_2, \ \ldots$$

and shall write x_n for the typical members of this sequence where n is an integer ≥ 0. The sequence's being *constant* means that x_n does not change with n and hence that

$$x_n = x_0$$

for all n. Equivalently

$$x_{n+1} = x_n \qquad (n = 0, 1, 2, \ \ldots).$$

To find constant solutions of a difference equation, either of these two criteria may be used.

Example 1. *Find the constant solutions of the difference equation*

$$x_{n+1} = x_n^2 \qquad (n = 0, 1, 2, \ \ldots). \tag{1}$$

Solution. *If the solution is constant, then all members of the sequence have the same value, which we denote by s. Hence $x_{n+1} = x_n = s$ for all $n \geq 0$. The difference equation (1) implies that*

$$s = s^2$$

and hence $s = 0$ or 1. Thus the only possible constant solutions of the difference equation are the sequences

$$x_0 = 0, \quad x_1 = 0, \quad x_2 = 0, \ \ldots \qquad and \qquad x_0 = 1, \quad x_1 = 1, \quad x_2 = 1, \ \ldots.$$

Conversely, these two constant sequences clearly satisfy the difference equation (1).

Constant solutions of difference equations will now be interpreted as fixed points of functions. This leads to a useful interpretation in terms of graphs.

The difference equations to be considered here are first-order ones having the form

$$x_{n+1} = g(x_n) \qquad (n = 0, 1, 2, \ldots) \qquad (2)$$

for some function $g: I \to \mathbb{R}$ with $I \subseteq \mathbb{R}$. The function g is said to *correspond* to the difference equation (2). It is important to be able to recognize the function g when it appears in specific examples.

Example 2. *Find the function g corresponding to the difference equation*

$$x_{n+1} = x_n^2 \qquad (3)$$

and describe its graph.

Solution. *Comparison of (2) and (3) shows that the function g corresponding to this difference equation is given by*

$$g(x_n) = x_n^2$$

for all x_n. In describing this function, however, we would normally write

$$g(x) = x^2$$

for all x. Hence g is just the squaring function, whose graph is a parabola.

For the difference equation (2), an initial condition $x_0 = s$ will give a constant solution

$$x_0 = s, \quad x_1 = s, \quad x_2 = s, \ldots$$

if and only if the number s satisfies the equation

$$s = g(s).$$

A number s which satisfies this equation is called a *fixed point* of the function g. The significance of this terminology is apparent from the mapping diagram, Figure 7.3.1, where g, when applied to s, leaves s fixed.

In the following example, the constant solutions of the difference equation found in Example 1 reappear in the guise of fixed points of the corresponding function.

Example 3. *Find all the fixed points s of the function g with $g(x) = x^2$.*

Solution. *The fixed points s are the solutions of $s = g(s)$, that is*

$$s = s^2.$$

Hence the fixed points of g are the numbers 0 and 1.

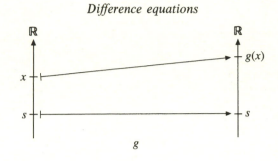

Fig. 7.3.1. Mapping diagram showing a fixed point s of the function g.

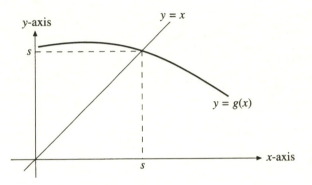

Fig. 7.3.2. Graph showing fixed point s of the function g.

Fixed points have the advantage of a simple graphical interpretation, which often provides information about fixed points even in cases where we cannot solve the equations exactly. This interpretation is shown in Figure 7.3.2: *a number s is a fixed point of g if and only if the point* $(s, g(s))$ *is a point of intersection of the graphs of* $y = g(x)$ *and* $y = x$.

A typical application of this graphical interpretation is given in the following example.

Example 4. *Show that the cosine function has just one fixed point s which lies in the interval* $0 < s < \pi/2$.

Solution. *The graphs of* $y = \cos(x)$ *and* $y = x$ *are shown in Figure 7.3.3. They intersect in just one point, whose x-coordinate lies between 0 and* $\pi/2$. *Hence there is only one fixed point s and it satisfies* $0 < s < \pi/2$.

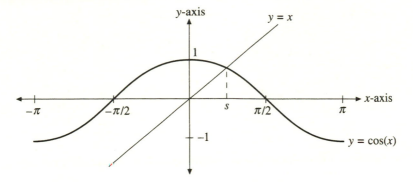

Fig. 7.3.3. Graph showing the single fixed point of the cosine function.

Exercises 7.3

1. In each case find all steady-state solutions of the difference equation using the method of Example 1 in the text.

(a) $x_{n+1} = 2x_n$ $(n = 0, 1, 2, \ldots)$

(b) $x_{n+1} = 2x_n + 1$

(c) $x_{n+1} = \frac{1}{2}(x_n^2 - 3)$.

2. How many constant solutions does the following difference equation have?

$$x_k = x_{k-1} \qquad (k = 2, 3, 4, \ldots)$$

3. The function g corresponding to a certain first-order difference equation is given by the formula $g(x) = 3x + 2$. What is the difference equation? [Give x_{n+1} as the appropriate function of x_n, where $n = 0, 1, 2, \ldots$.]

4. In each case find all of the fixed points of the function g.

(a) $g(x) = 2x - 2$

(b) $g(x) = x^2 - 4$

(c) $g(y) = \frac{1}{2}(y + \frac{1}{y})$ $(y > 0)$.

5. Suppose that a function $g : \mathbb{R} \to \mathbb{R}$ assumes only positive values. What follows about the fixed points (if any) of g?

6. Let g be the function corresponding to the difference equation

$$x_{n+1} = (x_n)^2 - 3 \qquad\qquad (n = 0, 1, 2, \ldots).$$

(a) Write down a formula giving $g(x)$.

(b) Find the fixed points of g.

(c) Hence find the steady-state solutions of the difference equation.

7. Repeat Exercise 6, but with the difference equation

$$x_{n+1} = \alpha(x_n)^2 \qquad (n = 0, 1, 2, \ldots.)$$

where $\alpha \neq 0$ is a constant.

8. In each case give a formula for $g(x)$ where g is the function corresponding to the difference equation.

(a) $N_{k+1} = N_k(1 - N_k)$ \qquad $(k = 0, 1, 2, \ldots)$,

(b) $Y_{n+1} = e^{Y_n} + 3$ \qquad $(n = 0, 1, 2, \ldots)$.

9. In each case sketch graphs of $y = g(x)$ and $y = x$ on the same axes. Hence decide whether the function g has any fixed points.

(a) $g(x) = x^2 - 1$ \quad (d) $g(x) = e^{-x}$

(b) $g(x) = x^2 + 1$ \quad (e) $g(x) = \ln(x)$ \qquad $(x > 0)$

(c) $g(x) = e^x$ \qquad (f) $g(x) = \tan(x)$ \qquad $(-\tfrac{1}{2}\pi < x < \tfrac{1}{2}\pi)$.

7.4 Iteration and cobweb diagrams

Cobweb diagrams are used in order to provide a simple graphical interpretation of the iteration process used to solve difference equations. They provide valuable insight into the possible ways in which solutions of difference equations can behave.

The difference equations to which such diagrams are relevant are first-order equations of the form

$$x_{n+1} = g(x_n) \qquad (n = 0, 1, 2, \ldots) \qquad (1)$$

Here g is the function corresponding to the difference equation, as explained in the previous section.

The main idea behind the construction of a cobweb diagram is to interpret equation (1) on a graph. To start with, graphs of both $y = g(x)$ and $y = x$ are sketched on the same set of axes. It is assumed also that the number x_n is given, and plotted on the x-axis. Figure 7.4.1 (a) shows how things might appear at this stage for a typical function g.

Given this set-up and given that (1) is satisfied, the question now is: *where on the x-axis should we put* x_{n+1}? The answer is shown in Figure 7.4.1 (b). To sketch the arrows in Figure 7.4.1. and thence to get the point x_{n+1} we perform the following steps.

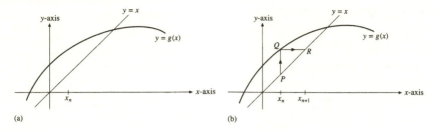

Fig. 7.4.1. Going from x_n to x_{n+1} via the graph.

STEP 1: *Start at the point P on the line $y = x$ directly above x_n and then project vertically to get the point Q on the graph of $y = g(x)$.*

STEP 2: *From Q project across horizontally until the point R is reached on the line $y = x$.*

STEP 3: *Choose the next point x_{n+1} to be the x-coordinate of R.*

We call this the *next-point procedure*. The reasons why it gives the correct point for x_{n+1} are as follows:

(a) The y-coordinate of Q is $g(x_n)$ (because Q lies on the graph of $y = g(x)$).

(b) Hence R has $g(x_n)$ for its y-coordinate also (as it is level with Q).

(c) Hence R has $g(x_n)$ for its x-coordinate (as it lies on the line $y = x$).

Thus the procedure gives $x_{n+1} = g(x_n)$, as required.

To construct a complete solution of the difference equation (1) we now choose an initial value x_0 and mark it on the x-axis. By repeatedly applying the next-point procedure we obtain from x_0 a sequence of points

$$x_0, \quad x_1, \quad x_2, \quad \ldots$$

lying on the x-axis and forming a solution of the difference equation (1). The pattern of arrows arising from use of the next-point procedure will typically form a cobweb (although it might sometimes be better called a zig-zag) path.

The following example gives a number of difference equations, whose solutions illustrate a variety of possible types of behaviour.

Example 1.   *For each of the following difference equations construct a cobweb diagram showing the solution of the difference equation which satisfies the initial condition $x_0 = 1$:*

(a) $x_{n+1} = 2x_n$ $(n = 0, 1, 2, \ldots)$

(b) $x_{n+1} = \dfrac{1}{2}x_n$

(c) $x_{n+1} = -2x_n$

(d) $x_{n+1} = -\dfrac{1}{2}x_n$.

Solution. *In each case the first step is to write down the function g for which the difference equation has the form*

$$x_{n+1} = g(x_n).$$

The graphs of $y = g(x)$ and $y = x$ are then plotted relative to the same pair of axes and the initial value x_0 is plotted on the x-axis. Repeated application of the next-point procedure then produces the cobwebs shown in Figure 7.4.2 and the values for

$$x_0, \quad x_1, \quad x_2, \quad x_3, \quad \ldots$$

(which check against those obtained directly from the relevant difference equation by iteration).

The cobweb diagrams exemplify some important types of long-term behaviour for solutions of difference equations. One of the most important questions in this respect is whether the solution converges to a limit. The answers, in the various cases, are as follows.

Case (a): no. We can make x_n as large as we like by making n sufficiently large. Hence we say that the sequence *diverges to ∞.*

Case (b): yes. We can make x_n as close as we like to 0 by making n sufficiently large. Hence we say that the sequence *converges to the limit* 0.

Case (c): no. As n increases, x_n oscillates from one side of the origin to the other with ever-increasing amplitude. Hence we say that the sequence *does not converge to any limit.*

Case (d): yes. Although x_n oscillates as n increases, the amplitude dies away and approaches 0. We can thus make x_n as close as we like to 0 by making n sufficiently large. Hence we say that the sequence *converges to the limit* 0.

In the preceding example, the functions corresponding to the difference

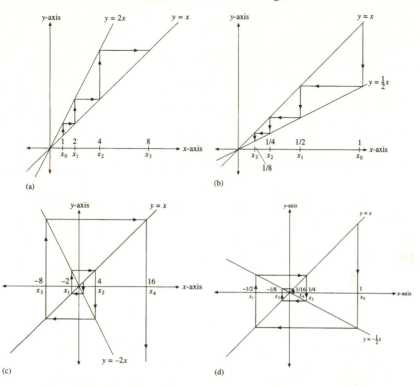

Fig. 7.4.2. Cobweb diagrams for Example 1. Cases (b) and (d) converge to a fixed point whereas (a) and (c) do not.

equations are all *linear*, in the sense of having straight-line graphs. While this makes the graphs easy to draw, it also makes the behaviour patterns for the solutions rather special. Thus, for example, if the function is linear, it cannot have exactly two fixed points. The following example provides a contrast with the linearity of the previous example.

Example 2. *Use a zig-zag diagram to investigate the long-term behaviour of the solutions of the difference equation*

$$x_{n+1} = x_n^2 \qquad (2)$$

for each of the following initial conditions: (a) $x_0 = 0.9$ and (b) $x_0 = 1.1$.

Solution. *The function g corresponding to the difference equation (2) is given by the formula $g(x) = x^2$. Hence the cobweb diagrams are as in Figure 7.4.3. The solution in case (a) converges to the limit 0 whereas the solution in case (b) diverges to ∞.*

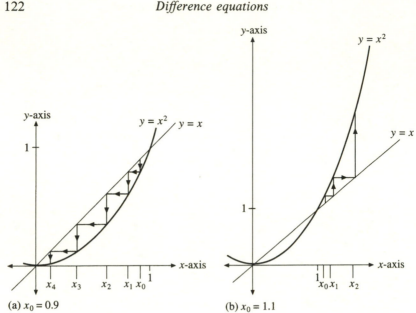

(a) $x_0 = 0.9$ (b) $x_0 = 1.1$

Fig. 7.4.3. Cobweb diagrams for Example 2 show different behaviour for different initial conditions.

Our final example illustrates how a cobweb diagram may, in exceptional cases, collapse to a square and thereby give rise to periodic behaviour.

Example 3. *Use a cobweb diagram to discuss the behaviour of the solution of the difference equation*

$$x_{n+1} = -x_n \tag{3}$$

which satisfies the initial condition $x_0 = 1$.

Solution. *The function g corresponding to the difference equation (3) is given by the formula $g(x) = -x$. The graphs of $y = g(x)$ and $y = x$ are shown in Figure 7.4.4. Because of the symmetry between the two graphs, two applications of the next-point procedure give back the starting point. Hence the cobweb diagram closes up to give a square described infinitely many times.*

The solution with initial condition $x_0 = 1$ is obtained by starting at the upper right corner of the square and following it around anti-clockwise. The values of the x-coordinate at successive crossings of the x-axis are then recorded to give

$$x_0 = 1, \quad x_1 = -1, \quad x_2 = 1, \quad x_3 = -1, \ldots.$$

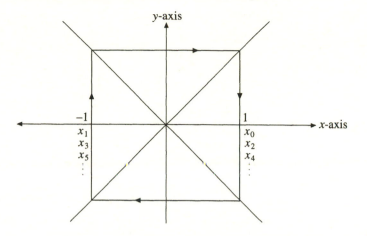

Fig. 7.4.4. Cobweb diagram which illustrates a 2-cycle.

The above sequence has the interesting feature that every second member has the same value, even though the sequence is not constant. This property may be expressed by saying that the solution has *period* 2 or is a *2-cycle*. More generally, we say that a sequence

$$x_0, \quad x_1, \quad x_2, \ldots$$

is *periodic* if there is a positive integer p such that

$$x_{n+p} = x_n \qquad (n = 0, 1, 2 \ldots)$$

The smallest such p is then called the *period* of the sequence and the sequence is called a *p-cycle*. In particular, a sequence of period 1 is a constant solution.

Exercises 7.4

1. Sketch graphs $y = x$ and $y = x + 1$, for $0 \le x \le 5$, relative to the same pair of axes. Hence draw a cobweb diagram for the solution of the difference equation

$$x_{n+1} = x_n + 1 \qquad (n = 0, 1, 2, \ldots)$$

which satisfies the initial condition $x_0 = 0$. Show $x_0, x_1, \ldots, x_5$ on the x-axis.

2. In each of the cases shown below, sketch a cobweb diagram for the solution of the difference equation

$$x_{n+1} = g(x_n) \qquad (n = 0, 1, 2, \ldots)$$

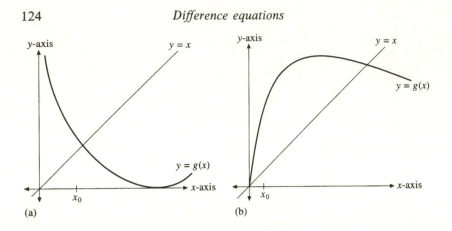

(a)   (b)

3. For each of the difference equations in Example 1 in the text, use iteration to find x_1, x_2, x_3, given that $x_0 = 1$. Hence check the statements made in Figure 7.4.2.

4. Repeat Example 1 in the text, but this time suppose the initial condition is $x_0 = -1$.

5. Repeat Example 2 in the text, but this time choose the initial conditions to be $x_0 = -0.9$ in case (a) and $x_0 = -1.1$ in case (b).

In Exercises 6 and 7 you will need to use graph paper.

6. Let g be the function with $g(x) = x^{\frac{1}{2}}$ for $x \geq 0$. Tabulate the values of g at 0, 0.1, 0.2,...,0.9, 1. Plot the corresponding points on the graph of $y = g(x)$ and connect them to form a smooth curve. Also draw the line $y = x$. From your graph, make the cobweb construction of the solution of the difference equation

$$x_{n+1} = (x_n)^{\frac{1}{2}} \qquad (n = 0,\ 1,\ 2,...)$$

which satisfies the initial condition $x_0 = 0.1$.
   To which limit does the solution appear to be converging?

7. Let g be the function with $g(x) = x/(1 + x)$ for $x \neq -1$. Tabulate the values of g at 0, 0.1,...,0.5. Plot the corresponding points on the graph of $y = g(x)$ and connect them to form a smooth curve. Also draw the line $y = x$. From your graph make the cobweb construction for the solution of the difference equation

$$x_{n+1} = x_n/(1 + x_n) \qquad (n = 0, 1, 2,...)$$

which satisfies the initial condition $x_0 = 0.5$.
   To which limit does the solution appear to be converging?

8. Consider the difference equation

$$x_{n+1} = 1/x_n \qquad (n = 0, 1, 2,...)$$

where it is assumed that each $x_n \neq 0$.

   (a) Draw a cobweb diagram for the solution which satisfies the initial condition $x_0 = 2$. Check your answers by iteration, directly from the difference equation.

(b) Find all the numbers $a \neq 0$ for which the initial condition $x_0 = a$ determines a constant solution.

(c) Deduce that all non-constant solutions are 2-cycles.

9. Consider the difference equation

$$x_{n+1} = g(x_n)$$

where g is one of the two functions shown below. By experimenting with cobweb diagrams corresponding to various initial conditions find:

(a) a solution of period 2 when g is given by the graph (i),

(b) a solution of period 3 when g is given by the graph (ii).

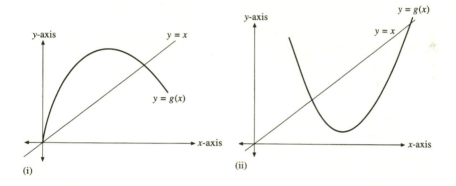

(i)   (ii)

8

Linear difference equations in finance and economics

This chapter begins with a study of linear difference equations, which are usually easier to solve in closed form than non-linear ones. A closed-form solution is found for a special type of linear difference equation that arises frequently in problems from finance and economics.

The applications to finance concern interest on loans and the repayment of debt. Hence they should prove just as useful to those who have money and wish to invest it as to those who don't have money and wish to borrow it.

Economists set up mathematical models with which they hope to predict movements in price, interest rates, level of unemployment, and so on. Although some of the economic concepts involved in these models are rather abstract, we have a rough idea of their meaning from everyday usage and this is all that is needed for the models studied in this chapter.

Economic models contain assumptions about the interdependence of the various quantities of interest to economists. A model is regarded as successful if these assumptions lead to reliable forecasts. The quantities studied in economics cannot always be measured with the same accuracy as those studied in the physical sciences and hence economists are more concerned with qualitative predictions — whether the prices will go up or down or whether they will oscillate about some equilibrium value.

Economists often simplify their mathematical models by assuming the functions involved are linear, in the sense of having straight-line graphs. This is particularly appropriate when the quantities involved stay near equilibrium values, because a sufficiently small piece of a smooth curve looks like a line segment. A quantity Y is said to be a linear function of another quantity X if $Y = mX + c$ for where m and c are constants. Y is said to be directly proportional to X if $Y = mX$.

126

Table 8.1.1. *Examples of first-order difference equations and their classifications.*

Difference equation	Linear	Homogeneous	Constant coefficient
$y_{k+1} = 3y_k + 2$	Yes	No	Yes
$y_{k+1} = 3y_k$	Yes	Yes	Yes
$y_{k+1} = (y_k)^2$	No	–	–
$y_{k+1} = k^2 y_k$	Yes	Yes	No

8.1 Linearity

The difference equations arising in this chapter are mainly *first-order linear* ones, which have the form

$$y_{k+1} = ay_k + b \qquad (k = 0, 1, 2, \ldots) \qquad (1)$$

where a and b are given functions of k. In searching for closed-form solutions later, we shall suppose that both a and b are constants.

The difference equation is said to be *homogeneous* if b is the constant 0 and to be *constant coefficient* if a is a constant. The *homogenized equation* for (1) is the difference equation

$$x_{k+1} = ax_k \qquad (k = 0, 1, 2, \ldots) \qquad (2)$$

which is obtained from (1) by putting $b = 0$. To avoid confusing solutions of the two equations, moreover, we have also replaced 'y' by 'x'. Table 8.1.1 illustrates these definitions.

The solutions of linear difference equations obey *superposition theorems*, analogous to those presented in Section 5.3 for differential equations, which show how solutions may be combined to form new ones. Such theorems lead to a useful procedure for solving linear difference equations in closed form. The procedure will be illustrated by an example.

Example 1. *Find the solution of the first-order linear difference equation*

$$y_{k+1} = 2y_k + 3 \qquad (k = 0, 1, 2, \ldots) \qquad (3)$$

which satisfies the initial condition $y_0 = 2$.

Solution.

STEP 1: Write down and solve the homogenized equation for (3). *The homogenized equation is*

$$x_{k+1} = 2x_k \qquad (k = 0, 1, 2, \ldots) \qquad (4)$$

Iteration suggests that its solutions are given by the formula

$$x_k = 2^k x_0 \qquad (5)$$

where the initial value x_0 can be chosen arbitrarily. Substitution of (5) back in (4) shows that (4) is satisfied.

STEP 2: Guess one particular solution of the original equation (3). *The RHS of (3) involves only constants (apart from y_k). This suggests trying a constant solution, in which $y_{k+1} = y_k$. The original equation (3) is then equivalent to*

$$y_k = 2y_k + 3$$

and hence to

$$y_k = -3. \qquad (6)$$

Substitution of (6) back into (3) shows that (3) is satisfied.

STEP 3: Add the solutions (5) of the homogenized equation to the particular solution (6) of the original equation — this gives all the solutions of the original equation. *Adding the solutions gives the formula*

$$y_k = -3 + x_0 2^k \qquad (7)$$

for the solutions of the original difference equation (3), in terms of the parameter x_0.

STEP 4: Choose x_0 to fit the given initial condition $y_0 = 2$. *Put $k = 0$ in (7) to get $y_0 = -3 + x_0$ and hence $x_0 = 3 + y_0 = 5$. Thus the required solution is given by*

$$y_k = -3 + 5 \times 2^k \qquad (k = 0, 1, 2, \ldots).$$

Finally, it is a good idea to check that this formula satisfies both the initial condition (put $k = 0$) and the original difference equation (substitute back in (3)).

A general formula for the solutions of (1), which includes the solution found in the above example as a special case, is given by the following proposition.

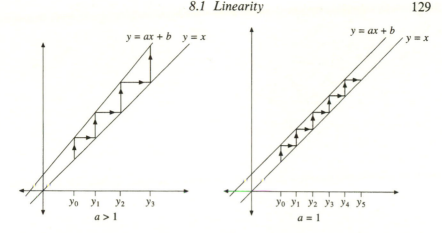

Fig. 8.1.1. Cobweb diagram for a linear constant-coefficient difference equation: exponential or linear growth depending on the value of a.

Proposition 1 *The linear first-order difference equation*

$$y_{k+1} = ay_k + b,$$

with a and b constant, has its solutions given by the formula

$$y_k = \begin{cases} a^k(y_0 - \frac{b}{1-a}) + \frac{b}{1-a} & \text{if } a \neq 1 \\ y_0 + kb & \text{if } a = 1 \end{cases}$$

where $k = 0, 1, 2, \ldots$ and where the initial value y_0 can be chosen as desired.

The derivation of this general formula in the case $a \neq 1$ involves the use of steps analogous to those used in Example 1 above. The derivation is even simpler in the case $a = 1$. The details are set as exercises later. In this regard, most of us would find it easier to remember the derivations than to remember the formulae!

The solutions do exhibit an interesting feature, however, which is worthy of comment. In the case $a \neq 1$ the solution gives y_k as an *exponential* function of k, but in the case $a = 1$ it is a *linear* function of k. The reason for the anomalous behaviour when $a = 1$ is revealed by the cobweb diagrams for the solutions in the two cases, shown in Figure 8.1.1.

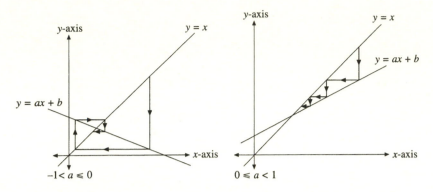

Fig. 8.1.2. Convergence of solutions to a limit.

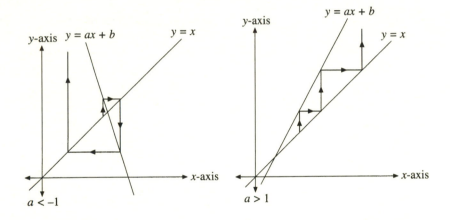

Fig. 8.1.3. Non-convergence of solutions.

As the parameter a approaches 1, the line $y = ax + b$ swings around and approaches a direction parallel to that of the line $y = x$; hence the diagram on the left becomes more and more like the one on the right. This gives rise to the solutions, for $a = 1$, which grow linearly rather than exponentially.

In the applications given later, it will be important to understand the long-term behaviour of the solutions given in Proposition 1. The possible cases which can arise, when $a \neq 1$, are illustrated in Figure 8.1.2 and Figure 8.1.3.

Although this section has been concerned exclusively with first-order linear difference equations, some of the ideas extend to *second-order* difference equations of the type

$$y_{k+2} = ay_{k+1} + by_k + c \qquad (k = 0, 1, 2, \ldots),$$

which are said to be *second-order linear*, to have *constant coefficients* if a and b are constants, and to be *homogeneous* if c is the constant 0. Use of appropriate superposition theorems makes it possible to find closed-form solutions, at least when a, b and c are constant. An example will be set in the exercises suggesting how this may be done.

Exercises 8.1

In the exercises which ask you to 'classify' a difference equation, state the order of the difference equation and whether it is linear, homogeneous, has constant coefficients.

1. In each case, classify the difference equation and state whether Proposition 1 in the text is applicable. If so, use it to get the closed-form solution of the difference equation satisfying the stated initial condition $y_0 = 3$. Assume $k = 0, 1, 2, \ldots$.

    (a) $y_{k+1} = 3y_k + 10$ (b) $y_{k+1} = 2(y_k)^3$

    (c) $y_{k+1} = y_k + 10$ (d) $y_{k+2} = 2y_{k+1} + 4y_k$

2. Classify each of the following difference equations and state whether Proposition 1 in the text is applicable.

    (a) $X_{k+2} = 2kX_k$ (b) $N_{k+1} = 3N_k$

    (c) $Z_{k+1} = \cos(Z_k)$ (d) $Y_{k+2} = -6Y_{k+1} + 7Y_k + 5$

3. For the difference equation

$$y_{k+1} = 1.1y_k \qquad (k = 0, 1, 2, \ldots)$$

    (a) write down in closed form the solution satisfying the initial condition $y_0 = 100$,

    (b) use your calculator to find the integer k such that

$$y_k < 200 < y_{k+1},$$

    (c) state the last integer k for which y_k is less than 200,

    (d) state the first integer k for which y_k exceeds 200.

    (e) To which of the cases illustrated in Figure 8.1.2 does the above difference equation belong?

4. For the difference equation

$$y_{k+1} = 0.9y_k \qquad (k = 0, 1, 2, \ldots)$$

    (a) write down in closed form the solution satisfying the initial condition $y_0 = 0.01$,

    (b) use your calculator to find the integer k such that

$$y_{k+1} < 0.005 < y_k,$$

(c) state the last integer for which y_k is greater than 0.005,

(d) state the first integer for which y_k is smaller than 0.005.

(e) To which of the cases illustrated in Figure 8.1.2 does the above difference equation correspond?

5. Find, in closed form, the solution of the difference equation

$$y_{k+1} = 3y_k + 2 \qquad (k = 0, 1, 2, \ldots)$$

which satisfies the initial condition $y_0 = 1$. Use steps similar to those given in Example 1 in the text. Check your answer.

6. Derive the solution given in Proposition 1 in the text in the case $a \neq 1$. Use steps similar to those given in Example 1 in the text. Check the answer.

7. Derive the solution given in Proposition 1 in the text in the case $a = 1$. Show how the solution may be guessed from the result of applying iteration. Check the answer.

8. Suppose that (1) and (2) in the text are satisfied. Show that if z_k denotes $y_k + x_k$ then

$$z_{k+1} = az_k + b \qquad (k = 0, 1, 2, \ldots).$$

[This shows that adding a solution of the homogenized equation (2) to a solution of the original equation (1) gives another solution of (1). This is the superposition theorem on which the steps in Example 1 in the text are based.]

9. Consider the difference equation

$$y_{k+1} = (y_k)^2 - 2 \qquad (k = 0, 1, 2, \ldots)$$

and the 'homogenized' version

$$x_{k+1} = (x_k)^2.$$

(a) Show that the original equation is satisfied by the constant solution $y_k = 2$.

(b) Show that the 'homogenized' equation is satisfied by the constant solution $x_k = 1$.

(c) Is the sequence defined by putting $z_k = y_k + x_k$ a solution of the original difference equation? Does this contradict the superposition principle proved in Exercise 8? Explain your answer.

10. This exercise illustrates how the methods of this section may be extended to *second-order* difference equations. Consider the Fibonacci equation

$$y_{k+2} = y_{k+1} + y_k \qquad (k = 1, 2, 3, \ldots),$$

which is second-order linear homogeneous with constant coefficients. The solution satisfying the initial conditions $y_1 = 1$ and $y_2 = 1$ will be found in closed form.

(a) *Guessing solutions.* The solutions for first-order equations suggest trying exponential solutions, of the form

$$y = \alpha^k \qquad (k = 1, 2, 3, \ldots)$$

where α is a constant. Show that this satisfies the Fibonacci equation only if α satisfies the quadratic equation $\alpha^2 - \alpha - 1 = 0$.

[Note that this gives two solutions, $y_k = (\alpha_1)^k$ and $y_k = (\alpha_2)^k$, where α_1 and α_2 are the roots of the quadratic equation.]

(b) *Superposing solutions.* Check that

$$y_k = c_1(\alpha_1)^k + c_2(\alpha_2)^k$$

is a solution of the Fibonacci equation where c_1 and c_2 are arbitrary real constants.

(c) *Fitting initial values.* Hence deduce that the solution of the Fibonacci equation satisfying the initial conditions $y_1 = 1$ and $y_2 = 1$ is given by

$$y_k = \frac{1}{\sqrt{5}}\left[\left(\frac{1+\sqrt{5}}{2}\right)^k - \left(\frac{1-\sqrt{5}}{2}\right)^k\right] \qquad (k = 1, 2, 3, \ldots).$$

[Note: It is surprising that this closed-form solution involving square roots produces only integers!]

11. Repeat Exercise 10 for the difference equation

$$y_{k+2} = y_{k+1} + 2y_k$$

with the initial conditions $y_0 = 1$, $y_1 = 2$.

12. Will the technique used in Exercise 10 work for the difference equation

$$y_{k+2} = y_{k+1} + (y_k)^2 \ ?$$

Give reasons.

8.2 Interest and loan repayment

A sum of money, when lent to a financial institution such as a bank, earns interest. The initial sum deposited, called the *principal*, thereby grows in value to an *amount* which is the sum of the principal and the interest earned. This may be expressed by writing

$$\{\text{amount on deposit}\} = \{\text{principal}\} + \{\text{interest}\}.$$

The amount which the lender finally gets back depends, among other things, on whether the money was lent at *simple interest* or at *compound interest*. The annual rate at which interest is earned is expressed as a percentage, say $p\%$ (or $p\%$ per annum). For compound interest, however, there is also an interest *conversion period*, which is usually a year or some fraction α of a year. How the interest is calculated and how the amount on deposit grows will now be explained for each type of loan.

Simple interest

For loans of this type, the interest earned in any year is obtained by applying the rate to the initial principle and thus it stays the same throughout the duration of the loan.

Thus, for an annual interest rate of $p\%$, the interest earned during any one year is $p\%$ of the principal, that is,

$$\{\text{interest}\} = \frac{p}{100} \times \{\text{principal}\}$$

and hence

$$\left\{ \begin{array}{l} \text{amount on} \\ \text{deposit after} \\ k+1 \text{ years} \end{array} \right\} = \left\{ \begin{array}{l} \text{amount on} \\ \text{deposit after} \\ k \text{ years} \end{array} \right\} + \frac{p}{100} \times \{\text{principal}\}.$$

To express this as a difference equation, let S_0 be the principal and for each k let S_k be the amount into which it grows after k years, giving

$$S_{k+1} = S_k + \frac{p}{100} S_0 \qquad (k = 0,\ 1,\ 2,\ldots).$$

This is a first-order linear difference equation of the type studied in Section 8.1 with $a = 1$ and $b = (p/100)S_0$. Hence, by Proposition 1 of Section 8.1, the solution is given by the formula

$$S_k = \left(1 + \frac{kp}{100} \right) S_0 \qquad\qquad (1)$$

which is called the *simple interest formula*.

The significance of this formula is that it gives the amount S_k into which the principal S_0 grows when it earns simple interest for k years at an annual rate of $p\%$.

Compound interest

This type of interest is more usual for loans made over longer periods. Interest is added to the principal at regular intervals, called *conversion periods*, and the new amount (rather than the principal) is used for calculating the interest for the next conversion period. The fraction of a year occupied by the conversion period is denoted by α so that conversion periods of 1 month, 1 quarter, 6 months and 1 year are given respectively by $\alpha = \frac{1}{12}, \alpha = \frac{1}{4}, \alpha = \frac{1}{2}$ and $\alpha = 1$. Instead of saying that the conversion period is 1 month, for example, we may say that the interest is *compounded* monthly.

For an interest rate of $p\%$ and conversion period equal to a fraction

α of a year, the interest earned for the period is $\alpha p\%$ of the amount on deposit at the start of the period, that is,

$$\{\text{interest}\} = \frac{\alpha p}{100} \times \left\{ \begin{array}{l} \text{amount on deposit at the} \\ \text{start of the conversion period} \end{array} \right\}$$

and hence

$$\left\{ \begin{array}{l} \text{amount on} \\ \text{deposit} \\ \text{after } k+1 \\ \text{conversion} \\ \text{periods} \end{array} \right\} = \left\{ \begin{array}{l} \text{amount on} \\ \text{deposit} \\ \text{after } k \\ \text{conversion} \\ \text{periods} \end{array} \right\} + \frac{\alpha p}{100} \times \left\{ \begin{array}{l} \text{amount on} \\ \text{deposit} \\ \text{after } k \\ \text{conversion} \\ \text{periods} \end{array} \right\}$$

To express this as a difference equation, for each k let S_k denote the amount on deposit after k conversion periods, giving

$$S_{k+1} = S_k + \frac{\alpha p}{100} S_k$$

$$= \left(1 + \frac{\alpha p}{100}\right) S_k \qquad (k = 0, 1, 2, \ldots).$$

This is a first-order linear homogeneous difference equation of the type studied in Section 8.1 with $a = (1 + \alpha p/100)$. Hence the solution is given by

$$S_k = \left(1 + \frac{\alpha p}{100}\right)^k S_0, \qquad (2)$$

which is called the *compound interest formula*.

This formula gives the amount S_k into which the principal S_0 grows when it earns compound interest for k conversion periods, each equal to a fraction α of a year, at an interest rate of p%.

The formulas (1) and (2) show that the amount S_k increases as a linear function of k when the interest is simple, but as an exponential function of k when the interest is compound. The cobweb diagrams in Figure 8.1.1 illustrate the difference between these rates of growth, the exponential growth being much greater than the linear one, at least in the long term.

When applying these formulae, recall that in (2) the letter k stands for the number of conversion periods (rather than the number of years, as in (1)). Thus to find the amount on deposit after 10 years at compound interest with a monthly conversion period, take $k = 120$ in (2) (rather than $k = 10$, as in equation (1)).

Loan repayments

Arguments similar to those used above may be applied to the study of loan repayments. The particular scheme considered here is the one normally used for the repayment of housing loans and is called *amortization*. Repayments are made at regular intervals, and usually in equal amounts, to reduce the principal (the amount borrowed) and to pay interest on the amount still owing.

It is supposed that compound interest at $p\%$ is charged on the outstanding debt, with conversion period equal to the same fraction α of the year as the period between repayments. Between payments, the debt increases because of the interest charged on the debt still outstanding after the last payment. Hence

$$\left\{\begin{matrix} \text{debt after} \\ k+1 \text{ payments} \end{matrix}\right\} = \left\{\begin{matrix} \text{debt after} \\ k \text{ payments} \end{matrix}\right\} + \left\{\begin{matrix} \text{interest on} \\ \text{this debt} \end{matrix}\right\} - \{\text{payment}\}.$$

To write this as a difference equation, let the initial debt to be repaid be D_0, for each k let the outstanding debt after the kth payment be D_k, and let the payment made after each conversion period be R. Hence

$$D_{k+1} = D_k + \frac{\alpha p}{100} D_k - R$$

$$= \left(1 + \frac{\alpha p}{100}\right) D_k - R.$$

This is a first-order linear difference equation of the type studied in Section 8.1 with $a = 1 + \alpha p/100$ and $b = -R$. Hence the solution is given by Proposition 1 in Section 8.1 as

$$D_k = \left(1 + \frac{\alpha p}{100}\right)^k \left(D_0 - \frac{100R}{\alpha p}\right) + \frac{100R}{\alpha p}, \tag{3}$$

which we shall call the *loan repayment formula*.

This formula gives the debt remaining on an initial debt D_0 after k payments of R made at the end of conversion periods of length equal to a fraction α of a year, the interest being compound interest at $p\%$.

Note that if $R = 0$ this reverts to the compound interest formula (with debt instead of credit).

Example 2. *A loan of $10 000 is to be repaid according to the amortization scheme. Calculate the monthly repayments needed to pay off the loan in five years if interest is charged at 15% compounded monthly.*

Solution. *Let R be the required monthly payment. In the loan repayment formula choose*

$$\alpha = \frac{1}{12} \qquad (since\ a\ month\ is\ \tfrac{1}{12}\ of\ a\ year)$$

$$p = 15 \qquad (the\ annual\ percentage\ rate)$$

$$D_0 = 10\,000 \qquad (the\ initial\ debt)$$

$$k = 60 \qquad (since\ 5\ years = 60\ months)$$

and hence get

$$D_{60} = 21\,071.81 - 88.57R.$$

But, for the loan to be repaid after 60 months, $D_{60} = 0$. Hence $R = 237.91$ and so the required monthly repayment is $237.91.

Difference equations are derived for interest and loan repayment problems in Goldberg (1958), pages 87–90. Technical details concerning a wide variety of financial calculations are given in Ayres (1963).

Exercises 8.2

The simple and compound interest formulae (1) and (2) in the text can be used wherever they are relevant.

1. Find p so that interest compounded annually at $p\%$ produces the same amount at the end of a year as 12% compounded quarterly.

2. Suppose that compound interest is earned on a certain investment, with an interest conversion period of six months. Calculate the interest rate $p\%$ necessary to double the initial principal in four years.

3. (a) Tabulate the amount on deposit at the end of each year for five years, for an initial principal of $1000 earning simple interest at an annual rate of 12.5%.

(b) Calculate the number of years it would take for the principal to double at this rate.

4. Repeat Exercise 3, but with a compound interest rate of 12.5% and an interest conversion period of one year.

5. Repeat Exercise 3, but with a compound interest rate of 12.5% and an interest conversion period of six months.

6. Two insurance companies offer investors insurance bonds earning compound interest at identical rates of interest. One company takes 5% of the initial principal as their up-front charge while the other company leaves the entire principal invested but takes 5% of the final amount as its withdrawal fee.

Does either scheme yield a better return to the investor? Give reasons for your answer. You may assume interest rates stay fixed.

7. (a) By making any necessary changes to the derivation given in the text of the compound interest formula, set up a difference equation to describe the following situation:

   An initial sum of money S_0 is deposited to earn compound interest at a rate of p% and the conversion period is a fraction α of a year. At the end of each conversion period a further sum S_0 is deposited, to give a total amount S_k at the end of the kth conversion period.

   (b) Find in closed form the solution of the difference equation found in (a).

8. Using Exercise 7 solve the following problem.

   A company deposits $2000 at the end of each quarter in a fund which earns 10% interest compounded quarterly. Calculate the amount in the fund 10 years after the initial deposit.

9. In the loan repayment formula in the text, find the value of the repayment R which just keeps the debt at its initial level. What happens if R is less than this value? greater than this value? Argue from the formula.

10. The average loan on an Australian home is often quoted to be $55 000. Calculate the monthly repayment necessary to have the loan repaid after 25 years if the interest rate is 7.5%. What is the total amount paid back on the loan?

11. Bernoulli's inequality states that, for each real number $x \geq -1$ and each positive integer n, $(1+x)^n \geq 1+nx$. Use this inequality to prove that, for the same principal and at the same rate of interest, compound interest always produces an amount at least as great as that for simple interest.

12. Show that $(1 + x/n)^n$ is an increasing function of $n > 0$ where x is a fixed real number ≥ 0. Hence show that, for a given principal and rate of interest, decreasing the conversion period can only increase the amount on deposit after a specified time.

13. Discuss how you would modify some of the models from this chapter to take account of inflation.

8.3 The cobweb model of supply and demand

Economics courses often begin with a discussion of the ideas of *supply* and *demand* and their effect on prices. Indeed, even people without formal training in economics have often heard of the 'law of supply and demand'. If pressed for an explanation of what it means they would probably say something like 'if the supply goes down or the demand goes up, then prices will rise'. The cobweb model, to be discussed in this section, is a refinement of this idea to include a time lag in the supply. This leads to a difference equation for the price.

The assumptions of the cobweb model, listed below, are suggested by the marketing of agricultural produce. Farmers, encouraged by high prices for a crop one season, will plant a larger area with that crop in anticipation of a similarly high price the following season. Thus supply lags one season behind the price of that particular crop. This provides the motivation for assumption (a) in our description below of the cobweb model. The assumption (b) relates demand to price in the same season, as would normally be expected. The assumption (c) is relevant to perishable goods like fruit and vegetables: the price will adjust to clear the market by making demand equal to supply.

The cobweb model. This model of supply and demand for a commodity makes the following assumptions relative to consecutive time periods:

(a) *The supply in period k (where $k = 1, 2, 3, \ldots$) is a linear function of the price in previous period $k - 1$, with the supply increasing when the price increases.*

(b) *The demand in period k is a linear function of the price in period k, with the demand decreasing when price increases.*

(c) *The market price is determined by the available supply, with the transaction taking place at the price which makes the demand equal to the supply.*

In the above model, the functions have been assumed to be linear merely for the sake of simplicity. Some economists use models in which the functions may be non-linear.

To illustrate these assumptions we shall use the following notation for the various quantities which are involved. For $k = 0, 1, 2, \ldots$ let

$S_k = \{$number of units of the commodity supplied in kth period$\}$,

$D_k = \{$number of units of the commodity demanded in kth period$\}$,

$p_k = \{$price of a unit of the commodity in the kth period$\}$.

Example 1. *Suppose that the supply and demand of potatoes are related to the price by the straight-line graphs shown in Figure 8.3.1. Verify that assumptions (a) and (b) of the cobweb model are satisfied. Show that, if (c) is assumed also, then the price satisfies the difference equation*

$$p_k = -\frac{1}{2}p_{k-1} + 1 \qquad (k = 1, 2, 3, \ldots).$$

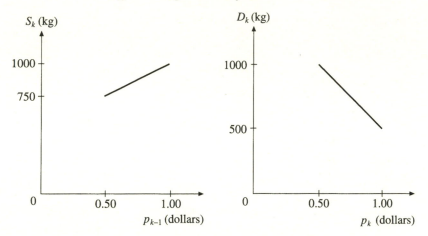

Fig. 8.3.1. Hypothetical supply and demand graphs for potatoes.

Solution. *The assumption (a) is satisfied because the supply S_k is a linear function of the price p_{k-1} and, from the graph, S_k increases when p_{k-1} increases. Similarly, assumption (b) is satisfied because D_k is a linear function of p_k and D_k decreases if p_k increases.*

It is easy to write down explicit formulae for the linear functions implicit in the graphs. In each case we read off the slope from the graph ('rise over run') and then add a constant to the RHS of the formula to make the graph pass through one of the end points. This gives

$$S_k = 500p_{k-1} + 500,$$
$$D_k = -1000p_k + 1500.$$

The assumption (c) translates to $D_k = S_k$ and hence from the last two equations

$$p_k = -\frac{1}{2}p_{k-1} + 1 \qquad (k = 1, 2, 3, \ldots).$$

This is the required difference equation for the price.

A description of the cobweb model of supply and demand may be found in Lipsey, Langley and Mahoney (1981), pages 149–152 (but note that, in the graph on page 150, mathematicians would regard the functions graphed as the *inverses* of the ones claimed). Lipsey *et al.* give some figures concerning acres planted and prices obtained by growers in the South Australian potato industry. Unfortunately, in this instance, the cobweb model does not give good agreement with the results.

The difference equation for the cobweb model is derived in Archibald and Lipsey (1973), pages 300–304, and in Goldberg (1958), pages 179–182.

Exercises 8.3

1. Suppose that in Example 1 in the text the price of potatoes is initially 90 cents.

(a) Sketch a cobweb diagram and deduce how the price behaves in the long term.

(b) Find the solution of the difference equation for the price in closed form and use it to check your answer in part (a) concerning the long-term behaviour of the price.

2. Repeat Example 1 in the text, but this time suppose the graphs of supply and demand against price are as shown below. Derive the difference equation for the price,

$$p_k = -2p_{k-1} + 5/2 \qquad (k = 1, 2, 3, \ldots).$$

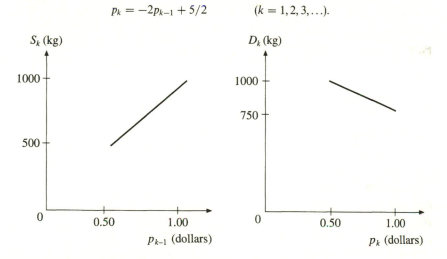

3. (a) Write the assumptions of the cobweb model as mathematical equations using the symbols S_k, D_k, p_k defined in the text together with any other notation you need to introduce.

(b) Hence derive a difference equation for the price of the form

$$p_k = -ap_{k-1} + b \qquad (k = 1, 2, 3, \ldots)$$

in which a and b are constants. Express a in terms of the notation you have introduced in part (a) and explain why $a > 0$. What happens if $b \le 0$?

(c) Describe the long-term behaviour of the price in each of the cases

(i) $0 < a < 1$ (ii) $a = 1$ (iii) $a > 1$.

Relate these inequalities to the slopes of the graphs of S_k against p_{k-1} and D_k against p_k.

8.4 National income: 'acceleration models'

The national income of a country is intended to measure the total level of economic activity within the country during a given period of time. A term often used synonymously with national income is GDP (gross domestic product). It provides a measure of living standards, dropping during a recession and rising during a boom. For this reason there is great interest in analyzing the factors which affect national income and in setting up models to predict its values in the future. One of the criteria used by economists to judge the effectiveness of these models is whether they predict regular fluctuations in the national income, thereby reflecting the business cycle.

The simple models studied here originated in 1939 with the American economist P. A. Samuelson (who won the Nobel Prize for Economics in 1970). The basis for these models is the *acceleration principle* according to which investment occurs not because of a high level of national income but rather because of an *increase* in the level of national income. Such an increase leads to a greater demand for goods and services, and hence to a need for greater capacity in the factories to produce more goods. Thus there is an increased opportunity for the investor.

The quantities to be used in the model, during the kth accounting period, are as follows:

$$Y_k = \{\text{national income}\}$$
$$C_k = \{\text{consumer expenditure}\}$$
$$I_k = \{\text{induced private investment}\}$$
$$G_k = \{\text{government expenditure}\}.$$

The consumer expenditure is the amount spent on consumer goods like food, clothing and appliances and the induced private expenditure is the amount invested in machinery, training programmes, etc.

The model will consist of four statements, expressed verbally, which relate these four quantities. It will be left as an exercise to translate these statements into mathematical equations involving the above symbols.

Model I for the national income consists of the four statements (a), (b), (c) and (d) listed below. Statement (a) is called the *accounting equation* and may be regarded as a definition of national income, while statement (c) expresses the *acceleration principle* explained above.

(a) *The national income is the sum of consumer expenditure, induced private investment and government expenditure.*

(b) *Consumer expenditure in each period is directly proportional to the national income in that period.*

(c) *Induced private investment in any period is directly proportional to the increase in the national income for that period above the national income for the preceding period.*

(d) *The government expenditure stays constant from one period to the next.*

These statements can be written as mathematical equations involving the symbols introduced above for the various quantities. It can then be shown that the national income satisfies a difference equation of the type

$$Y_k = aY_{k-1} + b \qquad (k = 1, 2, \ldots) \qquad (1)$$

where a and b are constants with $a > 1$ and $b < 0$. The details are left till later as a useful exercise. The difference equation (1) is first-order linear constant coefficient of the type studied in Section 8.1 and hence its solutions are given by

$$Y_k = a^k \left(Y_0 - \frac{b}{1-a} \right) + \frac{b}{1-a}. \qquad (2)$$

Because $a > 1$, the formula (2) shows that Y_k grows exponentially with k (at least if $Y_0 > b(1-a)$), but it does not oscillate. Hence Model I does not predict the regular fluctuations in the national income which we know occur in practice. To obtain more realistic predictions, some modifications to the model are needed.

An obvious point at which to try to improve Model I is at the interpretation (c) of the accelerator principle. Knowledge of the latest increase in national income comes too late for investors to take advantage of it. Hence it would be more realistic to introduce a time lag and to use the increase from one period back. This leads to the following model for the national income.

Model II consists of the statements (a), (b) and (d) from Model I together with the modified statement (c′) of the accelerator principle:

(c′) *Induced private investment in any period is directly proportional to the increase in the national income for the previous period above the national income for the period before that.*

This model leads to a *second-order* difference equation for the national income of the type

$$Y_k + aY_{k-1} + bY_{k-2} = c \qquad (3)$$

where a, b and c are constants satisfying certain inequalities. An example appears in the exercises which shows such an equation can have a solution showing cyclic (periodic) behaviour. Hence, in Model II, our objection to Model I has been removed.

Samuelson's models for the national income are described in Gandolfo (1971), pages 63–73, Kenkel (1974), pages 241–259, and in Pfouts (1972), pages 116–119. Kenkel gives some interesting comments as to how these models may be made more realistic.

Exercises 8.4

1.   (a) In Model I write each of the statements (a), (b), (c) and (d) in terms of the symbols introduced in the text for the various quantities. (Use the letters A and B for the constants of proportionality in statements (b) and (c) respectively.)

   (b) Deduce from your answers to part (a) of this exercise that the national income satisfies the difference equation

   $$(1 - A - B)Y_k = -BY_{k-1} + G_0 \qquad (k = 1, 2, 3, \ldots).$$

2. Let A and B be as in Exercise 1 (b).

   (a) Use the idea that consumer spending cannot exceed national income to derive an inequality for the constant A.

   (b) Use the idea that the investment to match a given increase in income must be greater than that increase to deduce an inequality for the constant B.

   (c) Hence show that the difference equation introduced in Exercise 1 (b) has the form

   $$Y_k = aY_{k-1} + b \qquad (k = 1, 2, 3, \ldots)$$

   where a and b are constants satisfying the inequalities claimed in the text.

3.   (a) In Model II write each of the statements (a), (b), (c') and (d) in terms of the symbols introduced in the text for the various quantities.

   (b) From your answers to part (a) show that the national income satisfies a difference equation of the form

   $$Y_k + aY_{k-1} + bY_{k-2} = c \qquad (k = 2, 3, 4, \ldots)$$

   where a, b and c are constants, as claimed in the text.

   (c) In the case $a = -1, b = 1, c = 100$ show that the solution of this difference equation satisfying the initial conditions $Y_0 = 100$ and $Y_1 = 101$ exhibits cyclic (periodic) behaviour.

4. Consider the model for the national income consisting of statements (a), (c) and (d) in the text together with the following assumption:

(b') Consumer spending in each period is directly proportional to the national income in the preceding period.

Write down each statement in terms of the symbols introduced in the text for the various quantities and hence find a difference equation for the national income.

5. Consider the model for the national income consisting of statements (a), (b) and (d) in the text together with the following assumption:

(c') Induced private investment in any period is directly proportional to the increase in consumer spending for that period above the consumer spending for the preceding period.

Write down each statement in terms of the symbols introduced in the text for the various quantities and hence find a difference equation for the national income.

9

Non-linear difference equations and population growth

Linear difference equations have the advantage that a closed-form solution can be easily obtained. But, in many cases, the behaviour of linear difference equation models is not consistent with observation. This is true in many areas of biology, and particularly in studies of populations, where non-linear models are better.

In this chapter non-linear models are developed which describe how a population of individuals grows over time. Difference equation models are appropriate when a species has a distinct breeding season. The simplest non-linear model, the 'logistic equation' is studied in detail. Similar ideas are also used to model a measles epidemic. This involves iterating a pair of simultaneous difference equations for the number of those who can infect others and for the number of those susceptible to being infected.

Closed-form solutions usually cannot be found for non-linear difference equations. Thus to interpret the models one has to resort to numerical simulation or devise approximate closed-form solutions to the equations. Both approaches are developed here. The ideas rely on concepts developed in Chapters 7 and 8.

9.1 Linear models for population growth

Many people are very interested in the way populations grow and in determining what factors influence their growth. Knowledge of this kind is important in studies of bacterial growth, wildlife management, ecology and harvesting. In this section a very simple model is formulated for a population which breeds at fixed time intervals. This model forms a starting point for the development of more realistic models.

While some models for population growth are very simple, others can be very sophisticated. A certain amount of insight into population growth can be gained by first looking at simple models which incorporate the most important features which affect population growth. Although simple models may not be accurate they are still valuable. The mathematical biologist, J. Maynard-Smith (1968), makes the following comments:

Any attempt to formulate the problem mathematically necessarily leaves out many relevant factors. The attempt is nevertheless illuminating — it provides a rapid way of discovering the kind of effect various features may have on the behaviour of a population and it suggests what needs to be measured before the behaviour of any particular species can be understood.

Many animals tend to breed only during a short, well-defined, breeding season. It is then natural to think of the population changing from *season* to *season*. Thus time is measured *discretely* with positive integers denoting each breeding season. Hence the obvious approach for describing the growth of such a population is to write down a suitable difference equation. Later in this book, differential equations will be used to study populations which breed continuously (for example, human populations).

Cell division

In nature, species typically compete with other species for food and are themselves sometimes preyed upon. Thus the populations of different species interact with each other. In the laboratory, however, a given species can be studied in isolation. We shall therefore concentrate, at first, on models for a *single* species. The first example we shall study is a population of yeast cells which reproduce by dividing into two. The rate at which yeast cells divide is governed by environmental factors such as nutrient availability and temperature. In the laboratory these factors may be controlled so that the rate of dividing is constant.

Example 1. *(Cell Division). Suppose a single cell divides every minute. Assuming that none of the cells die determine how many minutes it will take before there are more than one million cells.*

Solution. *We measure time in integer multiples of one minute. Let N_k be the number of cells after k minutes (where k is an integer). So N_{k+1} will be the number of cells one minute later. If each cell divides into two there will be twice this number of cells one minute later. Hence we may write down the difference equation*

$$N_{k+1} = 2N_k \qquad \text{with} \quad N_0 = 1. \tag{1}$$

This linear difference equation, which is of the type studied in Chapter 8, has the solution

$$N_k = 2^k, \tag{2}$$

obtained by direct iteration. Now we want k such that $N_k = 10^6$, so

$$2^k = 10^6. \tag{3}$$

To solve for k, take natural logarithms to get

$$\ln(2^k) = \ln(10^6),$$

which simplifies to

$$k = \frac{6\ln(10)}{\ln(2)} \simeq 19.93. \tag{4}$$

Thus in 19 minutes less than a million cells are produced and in 20 minutes more than a million cells are produced. So 20 minutes is the required time.

Births and deaths

In the previous model for cell division the population was affected only by two new members *replacing* an old member. We are also interested, however, in setting up models for populations of higher organisms (for example, insects, fish and mammals). In these populations (unlike those of cells) a number of individuals survive for at least several breeding seasons. To do this, we must take account of the births and the deaths which occur between the start of one breeding season and that of the next.

To make progress with the modelling it is expedient to make some simplifying assumptions about the population. We assume that we are dealing only with large populations. Thus we can treat the population as a whole and we do not have to deal with individuals. We then assume that the population growth is governed by the average behaviour of its individual members. With this in mind we make the following additional assumptions:

- Each member of the population produces the same number of offspring.
- Each member has an equal chance of dying (or surviving) before the next breeding season.

- The ratio of females to males remains the same in each breeding season.

We also assume

- Age differences between members of the population can be ignored.
- The population is isolated — there is no immigration or emigration.

It should be clear that the first two assumptions are reasonable only when dealing with *large* populations, where it is expected that differences between individuals are not significant. In the exercises models are formulated which relax the final two assumptions. We also note that, in certain populations where the ratios of females to males is not roughly the same (e.g. many insect populations), then it is more practical to count only the females and completely ignore the males.

Suppose that on average *each* member of the population gives birth to the same number of offspring, α, each season. The constant α is called the *per-capita birth rate* for an individual of the population. It is also possible to think of α as the probability that a given member of the population gives birth to a single offspring during the breeding season. We also define β as the probability that an individual will die before the start of the next breeding season. We call β the *per-capita death rate*. Thus

(a) *the number of individuals born in a particular breeding season is directly proportional to the population at the start of the breeding season, and*

(b) *the number of individuals who have died during the interval between the end of consecutive breeding seasons is directly proportional to to the population at the start of the breeding season.*

If we let N_k denote the number of individuals of the population at the start of the kth breeding season then

$$\left\{ \begin{matrix} \text{number} \\ \text{born in} \\ \text{breeding season} \end{matrix} \right\} = \alpha N_k \quad \text{and} \quad \left\{ \begin{matrix} \text{number} \\ \text{who die in} \\ \text{breeding season} \end{matrix} \right\} = \beta N_k \quad (5)$$

Experimental measurements of per-capita birth and death rates are usually expressed as average values. Because of this the value of N_k obtained from an equation involving these quantities will usually not be an integer. However, although N_k is not *calculated* as an integer it may be *interpreted as an integer* in the model by rounding N_k to the nearest integer. (We note that it is not always necessary to interpret the population as an integer. A practical way of measuring large populations is by counting the number of individuals in sample areas. This number, which is measured in individuals per unit area, is called the population density, and it will not usually be an integer.)

Example 2. *In a species of animals a constant fraction of the population $\alpha = 5.3$ are born each breeding season and a constant fraction $\beta = 4.97$ die. Formulate a difference equation for the population and find the number of individuals after two seasons given the initial number is $N_0 = 987$.*

Solution. *Let N_k be the number of individuals in the current breeding season. In the next breeding season*

$$N_{k+1} = \left\{ \begin{array}{c} \text{current number} \\ \text{of individuals} \end{array} \right\} - \left\{ \begin{array}{c} \text{number} \\ \text{who die} \end{array} \right\} + \left\{ \begin{array}{c} \text{number} \\ \text{born} \end{array} \right\}. \tag{6}$$

Using (5), (6) becomes

$$N_{k+1} = N_k - \beta N_k + \alpha N_k$$

or

$$N_{k+1} = (1 + \alpha - \beta) N_k. \tag{7}$$

Thus with $\alpha = 5.3$, $\beta = 4.97$ and $N_0 = 987$ we calculate

$$N_1 = 1312.71 \qquad N_2 = 1745.9043.$$

Thus we interpret the population after the first season as 1313 individuals and after the second season as 1746 individuals.

The difference equation (7) is of the linear type studied in the previous chapter. Its closed form solution, given the initial number N_0, is

$$N_k = (1 + \alpha - \beta)^k N_0, \qquad (k = 0, 1, 2, \ldots).$$

The model depends on the combination

$$r = \alpha - \beta. \tag{8}$$

This quantity is called the *growth rate*. Clearly if $r < 0$ (corresponding to the per-capita death rate exceeding the per-capita birth rate) then the population decreases towards extinction but if $r > 0$ then the population increases indefinitely. This is illustrated in Figure 9.1.1. This is not, however, what is often observed in practice, as will be seen in the next section where more realistic non-linear models are introduced.

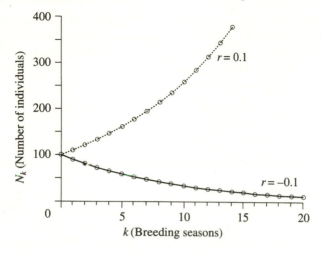

Fig. 9.1.1. Population growth as given by a linear difference equation with growth rate r.

Exercises 9.1

1. How many cells will a single cell produce after 10 divisions?

2. Suppose that a single yeast cell divides every 2 minutes. Also suppose 75% of yeast cells survive to divide in the next generation.

    (a) What is the growth rate?

    (b) How many yeast cells will there be after 3 hours?

3. A population of birds on an island has a constant per-capita birth rate α and a constant per-capita death rate β (per individual per year). Also, a constant number I of birds migrate to the island each year.

    (a) What is the appropriate time period to be used here?

    (b) Formulate a suitable difference equation.

    (c) Obtain the closed-form solution given the initial number of individuals, N_0.

4. In a certain type of female insect population *all* the adults die before the eggs hatch. *Each* adult contributes a constant number of eggs b, of which a fraction f survive and develop into adult females.

    (a) Set up a difference equation for the population after the kth hatching.

    (b) What is the minimum number of eggs that should be laid so that the population does not become extinct, given the fraction which survive is 20%.

5. Female houseflies produce approximately 120 eggs in one laying. Approximately *half* these develop into females. If all survive, starting with one female, what is the number of females after one year (seven layings)?

6. Modify the difference equation in the previous exercise given that all the flies die before the next hatching. Will the population be greater or smaller than that of the previous question given the same initial number?

7. A culture of bacteria grows from 2×10^6 cells to 3×10^8 cells in 2 hours. From this information deduce the time between successive cell divisions.

9.2 Restricted growth — non-linear models

The simple linear difference equation of the previous section is not generally suitable as a model of population growth since it predicts *unbounded* growth as the population increases. This is not what is observed in nature or in populations raised under controlled laboratory conditions. Rather than reject the model outright we try to build into it modifications so that it better approximates the observed behaviour. This leads to models described by *non-linear difference equations*.

The carrying capacity

Laboratory studies have shown that as a population increases the per-capita death rate goes up and the per-capita birth rate goes down. This is due to overcrowding and competition for food. In Figure 9.2.1 populations of laboratory-raised beetles are plotted against time. In each case there is a number which is representative of the number of beetles that a given environment can support. This is known as the *carrying capacity* of the environment.It corresponds to the number of individuals in the population when the birth rate and death rate are equal. In Figure 9.2.1(a) the population appears to be converging to a carrying capacity of approximately 1100. Figure 9.2.1(b) illustrates a population fluctuating above and below a carrying capacity of approximately 200. In Figure 9.2.1(c) the population does not appear to follow any simple pattern. We now introduce a non-linear model whose solutions exhibit all the different types of behaviour observed in Figure 9.2.1, when a parameter is varied.

Recall the linear difference equation of the previous section,

$$N_{k+1} = N_k + rN_k \tag{1}$$

where the constant r is the growth rate (the per-capita birth rate minus the per-capita death rate). To incorporate overcrowding, (1) is replaced by

$$N_{k+1} = N_k + R(N_k)N_k \tag{2}$$

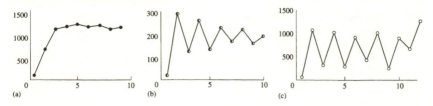

Fig. 9.2.1. Examples of populations of three different strains of stored product beetle, from May (1976).

where $R(N_k)$ is a growth rate which is a function of the size N_k of the population. It must satisfy certain requirements imposed on the growth of the population:

- Due to overcrowding, the per-capita death rate increases and the per-capita birth rate decreases; so $R(N_k)$ must decrease as N_k increases.
- When $N_k = K$, the carrying capacity, the growth rate is zero; so $R(K) = 0$.
- As $N_k \to 0$ the effects of overcrowding diminish and the growth rate tends towards a constant value r, which we call the *unrestricted* growth rate (measured per head of population per breeding season). Thus $R(0) = r$.

The unrestricted growth rate is determined approximately by measuring the growth of a population in a situation which is not affected by overcrowding.

The discrete logistic equation

There are many choices of $R(N_k)$, as a function of N_k, which satisfy the above conditions. In the spirit of mathematical modelling we choose the *simplest* such function. This has as its graph a straight line from $(0, r)$ to $(K, 0)$ (see Figure 9.2.2). The slope of the line is $-r/K$ and so the equation of the line is

$$R(N_k) = -\frac{r}{K} N_k + r. \tag{3}$$

Substituting (3) into (2) shows that the population satisfies

$$N_{k+1} = N_k + rN_k \left(1 - \frac{N_k}{K}\right). \tag{4}$$

This non-linear difference equation is called the *discrete logistic equation*.

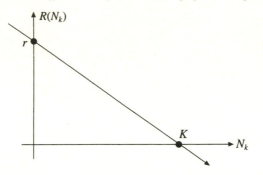

Fig. 9.2.2. Growth rate $R(N_k)$, plotted against N_k, for the discrete logistic equation.

Because the discrete logistic equation is non-linear it is not possible to use the general solution from Section 8.1. In fact a general closed-form solution has not yet been discovered. However, it is still quite simple to iterate numerically for given values of the parameters r (the unrestricted growth rate), K (the carrying capacity) and N_0 (the initial population), as is done in the next section.

General features of the discrete logistic equation

Before iterating numerically it is useful to explore the equations to see if some preliminary information can be extracted.

First we look for steady-state solutions. The steady-state solutions of the discrete logistic equation are determined by setting $N_{k+1} = N_k = s$, as explained in Section 7.3. Substitution into the discrete logistic equation (4) yields

$$s = s + rs\left(1 - \frac{s}{K}\right)$$

or

$$s\left(1 - \frac{s}{K}\right) = 0. \tag{5}$$

This equation has two solutions, $s = 0$ and $s = K$. The more interesting of these is $s = K$ since this gives the steady-state solution corresponding to the carrying capacity of the environment. Thus, if the initial population is given by $N_0 = K$, the population will remain at that same value.

Second, by writing the logistic equation (4) in the form

$$N_{k+1} - N_k = rN_k\left(1 - \frac{N_k}{K}\right) \tag{6}$$

we can discover some important features of the solutions. The LHS of
(6) is the *change* in the population between successive time periods. Thus
it follows that if $N_k < K$ then the total population will *increase* in the
next time interval, since $N_{k+1} - N_k > 0$. Conversely, if $N_k > K$ then the
population *decreases* in the next time interval.

Note that (6) shows the change $N_{k+1} - N_k$ is directly proportional to
the unrestricted growth rate r. Hence we might expect the population
to increase steadily towards the carrying capacity when r is small, but
to oscillate above and below the carrying capacity when r is large. It is
interesting to observe that these two types of behaviour are exhibited by
the population of beetles in Figures 9.2.1 (a) and (b).

A third feature of the discrete logistic equation, that can be found
directly from the equation, is that it can produce negative values of the
population if $r > 3$. This is a major limitation of this model.

Other models

There are many examples of populations raised in laboratories which
have growth rates greater than 3. Thus the fact that the logistic equation
produces negative population for $r > 3$ means that this model is not
suitable for such populations.

We must not forget that the discrete logistic equation is based on an
assumption of simplicity — namely the straight-line form of $R(N_k)$. In
practice the form of $R(N_k)$ can be more complicated. There are many
choices that we can make for the form of $R(N_k)$ based on experimental
evidence. One such model, which still has the advantage of being fairly
simple, is popular in the biological literature. In particular it has been
used for fish populations; for example, see Greenwell and Ng (1984).

This model uses an exponential curve, yielding a difference equation

$$N_{k+1} = N_k e^{a(1 - N_k/K)}, \qquad (7)$$

where a is a constant and K is the carrying capacity.

This model, along with others, is explored in the exercises in the same
way that we explored the discrete logistic equation in this section. In
each of these models the unrestricted growth rate r is defined as the limit
of $R(N_k)$ as $N_k \to 0$. The carrying capacity K, on the other hand, can
be regarded as the population for which the growth rate is zero; that
is, $R(K) = 0$. A discussion of some of these models can be found in
Chapter 3 of Edelstein-Keshet (1988).

Exercises 9.2

1. Use a calculator to find the population for the first five breeding seasons:

(a) using the discrete logistic equation with $K = 1000, r = 0.5, N_0 = 200$;

(b) using equation (7) in the text with $K = 1000, N_0 = 200$ and $\alpha = \ln(1.5)$.

2.   Define the variable $X_k = N_k/K$, where K is the carrying capacity. Note that X_k is a dimensionless form of the population N_k, scaled with respect to the carrying capacity K.

(a) Substitute into the discrete logistic equation and obtain

$$X_{k+1} = X_k + rX_k(1 - X_k).$$

This is the dimensionless form of the discrete logistic equation.

(b) What is the carrying capacity corresponding to this equation?

3. A model for insect populations (in which all adults are assumed to die before next breeding) leads to the difference equation

$$N_{k+1} = \frac{\lambda N_k}{1 + aN_k}$$

where λ and a are positive constants.

(a) Write the equation in the form $N_{k+1} = N_k + R(N_k)N_k$ and hence identify the growth rate.

(b) Show the general shape of the graph of $R(N_k)$ as a function of N_k, on a diagram similar to Figure 9.2.2.

(c) Express the unrestricted growth rate r and the carrying capacity K, for this model, in terms of the parameters a and λ.

(d) Find the steady-state solutions of this model and state briefly their biological significance.

(e) In this model, can the population ever switch from positive to negative values?

4. Another model for restricted population growth is given by

$$N_{k+1} = N_k e^{a(1-N_k/K)},$$

where K is the carrying capacity and a is a positive constant.

(a) Write the equation in the form $N_{k+1} = N_k + R(N_k)N_k$ and identify the variable growth rate $R(N_k)$.

(b) Sketch $R(N_k)$ in a diagram similar to Figure 9.2.2. Also sketch, with this, the growth rate for the discrete logistic equation.

(c) Show that $a = \ln(r + 1)$ where r, the unrestricted growth rate, is defined as the limit of $R(N_k)$ as $N_k \to 0$.

9.3 A computer experiment

In the absence of a closed-form solution to a non-linear difference equation, the next best thing is a numerical solution obtained from the difference equation by iteration. This is performed for a range of values of both the initial condition and any parameter in the difference equation. Any resulting changes in the way the population grows can then be observed. In this way we hope to obtain an overview of all possible types of behaviour which the equation predicts.

The process of carrying out numerical iterations for a range of parameter values is very tedious if done by hand. We recommend the use of a programmable calculator or a personal computer. A personal computer has the advantage that it can be programmed to display the results of the iteration graphically. For those who are not confident at writing programs the use of a 'spreadsheet program' is suggested. A spreadsheet program is used to manipulate rows and columns of data. It provides a convenient environment for numerical experimentation since one can change a parameter, recalculate the spreadsheet and view the graph with just a few keystrokes.

In the following we carry out a computer experiment on the discrete logistic equation (4) of Section 9.2,

$$N_{k+1} = N_k + rN_k \left(1 - \frac{N_k}{K}\right). \qquad (1)$$

The most interesting parameter to vary is r since an increase in r corresponds to an increase in fertility of a typical individual. Here r is varied in the range $0 < r \le 3$ while K and N_0 are held constant at $K = 1000$ and $N_0 = 100$. (Note that for $r > 3$ the population will become negative and the model ceases to apply.) The type of growth varies as r varies and we have attempted to classify below a number a different modes of growth: stable growth, cyclic growth and chaotic growth.

Stable growth $(0 < r \le 2)$

In Figure 9.3.1 populations corresponding to low values of r, $0 < r < 1$, are plotted. The populations, in each case, tend towards a carrying capacity $K = 1000$ as time increases. The population at first appears to grow exponentially, as in Figure 9.2.1(a), but then the effect of overcrowding becomes more pronounced as the population becomes larger, causing the population to level out. A gently sloping curve results. For the larger value of r the population growth initially is more rapid than for the

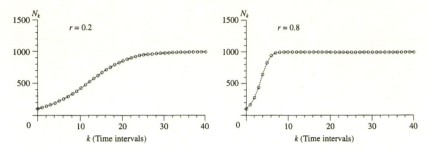

Fig. 9.3.1. Discrete logistic equation for $r = 0.2$ and $r = 0.8$. The population increases then levels out to an equilibrium value.

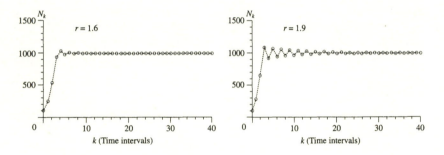

Fig. 9.3.2. Discrete logistic equation with $r = 1.6$ and $r = 1.9$. Damped oscillations.

smaller value of r, as we would expect. However overcrowding manifests itself more rapidly and the population tends toward the carrying capacity $K = 1000$ more quickly than for smaller values of r.

For $r > 1$ the character of the population growth changes as shown in Figure 9.3.2. In the early stages the growth of the population is so rapid that the population is able to *overshoot* the carrying capacity *before* the overcrowding effect is felt. Thus, from Figure 9.2.2 $R(N_k)$ is negative in the next time interval so the population decreases in the next time interval to below the carrying capacity. Next the population overshoots the carrying capacity again but this time it is closer to the carrying capacity. A 'damped oscillation' results, which converges to the carrying capacity. Note, from Figure 9.3.2, that for $r = 1.6$ the maximum amplitude of the oscillation is smaller compared with that of $r = 1.9$ and the damping of the oscillation is greater.

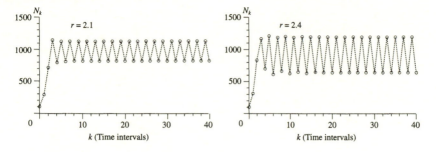

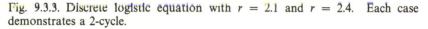

Fig. 9.3.3. Discrete logistic equation with $r = 2.1$ and $r = 2.4$. Each case demonstrates a 2-cycle.

Cyclic growth $(2 < r < 2.57)$

In Figure 9.3.3 populations corresponding to r in the range $2 < r < 2.57$ are plotted. A new type of behaviour occurs. The population tends towards an oscillation which is no longer damped but, at large times, fluctuates periodically above and below the carrying capacity $K = 1000$. In Figure 9.3.3 the populations oscillate, coming back every *second* breeding season. We call this a *2-cycle*. For $r = 2.1$ the population oscillates between the values 824 and 1129 whereas for $r = 2.4$ the population oscillates between 640 and 1193. (In Exercise 5 these values are derived analytically.)

As we increase r further to $r = 2.5$ the long-term behaviour of the population changes (see Figure 9.3.4). Instead of repeating itself after two time intervals, the population now repeats itself every *fourth* time interval, with two values below the carrying capacity and two values above. This is called a *4-cycle*. Beyond $r = 2.5$, 4-cycles become 8-cycles and then 16-cycles and so on up to $r \simeq 2.57$. This is called *period doubling*.

The American physicist Mitchell Feigenbaum noted values of the parameter at which these cycles first appeared (although he used a slightly different form of the discrete logistic equation; see Exercise 4). He was thereby led to the discovery of a famous number 4.669... which now bears his name.

Chaotic growth $(2.57 < r \le 3)$

As we increase the value of r past 2.57 some remarkable behaviour is noticed (see Figure 9.3.5). The pattern of the growth appears to be random, even though the population is predicted by a simple difference

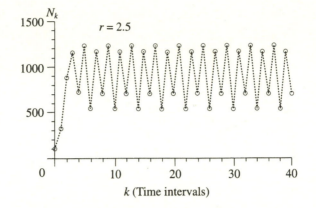

Fig. 9.3.4. Discrete logistic equation with $r = 2.5$: a 4-cycle.

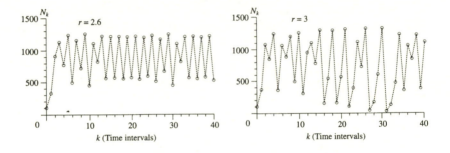

Fig. 9.3.5. Chaotic growth with $r = 2.6$ and $r = 3$.

equation, the discrete logistic equation. This random type of behaviour is called *chaotic*. One of the first people to realize how such simple models could lead to such complicated behaviour was Robert May who published several articles on this topic; see May (1975) and May (1976). It should be noted, however, that in the regime $2.67 \leq r \leq 3$ other types of behaviour can occur. For example, for a small range of values of r, 3-cycles occur which then go through the same period doubling phenomenon as we saw with 2-cycles.

The discovery that apparently random patterns of growth could arise from simple equations was very important in the study of populations. Before this discovery, observations from the field which did not exhibit a simple pattern (such as Figure 9.2.1(c)) were thought to be due to external random environmental effects. It is now clear, however, that random behaviour can be generated from within the system.

It is often difficult, in practice, to tell whether the behaviour is chaotic or merely periodic with a very long period. One feature of chaotic behaviour is that if the initial population is varied, ever so slightly, the population in subsequent times can change dramatically from that with the original value of N_0.

Discussion

We have now completed our numerical investigation of the effect of varying the parameter r since, when $r > 3$, N_k becomes negative and the model ceases to apply. Varying the parameter K does not provide any useful insight since this corresponds merely to a change of scale in the vertical axis. Varying the initial population N_0 does not normally affect the long-term behaviour of the population except when the the behaviour is chaotic. This is also investigated in the exercises.

Numerical experimentation has now become an accepted part of applied mathematics. The value of numerical experiment is illustrated by the results of the computer experiments, for the discrete logistic equation model, in Figures 9.3.1–9.3.5 since they raise some very interesting questions. For instance, we might ask how realistic the predictions of our model are compared with the results shown in Figure 9.2.1 for beetle populations. Also, what is the exact value of r for which the population first executes oscillations (2-cycles)? But most importantly, the results lead us to look for underlying causes of the oscillations and their biological significance.

The key to understanding why the population oscillates depends on two factors. Firstly, the population is *self-regulating* through the population-dependent growth rate. (Note that we found, in the previous section from studying the difference equation, that population decreases in the next time interval when it is greater than the carrying capacity and increases when it is less than the carrying capacity.) Secondly, the regulating effect is *felt* in the *next* time interval but actually *determined* by the population in the current time interval. In other words there is a natural *delay* in the population responding to overcrowding. As a result when the ideal growth rate r is sufficiently high the population responds by overcorrecting itself which leads to oscillations and sometimes chaos. Note that for continuously breeding populations (described by differential equations) the growth rate responds *instantaneously* to the population and oscillations do not normally occur, except through external seasonal factors.

One major limitation of the discrete logistic model is that it predicts negative populations when $r > 3$, which is clearly unrealistic since many real populations have growth rates exceeding 3. Despite this limitation the model is still useful since it predicts the qualitative range of behaviour as seen in Figure 9.2.1 and it does illustrate the importance of the growth rate. These considerations are certainly important to population biologists.

For general discussions on chaos see the books by Gleick (1987) and Stewart (1990). A more technical, but still very readable reference, is the book by Devaney (1986). Also the book by Tuck and De Mestre (1991), which discusses how to use the programing language BASIC to do computer experiments in population dynamics and ecology, is pitched at an introductory level.

Exercises 9.3

1. (Computer Simulation). Use a computer to iterate numerically the equation

$$N_{k+1} = N_k e^{a(1-N_k/K)},$$

where K is the carrying capacity and a is a positive constant. You are given $K = 1000, N_0 = 100$ and you should use your own choice of values of a. Describe the types of behaviour which appear for different values of a. In particular:

(a) When do 2-cycles first appear?

(b) When do 2-cycles become 4-cycles?

(c) Does the model exhibit chaos?

2. (Computer Simulation). A model which has been used to analyse insect populations is a modification of the one introduced in Exercises 9.2, question 3. The model leads to the difference equation

$$N_{k+1} = \frac{\lambda N_k}{(1 + N_k)^b}$$

for the insect population N_k. Using a computer, sketch solutions for the following values of the parameters:

	λ	b
Moth	1.3	0.1
Mosquito	10.6	1.9
Potato Beetle	75.0	3.4

3. (Computer experiment). For the discrete logistic equation with $K = 1000$ examine N_{100} as you vary N_0 slightly in each of the following cases:

(a) $r = 0.5$,

(b) $r = 2.5$,

(c) $r = 3$.

Start with $N_0 = 100$ and then try $N_0 = 101$ and $N_0 = 99$. What happens for each value of r?

4. The discrete logistic equation can be put into another form. The following shows how to do this.

(a) Write the discrete logistic equation

$$N_{k+1} = N_k + rN_k \left(1 - \frac{N_k}{K}\right)$$

in the form

$$N_{k+1} = \lambda N_k \left(1 - \frac{N_k}{K_1}\right)$$

and identify the constants λ and K_1 in terms of r and K.

(b) Make the substitution $X_k = N_k/K_1$ and show that the above difference equation becomes

$$X_{k+1} = \lambda X_k (1 - X_k).$$

(c) Find all the steady-state solutions for the difference equation in (b).

(d) Draw a cobweb diagram for the difference equation in (b). Use a few typical values of r.

[Note: The advantage of the first form of the discrete logistic equation is that the parameters have a direct biological meaning, but the form involving X_k is better to analyse mathematically because it has a simpler cobweb diagram.]

5. Consider the dimensionless discrete logistic equation

$$X_{k+1} = X_k + rX_k(1 - X_k).$$

(a) Show that every *second* term in the sequence $X_0, X_1, X_2, X_3, \ldots$ satisfies the difference equation

$$X_{k+2} = (1 + 2r + r^2)X_k - (2r + 3r^2 + r^3)X_k^2 + (2r^2 + 2r^3)X_k^3 - r^3 X_k^4.$$

(b) Put $X_{k+2} = X_k = s$ and hence show that the steady-state solutions satisfy a quartic equation.

(c) Explain why $s = 0$ and $s = 1$ must be solutions of this quartic equation. Hence show that the other two real solutions are

$$s = \frac{(2 + r) \pm \sqrt{r^2 - 4}}{2r}, \qquad \text{if } r \geq 2.$$

[Note: This shows that 2-cycles *first appear* when $r > 2$. The two values of s give the values that the 2-cycle fluctuates between (check this with the numerical simulation in the text).]

9.4 A coupled model of a measles epidemic

In nature, populations of different species interact with each other. For example, one species may be the food for another species or two species may be in direct competition for the same food supply. Even populations of a single species may be divided into several groups which interact with each other. An example of this is the study of infectious diseases where the population can be divided into several groups: those who have recovered and those who are susceptible to catching the disease. In this section, we illustrate this with a model which describes the spread of measles amongst children. This leads to *coupled* difference equations which have as their solution a *pair* of infinite sequences. The approach taken here to solve the difference equations is that of numerical investigation using a computer since we are unable to find a closed-form solution.

Measles epidemics

Measles is a highly contagious disease, caused by a virus and spread by effective contact between individuals. It tends to affect mainly children. Epidemics of measles have been observed in Britain and the United States roughly every two to three years. They occur more frequently in developing countries. An important problem is to understand what factors affect the timing and severity of measles epidemics.

Let us now look at the duration of the disease for a single child. A child who has not yet been exposed to measles is called a *susceptible*. Immediately after the child first catches the disease there is a latent period where the child is not contagious and does not exhibit any symptoms of the disease. This is because the virus has not yet multiplied sufficiently. The latent period lasts, on average, 5 to 7 days. After this the child enters the contagious period. The child is now called an *infective* since it is possible for another child who comes in contact with the infective to catch the disease. This period lasts approximately one week. After this time red spots appear on the skin of the child for a few days after which the child recovers. During this period, and subsequently, the child is *immune* to the disease and cannot be reinfected. Note that an individual cannot become an infective until a week after they have been infected, due to the latent period.

Formulating the model

For simplicity let us assume that both the latent period and contagious period last one week. Furthermore, we also assume that all contact between infectives and susceptibles occurs only on weekends, so that the number of infectives and susceptibles remains constant over the rest of the week.

There is then a time delay of one week from when a susceptible first catches the disease until that person becomes an infective; and a further delay of one week before the infective recovers from the disease. This suggests a model in the form of a difference equation where time is measured discretely in intervals of one week. We thus define

$$I_k = \left\{ \begin{array}{c} \text{number of} \\ \text{contagious infectives present} \\ \text{in the } k\text{th week} \end{array} \right\}$$

$$S_k = \left\{ \begin{array}{c} \text{number of} \\ \text{susceptibles present} \\ \text{in the } k\text{th week} \end{array} \right\}.$$

Next we need to write down equations which determine the number of infectives and susceptibles one week later. The following numerical example illustrates how to do this.

Example 1. *Suppose there are 10 infectives and 1000 susceptibles present in the kth week. Suppose that each infectives infects two susceptibles. How many infectives and susceptibles are there in the (k + 1)th week if we ignore births and deaths?*

Solution. *We have* $S_k = 1000$, $I_k = 10$, *and* $I_{k+1} = 20$. *So* $S_{k+1} = 1000 - 20 = 980$.
Note that none of the 10 original infectives is counted in the (k+1)th week since after one week they are no longer contagious.

To develop an equation for the number of infectives we consider the number of infectives one week later, in the $(k + 1)$th week. Now

$$\left\{ \begin{array}{c} \text{number of infectives} \\ \text{in } (k+1)\text{th week} \end{array} \right\} = \left\{ \begin{array}{c} \text{number of susceptibles} \\ \text{who caught measles} \\ \text{at beginning of } k\text{th week} \end{array} \right\}$$

which we write as

$$I_{k+1} = \left\{ \begin{array}{c} \text{number of susceptibles} \\ \text{who caught measles} \\ \text{at beginning of } k\text{th week} \end{array} \right\}. \tag{1}$$

It is generally thought that the number of new births is an important factor in measles epidemics. We also have to account for the susceptibles who get infected. So

$$
S_{k+1} = \left\{\begin{array}{c}\text{number of}\\\text{susceptibles}\\\text{in } k\text{th week}\end{array}\right\} - \left\{\begin{array}{c}\text{number of}\\\text{susceptibles who}\\\text{caught measles at}\\\text{beginning of } k\text{th week}\end{array}\right\} + \left\{\begin{array}{c}\text{number of}\\\text{births during}\\k\text{th week}\end{array}\right\}.
$$

$$(2)$$

It is further assumed that the number of births each week is a constant B. To calculate the number of susceptibles infected in a week it is *assumed* that a *single* infective infects a constant fraction f of the total number of susceptibles. Thus fS_k is the number of susceptibles infected by a single infective so, with a total I_k infectives, then

$$
\left\{\begin{array}{c}\text{number who caught measles}\\\text{at beginning of } k\text{th week}\end{array}\right\} = fS_kI_k.
$$

Hence (1) and (2) become

$$
\begin{aligned}
I_{k+1} &= fS_kI_k\\
S_{k+1} &= S_k - fS_kI_k + B
\end{aligned}
$$

$$(3)$$

where B and f are constant parameters of the model. Note that the higher the value of f the more easily the disease is spread between individuals.

Computer results

In an article, Anderson and May (1982) use this model to discuss measles epidemics in a typical city in Britain and in Nigeria. They assume initial values of infectives, and susceptibles of $I_0 = 20$ and $S_0 = 30\,000$ respectively. They also choose $f = 0.3 \times 10^{-4}$ — it can be shown that this value of f corresponds approximately to one infective infecting a single susceptible, during one week. In their article the number of new births, in one week in the typical British city, is given as $B = 120$.

Numerical iteration of (3) with these values gives the results shown in Figure 9.4.1. This shows that there is a dramatic increase in the number of infectives every 130 weeks (roughly 2–3 years). This corresponds to the epidemics observed every 2–3 years in Britain.

Now let us examine what happens when we change the parameter B to 360 (three times its previous value) corresponding to the birth rate of a developing country such as Nigeria. Here the birth rate is much higher. The results are shown in Figure 9.4.2. In this case the model

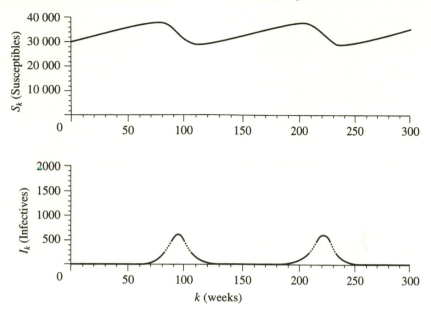

Fig. 9.4.1. Numerical simulation of a measles epidemic for a typical British city, with $B = 120$ births per week.

predicts epidemics every year and the epidemics are much more severe. This result is consistent with observation.

Steady-state solutions

Steady-state solutions are sometimes useful for understanding simple models. Finding steady-state solutions for coupled difference equations is the same, in principle, as for single difference equations. Now, however, both the quantities I_k and S_k are to assume constant values $\hat{I}$ and $\hat{S}$, say. This leads to a pair of simultaneous equations for $\hat{I}$ and $\hat{S}$. Care must be taken that all solutions are found. It is also a good idea to verify results by substitution back into the original equations.

Example 2. *Find all steady-state solutions of (3).*

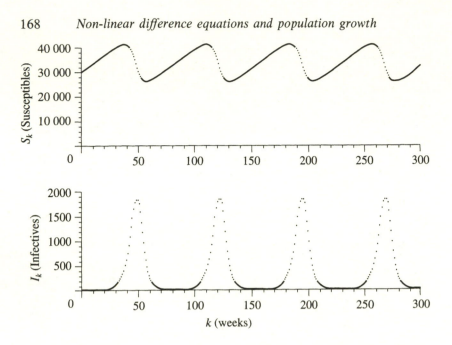

Fig. 9.4.2. Computer simulation of a measles epidemic in a typical city in Nigeria, with $B = 360$ births per week.

Solution. *Let us denote the steady-state number of infectives by $\hat{I}$ and the steady-state number of susceptibles by $\hat{S}$, where $\hat{I}$ and $\hat{S}$ are constants. Then,*

$$I_{k+1} = I_k = \hat{I} \quad and \quad S_{k+1} = S_k = \hat{S} \tag{4}$$

are the steady-state solutions. Substituting (4) into (3) we obtain

$$\hat{I} = f\hat{S}\hat{I},$$
$$\hat{S} = \hat{S} - f\hat{S}\hat{I} + B,$$

or

$$\hat{I}(1 - f\hat{S}) = 0, \tag{5a}$$
$$f\hat{S}\hat{I} - B = 0. \tag{5b}$$

Our aim is to solve for $\hat{I}$ and $\hat{S}$. From equation (5a) there are two cases to be considered: $\hat{I} = 0$ or $\hat{S} = 1/f$.

Case (a): $\hat{I} = 0$. *Substituting $\hat{I} = 0$ into (5b), to determine $\hat{S}$, yields $B = 0$. But this contradicts the fact that B is a positive constant. Thus, there is no solution of both (5a) and (5b) corresponding to $\hat{I} = 0$.*

Case (b): $\hat{S} = 1/f$. *Substituting $\hat{S} = 1/f$ into (5a), to determine $\hat{I}$, yields $\hat{I} = B$. Hence the pair, $\hat{S} = 1/f$, $\hat{I} = B$, is a solution of the coupled system (5a) and (5b), which is easily verified by substitution.*

One can also try to find solutions of (5b) first and then substitute into (5a) but this does not yield any additional solutions.

Thus there is only one steady-state solution;

$$I_k = B, \qquad S_k = 1/f,$$

which corresponds to all of the new born being infected.

Discussion

The first of the equations (3) shows that if S_k is below $1/f$ then

$$\frac{I_{k+1}}{I_k} < 1$$

so I_k decreases as k increases. Now the second of the equations (3) shows that for small I_k (after an epidemic is over), S_k increases due to births until S_k eventually becomes greater than $1/f$ and the number of infectives begins to increase again. This signifies the important part played by the birth rate, as seen in the numerical simulations. Also one can easily see the advantage in trying to keep the number of susceptibles down below $1/f$, through vaccination, to prevent epidemics.

A comprehensive discussion of this model is given in the article by Anderson and May (1982). This article is certainly accessible to students. Further models of epidemics will be discussed in Chapter 19 of this book.

Exercises 9.4

1. In some diseases infected individuals do not become immune but return to the susceptible class. Ignoring births, put together an argument which gives

$$I_{k+1} = fS_kI_k$$
$$S_{k+1} = S_k + I_k - fS_kI_k.$$

2. Find all steady-state solutions of the difference equations in Exercise 1.

3. Deduce from the equations in Exercise 1 that

(a) $S_k + I_k = M$ where M is a constant.

(b) Hence deduce that $I_{k+1} = fI_k(M - I_k)$. Find the steady-state solutions.

(c) Relate the equation in (b) to the discrete logistic equation.

4. Modify the measles model in the text given that a constant fraction γ of those recovered are reinfected. You will need to introduce an additional variable.

5. Modify the measles model in the text given that a constant fraction β of the new births are vaccinated.

6. In the model in the text we assumed that the number of new infected individuals per week was given by fI_kS_k. Another model assumes a constant probability p of contact between two individuals selected at random.

(a) Given the probability p above, what is the probability that a given susceptible does *not* have contact with a single infected, picked at random? Hence find the probability that a given susceptible does not have contact with any of the I_k infected and hence argue that the expected number of new infected individuals during week k is

$$S_k - S_k(1-p)^{I_k}.$$

(b) Put $e^{-\gamma} = 1 - p$ and show that the expected number of new infections is given by

$$I_{k+1} = S_k \left[1 - e^{-\gamma I_k}\right].$$

Hold S_k constant and sketch a graph of the number of new infections as I_k varies. Is this model better than the one used in the text when the number of infectives is large?

7. Using the result of Exercise 6 show that the modified model from Exercise 1 becomes

$$I_{k+1} = S_k \left[1 - e^{-\gamma I_k}\right]$$
$$S_{k+1} = I_k + S_k e^{-\gamma I_k}.$$

Hence show that

$$I_{k+1} = (M - I_k)\left[1 - e^{-\gamma I_k}\right]$$

where M is a constant.

8. (Computer Simulation). Modify the model in the text for a measles epidemic, using the result of Exercise 6. Use $\gamma \simeq 0.3 \times 10^{-4}$. What differences do you observe?

9. In a host-parasite system, a parasite searches for a host on which to deposit its eggs. Define

$N_k = \{\text{number of host species in } k\text{th breeding season}\}$

$P_k = \{\text{number of parasite species in } k\text{th breeding seasons}\}$

$f = \{\text{fraction of hosts not parasitized}\}$

$c = \{\text{average number of eggs laid by parasite which survive}\}$

$\lambda = \{\text{host rate, given that all adults die before their offspring can breed}\}.$

Argue that N_k and P_k satisfy

$$N_{k+1} = \lambda f N_k \qquad \text{and} \qquad P_{k+1} = c N_k[1 - f].$$

10. (Computer experiment). The Nicholson-Bailey model is a model for host-parasite systems (see previous question). It uses probability theory to argue that $f = e^{\gamma P_k}$. Numerically iterate for $\gamma = 0.068$, $c = 1$ and $\lambda = 2$. What happens?

9.5 Linearizing non-linear equations

In the previous two sections, non-linear difference equations have been studied via numerical iteration. The reason for this choice of method is that no closed-form solution is in general possible for non-linear equations. However it will be explained in this section that, if the initial value is close to a fixed point, then it is possible to obtain a linear difference equation which approximates the non-linear equation. The linear equation can be solved exactly and some properties of the solution can therefore be discussed for all values of the parameters.

Linearizing the discrete logistic equation

This technique of *linearization about a fixed point* is applicable to both single and coupled non-linear difference equations. However, the details of the method are simplest when applied to single equations and so for this reason the linearization of the discrete logistic equation of Section 9.2 will be our principal example. The following question is addressed: *if a population is described by the discrete logistic equation*

$$N_{k+1} = N_k + rN_k \left(1 - \frac{N_k}{K}\right), \tag{1}$$

for what values of the growth rate r will the population tend to the carrying capacity K if the initial population is close to K?

To answer this question the scaled variable

$$X_k = \frac{N_k}{K} \tag{2}$$

is introduced, which is the ratio of the population to the carrying capacity. Hence changes in the population which are small compared with the carrying capacity correspond to changes in X_k which are *small compared with 1*. This feature plays a crucial rôle in the linearization procedure.

In terms of the scaled variable, the discrete logistic equation becomes

$$X_{k+1} = X_k + rX_k(1 - X_k).$$

The fixed point of this equation corresponding to $N_k = K$ is $X_k = 1$. The equation can be studied close to this fixed point by introducing the small variable Y_k such that

$$\begin{aligned} X_k &= \{\text{steady-state solution}\} + Y_k \\ &= 1 + Y_k. \end{aligned} \tag{3}$$

The new variable Y_k has the advantage of being equal to 0 when X_k is at the fixed point and being small when X_k is close to the fixed point.

Substituting (3) into the difference equation and rearranging gives

$$Y_{k+1} = (1-r)Y_k - r(Y_k)^2. \tag{4}$$

Although we cannot solve this difference equation in closed form, it is easy to find a simple approximation to the RHS, valid when Y_k is small. To see this note that *if Y_k is small compared with 1 then Y_k^2 is very much smaller.* For example, if $Y_k = 10^{-3}$ then $Y_k^2 = 10^{-6}$, so that Y_k^2 is not only small, but *small in comparison with Y_k.* Hence, as a good approximation, the term $r(Y_k)^2$ in (4) can be ignored since it represents only a small correction to the RHS. We are therefore left with the linear difference equation

$$Y_{k+1} = (1-r)Y_k. \tag{5}$$

The solutions of the linear difference equation (5) are only approximate solutions of (4) but they have the advantage that they can be found in closed form.

The closed-form solution of equation (5) is

$$Y_k = (1-r)^k Y_0$$

where Y_0 can be calculated from the initial population N_0 using equations (2) and (3). Hence an approximate solution to the original equation (1) is given by

$$N_k \simeq K + K(1-r)^k Y_0. \tag{6}$$

Now for $|1-r| < 1$, the term $(1-r)^k$ tends to zero as k approaches infinity. Thus the approximate solution converges to the steady-state solution. This means that for $0 < r < 2$ the approximate solution is attracted to the steady-state, while for $r > 2$ the solution is repelled. This is in precise agreement with numerical experiments on the original non-linear difference equation, provided that the initial population N_0 is sufficiently close to the carrying capacity K.

The types of behaviour just discussed occur not only for the discrete logistic equation, but for many other difference equations as well. In general, given a non-linear difference equation, a steady-state solution s is called an *attractor*, or is said to be stable, if all the approximate solutions obtained from the linearized equation converge to the steady-state. If, on the other hand, all these solutions (apart from the steady-state solution) diverge away from the steady-state, then it is said to be a *repellor*, or is said to be unstable.

Linearization by differentiation

The above discussion illustrates the main ideas involved in the linearization of a difference equation. The method used, however, is only suitable when the function on the RHS of the difference equation is a polynomial of low degree. A more generally applicable technique uses a formula involving the derivative to approximate the RHS. This new technique will now be explained.

Consider a general first-order non-linear difference equation

$$X_{k+1} = g(X_k) \qquad (k = 0, 1, \ldots) \qquad (7)$$

where the sequence members X_k are real numbers and g is a sufficiently smooth function. Our interest is in the following problem: *obtain an approximate solution of the non-linear equation (7) if the initial value X_0 is close to a steady-state solution $X_k = s$.*

The first step is to introduce a new variable Y_k such that

$$X_k = s + Y_k. \qquad (8)$$

The new variable Y_k equals 0 when X_k is at the fixed point s and is small when X_k is close to the fixed point. We shall assume that Y_k stays small enough to ensure that

$$Y_k^2 \text{ is very small compared with } Y_k. \qquad (9)$$

By (8) the equation (7) becomes

$$Y_{k+1} + s = g(s + Y_k). \qquad (10)$$

The function $g(s + Y_k)$ can be approximated as

$$g(s + Y_k) \simeq g(s) + g'(s)Y_k. \qquad (11)$$

A theorem of calculus (the Taylor series expansion) tells us that the magnitude of the term ignored on the RHS of (11) is less than a constant times Y_k^2. By the assumption (9), this term is very small relative to Y_k and so can be ignored while Y_k stays small. The approximation (11) has the graphical interpretation of replacing the function $g(s + Y_k)$ by its tangent at the fixed point s. See Figure 9.5.1.

If we regard (11) as an exact equality (rather than as an approximate one) and then substitute into (10) we get

$$Y_{k+1} + s = g(s) + g'(s)Y_k.$$

But

$$s = g(s),$$

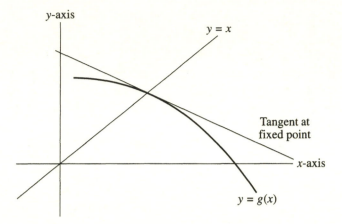

Fig. 9.5.1. Linear approximation via tangent to graph.

since s is a fixed point, and so the linear difference equation

$$Y_{k+1} = g'(s)Y_k$$

is obtained. This difference equation is not exactly the same as (11), but only approximately so. We nevertheless expect its solutions will stay close to those of (11) provided Y_k stays small.

The solution of this linear equation is

$$Y_k = (g'(s))^k Y_0$$

and hence, after substituting for X_k using (8), we see that the desired approximate solution of the non-linear equation

$$X_{k+1} = g(X_k),$$

when X_0 is close to the fixed point s, is

$$X_k = s + \left((g'(s))^k (X_0 - s)\right). \tag{12}$$

It is simple to use this approximate solution to determine under what conditions the fixed point is an attractor (see Exercise 4). The method of linearization by differentiation is summarized in Table 9.5.1.

Example 1. *Find the non-zero fixed point of the difference equation*

$$X_{n+1} = (X_n)^2$$

and determine if the fixed point is an attractor.

Table 9.5.1. *Linearization by differentiation.*

Aim: To obtain an approximate solution of the non-linear difference equation

$$X_{k+1} = g(X_k)$$

when the initial value X_0 is close to a fixed point s.

Method and solution: Introduce the small variable Y_k by defining

$$X_k = s + Y_k.$$

The solution of the linear equation

$$Y_{k+1} = g'(s)Y_k,$$

when added to s, then approximates the solution of the original non-linear equation.

Key approximation: The method relies on Y_k being small enough to ensure that

$$Y_k{}^2 \text{ is very small compared with } |Y_k|.$$

Properties of the solution: If $|g'(s)| < 1$ then the fixed point is an attractor and the solution converges to s, while if $|g'(s)| > 1$ then the fixed point is a repellor and the solution does not converge to s.

Solution. *The difference equation is of the form*

$$X_{n+1} = g(X_n)$$

with

$$g(x) = x^2.$$

The fixed points s are given by the solutions to the equation

$$s = g(s) = s^2$$

and so are $s = 0$ and $s = 1$. Thus the non-zero fixed point is $s = 1$.
  Now

$$|g'(1)| = 2$$

and thus

$$|g'(s)| > 1.$$

Hence the fixed point is a repellor rather than an attractor.

A thorough discussion of the mathematics involved in linearizing a difference equation is given in Devaney (1986).

Exercises 9.5

1. The approximation used in the linearization procedure (equation (11) in the text) is

$$g(s + y) \simeq g(s) + g'(s)y$$

where y is assumed small. Use your calculator to test the accuracy of this approximation for

(a) $g(x) = x^2$, $s = 1$, $y = 0.1$,

(b) $g(x) = \sin(x)$, $s = 0$, $y = 0.1$.

by comparing right and left hand sides of the approximation. Repeat (a) and (b) for $y = 0.05$. Does the accuracy improve?

2. Graphically, the approximation

$$g(s + y) \simeq g(s) + g'(s)y$$

corresponds to replacing the function $y \to g(s + y)$ by its tangent at the point s. Tabulate, as a function of y,

$$g(1 + y) = (1 + y)^2$$

for y in the interval $[-0.1, 0.1]$ at spacings of 0.02 and plot the points on graph paper. Now plot, as a function of y,

$$g(s) + g'(s)y = 1 + 2y$$

on the axes. Are the two graphs similar?

3. Use differentiation to linearize the discrete logistic equation, in scaled form,

$$X_{k+1} = X_k + rX_k(1 - X_k)$$

about the fixed point $s = 1$ and thus re-derive equation (5) in the text.

4. Use the approximate solution (12) in the text to deduce the 'properties of the solution' listed in Table 9.5.1: if $|g'(s)| < 1$ the fixed point is an attractor and the sequence converges to s, while if $|g'(s)| > 1$ then the fixed point is a repellor and the sequence will not converge to s.

5.   (a) Show that $N_k = K$ is a fixed point of the population model described by the difference equation

$$N_{k+1} = N_k e^{a(1 - N_k/K)}.$$

Introduce the scaled variable

$$X_k = N_k/K$$

to rewrite this difference equation.

(i) Use the method of linearization by differentiation to obtain the linearized form of the scaled equation and thus determine an approximate solution to the original equation.

(ii) An initial population of 20 000 is well described by the above difference equation with $a = 1.5$ and $K = 19\,800$. Calculate the subsequent evolution of the population in the next three time periods using the approximate solution of (b). Does the population appear to be stable? (In other words, is the fixed point an attractor?)

(iii) Show that the fixed point is an attractor if $0 < a < 2$.

10

Models for population genetics

In the previous chapter, difference equations were used to predict the change in the total population of a species from generation to generation. This leads naturally to the question of predicting the change in a particular characteristic of the individuals in the population. Since the heredity units which determine characteristics are the genes, this question comes under the heading of population genetics.

The formulation of models in population genetics requires a knowledge of the fundamentals of the theory of genetics. This chapter begins by presenting the required theory, before formulating particular models in terms of difference equations. These models differ from those obtained in the previous two chapters in that they are based on some firmly established laws. In this respect they are similar to the models obtained in mechanics. The laws of genetics, however, are expressed as probabilities and hence are only relevant, in practice, to populations which are sufficiently large.

The models in population genetics give rise to non-linear difference equations. This chapter assumes the elements of probability theory.

10.1 Some background genetics

An ability to predict the most probable characteristics of offspring from knowledge of the characteristics of the parents is a skill of prime importance to plant and animal breeders as well as to medical scientists. In many cases this problem can be reduced to the study of difference equations — but only after the fundamentals of the theory of genetics have been learned. This section presents the necessary background theory.

As in many of the physical sciences, there are some basic principles (or laws) by which the results of a large number of experiments can be explained. In genetics, the laws are due to the monk, amateur scientist and plant breeder Gregor Mendel, who performed a series of famous experiments in the mid 1800s.

Alleles and genotypes

For our purposes, the main result from Mendel's experiments is that some characteristics of plants and animals are determined by just two genes. Fur colour and fur length in hamsters, eye colour and wing length in flies, and the colour of flowers in many plants are a few examples of characteristics controlled by pairs of genes. Two genes responsible for the same characteristic, say eye colour, are called *alleles* of each other, and the two genes are said to form a *pair of alleles*.

Different pairs of alleles, with different locations along the chromosome, are responsible for different characteristics. Once a particular characteristic has been selected, say the colour of the flower in a pea plant, it is convenient to denote the pair of alleles by, say, the letters

$$A \quad \text{and} \quad \alpha.$$

In any given individual in the species, the alleles can occur in just one of the combinations

$$AA, \qquad A\alpha \quad (which \ is \ equivalent \ to \quad \alpha A) \qquad \text{or} \qquad \alpha\alpha$$

which are called *genotypes*. Each individual can therefore be classified as being of a particular genotype. A given genotype determines a physical characteristic of the individual.

For example, in pea plants there is a pair of alleles responsible for flower colour consisting of A which causes white flowers and α which causes pink flowers. As to the genotypes (each of which contains two of these genes), the genotypes AA will, of course, have white flowers and the genotypes $\alpha\alpha$ pink flowers. It turns out that the genotypes $A\alpha$ have white flowers. For this reason, the A-allele, which causes the white flowers, is said to *dominate* the α-allele, which causes the pink flowers. We also say the gene corresponding to α is *recessive*. In some cases, however, the $A\alpha$ genotype is different from both the AA genotype and the $\alpha\alpha$ genotype so the concept of dominant and recessive genes has no meaning. It is also noted that many physical characteristics are due to the expression of several different genes. Here, however, we will assume that only one gene is responsible, for the sake of simplicity.

Fig. 10.1.1. Genotypes of parents.

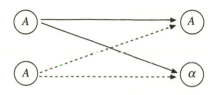

Fig. 10.1.2. Alleles from first and second parents.

Reproduction

Our aim is to determine how the proportions of the three genotypes in a population vary with the time. To achieve this aim, it is necessary to know how the genes are transmitted during reproduction. The only type of reproduction we shall consider is the most common one in which the male and female both contribute one *gamete* (an egg or a sperm) to the offspring. The gamete contains only one of the alleles A or α, the allele being chosen in accordance with the following law.

Mendel's first law. *The allele in the gamete is chosen at random from the two alleles in the genotype of the parent.*

This law is sometimes called the *law of segregation*. The following example illustrates the use of this law in predicting the genotypes of the offspring.

Example 1. *If an AA genotype mates with an Aα genotype, what are the possible genotypes of the offspring and what are the probabilities of occurrence of each genotype?*

Solution. *The genotypes of the parents are shown in Figure 10.1.1.*

According to Mendel's first law, all the gametes of the first parent will contain the allele A, while on average half the gametes of the second parent will contain the allele A and the other half of the allele α. These alleles are shown in Figure 10.1.2.

The possible combinations of these alleles, one from each parent, are

$$AA, \quad A\alpha, \quad AA, \quad \alpha A.$$

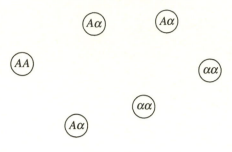

Fig. 10.1.3. Genotypes of a hypothetical population of six individuals.

Thus the possible genotypes of the offspring are AA and Aα, noting that αA is the same as Aα. Thus both AA and Aα occur with the same probability 1/2, in this example.

The above example provides a simple illustration of how Mendel's law can be used to predict the genotypes of the offspring. To apply this law to more complicated examples, it is first desirable to introduce some notation and formulae for the proportions of the different genotypes and alleles in a population. This will provide an important step towards our ultimate goal of modelling how these proportions change as the population increases.

Proportions of genotypes and alleles

As an illustration of how the proportions of genotypes are calculated, consider the population of six individuals shown in Figure 10.1.3.

There is one individual of genotype AA, three of genotype $A\alpha$ and two of genotype $\alpha\alpha$. Thus 1/6 of the population has genotype AA, 1/2 has genotype $A\alpha$ and 1/3 has genotype $\alpha\alpha$.

The general procedure for calculating the proportion of a population with a given genotype is to divide the number of individuals with that genotype by the total number of individuals in the population. Thus

$$G(AA) = \frac{\mathcal{N}(AA)}{N} \tag{1a}$$

where $G(AA)$ is the *proportion* with genotype AA, $\mathcal{N}(AA)$ is the *number* with genotype AA, and N is the *total number* in the population. Similar

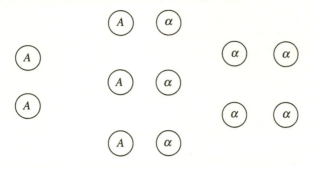

Fig. 10.1.4. Gene pool for the population in Figure 10.1.3.

formulae hold for the proportions with genotypes $A\alpha$ and $\alpha\alpha$:

$$G(A\alpha) = \frac{\mathcal{N}(A\alpha)}{N} \qquad (1b)$$

$$G(\alpha\alpha) = \frac{\mathcal{N}(\alpha\alpha)}{N} \qquad (1c)$$

Note also that

$$N = \mathcal{N}(AA) + \mathcal{N}(A\alpha) + \mathcal{N}(\alpha\alpha).$$

As an illustration, we apply these formulae to the population shown above in Figure 10.1.3 for which

$$\mathcal{N}(AA) = 1, \quad \mathcal{N}(A\alpha) = 3 \quad \text{and} \quad \mathcal{N}(\alpha\alpha) = 2,$$

while $N = 6$. Thus the formulae (1) give

$$G(AA) = 1/6, \quad G(A\alpha) = 1/2 \quad \text{and} \quad G(\alpha\alpha) = 1/3,$$

in agreement with the results noted earlier.

Two quantities which will be needed to predict the genotypes of the offspring are the proportions of the alleles A and α in the population. To clarify the meaning of these quantities it is convenient to introduce the concept of a *gene pool* to which each member of the population contributes the two alleles in its genotype. Thus an individual of genotype AA contributes two A-alleles, and so on.

When applied to the population in Figure 10.1.3, this procedure gives the gene pool shown in Figure 10.1.4. This gene pool contains five A-alleles and seven α-alleles. Hence we say that the proportions of the alleles A and α in the population are 5/12 and 7/12 respectively.

For an arbitrary population let

$$P(A) = \left\{ \begin{array}{c} \text{proportion of } A\text{-alleles} \\ \text{in the gene pool} \end{array} \right\}$$

$$P(\alpha) = \left\{ \begin{array}{c} \text{proportion of } \alpha\text{-alleles} \\ \text{in the gene pool} \end{array} \right\}. \tag{2}$$

To express these proportions in terms of the numbers (1) of each geno-type, note that each individual contributes *two* alleles to the gene pool. Hence the number of A-alleles in the gene pool is $2\mathcal{N}(AA) + \mathcal{N}(A\alpha)$ while the total number of genes is $2N$. This gives

$$P(A) = \frac{2\mathcal{N}(AA) + \mathcal{N}(A\alpha)}{2N}. \tag{3a}$$

Similarly

$$P(\alpha) = \frac{2\mathcal{N}(\alpha\alpha) + \mathcal{N}(A\alpha)}{2N}. \tag{3b}$$

As an illustration, we apply the formulae (3) to the population in Figure 10.1.3, for which $\mathcal{N}(AA) = 1$, $\mathcal{N}(A\alpha) = 3$ and $\mathcal{N}(\alpha\alpha) = 2$. This gives

$$P(A) = \frac{2 \times 1 + 3}{2 \times 6} = \frac{5}{12}$$

and

$$P(\alpha) = \frac{2 \times 2 + 3}{2 \times 6} = \frac{7}{12}$$

in agreement with the answers stated earlier.

Proportions as probabilities

The proportions of genotypes and alleles in the population can be re-garded as the probabilities of certain events. For example, since $G(AA)$ is the proportion of AA genotypes in the population, we may also write

$$G(AA) = \left\{ \begin{array}{c} \text{probability that an individual selected} \\ \text{at random has genotype AA} \end{array} \right\}. \tag{4}$$

Again, since $P(A)$ is the proportion of A-alleles in the gene pool, it is also the probability that an allele selected at random from the gene pool happens to be an A-allele. By Mendel's first law this implies that

$$P(A) = \left\{ \begin{array}{c} \text{probability that a gamete from a randomly} \\ \text{selected individual in the population} \\ \text{contains an } A\text{-allele} \end{array} \right\}. \tag{5}$$

Interpretations similar to (4) or (5) also apply to $G(A\alpha)$, $G(\alpha\alpha)$ and $P(\alpha)$.

During reproduction a gamete is needed from each of the male and female parents. Provided that the gene A is not *sex linked* (contained on the X or Y chromosome), the probability on the RHS of (5) will be the same when calculated for male and female individuals separately. Hence $P(A)$ gives this probability for each parent, as we assume from now on.

The formulae given above can be used to predict the probability of a particular genotype occurring in the offspring in more complicated examples than Example 1.

Example 2. *Suppose that 40% of a population are of genotype AA, 40% are of genotype Aα, and 20% of genotype αα. Find the probability that a gamete from a randomly selected individual in the population contains an A-allele. Find also the probability that it contains an α-allele.*

Solution. *Let N be the total number of individuals in the population, so that*

$$\mathcal{N}(AA) = 0.4N, \quad \mathcal{N}(A\alpha) = 0.4N, \quad \mathcal{N}(\alpha\alpha) = 0.2N.$$

From (3) the proportions of alleles are given by

$$P(A) = \frac{2 \times 0.4N + 0.4N}{2N} = 0.6,$$

and

$$P(\alpha) = \frac{2 \times 0.2N + 0.4N}{2N} = 0.4.$$

Note that $P(A) + P(\alpha) = 1.$

Example 3. *For the problem in Example 2 state the possible genotypes of the offspring and find the probability of each genotype occurring, given that the two parents are chosen at random.*

Solution. *The possible genotypes for the offspring are AA, Aα and αα. For offspring of genotype AA, each parent must contribute a gamete with the allele-A. Whether this occurs for the male parent is independent of whether it occurs for the female, the probability of occurrence being P(A) in either case. Hence the probability of both occurring — and giving offspring of genotype AA — is the product*

$$P(A)P(A) = [P(\alpha)]^2 = 0.36.$$

Similarly, for offspring of genotype αα, the probability is

$$P(\alpha)P(\alpha) = [P(\alpha)]^2 = 0.16.$$

Finally, offspring of genotype Aα may be obtained by either the male parent contributing the A-allele and the female the α-allele, or vice versa. Hence the probability of offspring having genotype Aα is

$$P(A)P(\alpha) + P(\alpha)P(A) = 2P(A)P(\alpha) = 0.48.$$

Thus we have obtained the probabilities of all the possible genotypes.

Once the offspring are born, the genotype composition of the population alters and this must be taken into account in predicting the genotypes of those born subsequently. Various assumptions which enable us to do this will be considered in the following sections.

Finally, recall that the sum of probabilities of mutually exclusive events, exhausting all possible cases, is 1. This leads to two useful identities:

$$G(AA) + G(A\alpha) + G(\alpha\alpha) = 1, \tag{6}$$

$$P(A) + P(\alpha) = 1. \tag{7}$$

You should also check that these identities hold in the previous examples.

This section has covered all the background theory from genetics needed for the task of predicting the changes in genotype proportions. You may well be interested, however, in learning more about genetics. If so, you should consult one of the more specialized books on genetics, such as Hexter and Yost (1976) or Hartl (1980).

Exercises 10.1

1.   Match each of the symbols on the left, which were introduced in the text, with the appropriate description on the right:

(a) $G(AA)$ (a′) *total population*

(b) $P(A)$ (b′) *number with genotype AA*

(c) $\mathcal{N}(AA)$ (c′) *proportion of A-alleles*

(d) N (d′) *proportion with genotype AA*

Interpretation as probabilities were given in the text for $G(AA)$ and $P(\alpha)$. Write down similar interpretations as probabilities for $G(A\alpha)$, $G(\alpha\alpha)$ and $P(\alpha)$.

2.   Verify the identities stated in the text,

$$G(AA) + G(A\alpha) + G(\alpha\alpha) = 1 \quad \text{and} \quad P(A) + P(\alpha) = 1,$$

by using the definitions (1) and (3) in the text.

3.   A population consists of seven individuals: two of genotype AA, two of genotype $A\alpha$ and three of genotype $\alpha\alpha$.

(a) Draw a diagram like that in Figure 10.1.3 showing the genotypes and then write down the proportions of each genotype in the population.

(b) Draw a diagram like that in Figure 10.1.4 showing the gene pool for the population and then write down the proportions of each of the alleles A and α in the gene pool.

4.   A population consists entirely of genotypes $A\alpha$.

(a) What answers would you expect for the allele proportions $P(A)$ and $P(\alpha)$?

(b) Verify your answer to part (a) by using the formulae (3) in the text.

5. A population consists of 20% genotypes AA, 40% genotypes $A\alpha$, and 40% genotypes $\alpha\alpha$. Calculate the allele proportions $P(A)$ and $P(\alpha)$.

6. In humans there is a blood group known as the 'MN group'. It is composed of individuals with the genotypes MM, MN or NN, each genotype consisting of a combination of the genes M and N.

    (a) What are the possible genotypes of the children of an MM and an NN parent? What is the probability of occurrence of each genotype?

    (b) As in (a), but now suppose both parents are of genotype MN.

7. Suppose that, in a survey of people from the MN blood group, it was found that 1/5 were of genotype MM, 1/5 were of genotype MN, and 3/5 of genotype NN.

    (a) Calculate the proportions of the genes M and N respectively, in the gene pool.

    (b) Give the probabilities for each of the possible genotypes occurring in children of those surveyed, assuming the mating is random.

8. The colouring of fur in rabbits is determined by a single pair of genes, to be denoted by A and α. The allele A is responsible for the pigmentation of the fur, while the allele α is responsible for white fur. The allele A is dominant to the allele α.

    (a) What genotypes have coloured fur?

    (b) Suppose that, in a population of rabbits, 25% of genotypes are AA, 50% of genotypes are $A\alpha$, and 25% are $\alpha\alpha$. Calculate $P(A)$ and $P(\alpha)$.

    (c) Deduce the probability that a particular offspring is a white rabbit, assuming the mating is random.

9. Show from appropriate formulae in the text that

$$P(A) = G(AA) + \frac{1}{2}G(A\alpha),$$

$$P(\alpha) = \frac{1}{2}G(A\alpha) + G(\alpha\alpha).$$

10.2 Random mating with equal survival

This section begins by discussing some assumptions from which the likely changes in the genotype proportions can be predicted over a period of time. All the assumptions of the previous section will be retained. The main objective is to analyse how the proportion of a recessive gene in the gene pool changes over time. To make some progress we must make some simplifying assumptions about the way in which genes may be combined and also on how many of the various genotypes survive to produce offspring.

Basic model I

In this section we will assume *random mating* between individuals:

> (a) *mating occurs at random: the choice of mate does not depend on the mate's genotype.*

This particular assumption is mentioned explicitly so as to emphasize its rôle in this section. It is applicable whenever the different genotypes within the population are either unaware of or indifferent to the genotypes of their partners. For example, mating in humans is typically random with respect to the genes responsible for blood group type, whereas it may not be completely random with respect to the genes controlling height. In the latter case there may be a bias towards tall people's marrying tall people — the choice of partner being thus influenced by the partner's genotype. This modification of assumption (a) is dealt with in Section 10.4.

Besides the random mating assumption, further modelling of the population is necessary before it is possible to predict the changes in genotype proportions. These models rely on the idea of a *generation*, which will now be explained.

In certain populations, breeding occurs during short well-defined periods, equally spaced. Crops sown by farmers, at the same time every year, fall into this pattern as do many species of migratory animals. The rôle of this assumption here, as in Chapter 9, is to yield difference equations (rather than differential equations). Thus the population divides into discrete generations. For each integer $k \geq 0$, the kth generation consists of all the individuals born during the kth breeding period.

To simplify our models, moreover, it will be assumed that the generations do not overlap and that different generations do not interbreed. Each generation begins with the fertilized egg (or zygote) birth and ends with reproduction to form the next generation, as illustrated in Figure 10.2.1. This assumption is certainly applicable for many animal and plant species which only breed at a specific time and where all the adults die after breeding. It is not, strictly speaking, correct for human populations. It may be a useful approximation, however, if we take a generation to be 25 years and only bother to measure the population every generation.

The expected proportions of the different *genotypes* in the population at the *end* of the kth generation will be denoted respectively by

$$G_k(AA), \quad G_k(A\alpha) \quad \text{and} \quad G_k(\alpha\alpha).$$

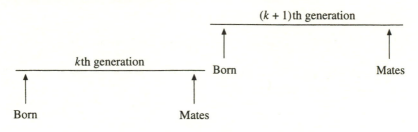

Fig. 10.2.1. Generations.

Similar notation will be used for the values of the other quantities associated with the population, at that time. The first of our models will now be described.

The other major assumptions, to be used in this section, are *equal survival* and *equal fertility*. These assumptions are as follows.

(b) *Equal survival: each genotype has the same chance of surviving from the fertilized egg to the end of the generation, where mating occurs.*

(c) *Equal fertility: each couple produces on average, the same number of viable sperm and eggs, regardless of the genotypes in the couple.*

The equal survival assumption means that we can assume that a constant fraction, r say, of the number of *each* genotype of offspring survives to the end of the generation. Later, in Section 10.3, we shall see how to modify assumption (b) for when different genotypes have different chances of surviving. We shall refer to the model encompassing (a) random mating, (b) equal survival, and (c) equal fertility, as *Model I* for population genetics.

Offspring genotype proportions from random mating

Our overall objective is to obtain a difference equation for one of the allele proportions. First we need to determine how to get from the end of one generation to the beginning of the next generation, under the assumption of (a) random mating.

We introduce the notation $P_k(A)$ and $P_k(\alpha)$ for the two allele proportions at the *end* of the kth generation, and $G_k(AA)$, $G_k(A\alpha)$ and $G_k(\alpha\alpha)$ for the three genotype proportions, at the *end* of the kth generation.

Let us use an asterisk to denote quantities at the *beginning* of the generation. Thus

$$G^*_{k+1}(AA), \quad G^*_{k+1}(A\alpha) \quad \text{and} \quad G^*_{k+1}(\alpha\alpha)$$

denote the various genotype proportions of the fertilized eggs, which give rise to the $(k+1)$th generation.

Example 1. *Find an expression for $G^*_{k+1}(AA)$ in terms of the the allele proportions from the kth generation.*

Solution. *Recall that, by definition,*

$$G^*_{k+1}(AA) = \left\{ \begin{array}{l} \text{probability that an individual at the beginning} \\ \text{of the } (k+1)\text{th generation has genotype } AA \end{array} \right\}$$

Hence, by assumption (c), equal fertility,

$$G^*_{k+1}(AA) = \left\{ \begin{array}{l} \text{probability that an offspring from a given} \\ \text{couple in the kth generation has genotype } AA \end{array} \right\}.$$

This, in turn, is equal to the product of the separate probabilities that each parent contributes an A-allele. Each of these latter probabilities is equal to the proportion of A-alleles in the gene pool at the time of mating (by assumption (a)) and hence to $P_k(A)$. Thus

$$G^*_{k+1}(AA) = P_k(A)P_k(A) = [P_k(A)]^2.$$

Similarly, results are easily obtained for $G^*_{k+1}(A\alpha)$ and $G^*_{k+1}(\alpha\alpha)$. These are summarized as

$$\begin{aligned} G^*_{k+1}(AA) &= [P_k(A)]^2, \\ G^*_{k+1}(A\alpha) &= 2P_k(A)P_k(\alpha), \\ G^*_{k+1}(\alpha\alpha) &= [P_k(\alpha)]^2) \end{aligned} \tag{1}$$

These equations are always valid for random mating.

The British geneticist R.C. Punnet deduced the formulae (1) from a diagram which is now called the *Punnet square*. In the Punnet square, shown in Figure 10.2.2, the alleles present in the gametes of the male parents and their expected proportions are arranged along the top of the square, and those of the female along the side. The genotypes of the offspring and their expected proportions are contained within the square.

Thus the genotype proportions of the *offspring*, just after birth, can be read from the Punnet square as $[P_k(A)]^2$ for the AA genotype, $[P_k(\alpha)]^2$ for the $\alpha\alpha$ genotype and $2P_k(A)P_k(\alpha)$ for the $A\alpha$ genotype. Note that the cross terms in the Punnet square add to give the $A\alpha$ genotype since $A\alpha$ is the same as αA.

♂ / ♀	$A \qquad P_k(A)$	$\alpha \qquad P_k(\alpha)$
A $P_k(A)$	AA $G^*_{k+1}(AA) = (P_k(A))^2$	$A\alpha$ $G^*_{k+1}(A\alpha) = P_k(A)\,P_k(\alpha)$
α $P_k(\alpha)$	αA $G^*_{k+1}(\alpha A) = P_k(\alpha)\,P_k(A)$	$\alpha\alpha$ $G^*_{k+1}(\alpha\alpha) = (P_k(\alpha))^2$

Fig. 10.2.2. The Punnet square.

Obtaining difference equations

We now have enough relations between the various quantities to enable us to derive a difference equation, giving say $P_{k+1}(\alpha)$ in terms of $P_k(\alpha)$. Solving this difference equation will then tell us how the proportion of the recessive gene α changes over time.

The procedure for deriving a difference equation for allele proportions is essentially a two-step process. This is described below.

STEP 1: *Using (1) and the fraction of each genotype surviving, determine the numbers of each genotype at the end of the $(k+1)$th generation.*

STEP 2: *Now count up one of the allele proportions and eliminate the other using $P_k(A) + P_k(\alpha) = 1$.*

The following example shows how to use these steps.

Example 2. *Find a difference equation for the recessive gene α given that a constant fraction r of each genotype of fertilized egg survives to the end of each generation.*

Solution. *We now carry out the two steps in detail for $P_{k+1}(A)$ in terms of $P_k(A)$.*

STEP 1: *Define N^*_{k+1} as the total number of fertilized eggs giving rise to the $(k+1)$th generation and let $\mathcal{N}^*_{k+1}(AA)$, $\mathcal{N}^*_{k+1}(A\alpha)$ and $\mathcal{N}^*_{k+1}(\alpha\alpha)$ denote the number of each genotype of the fertilized eggs. Now $\mathcal{N}^*_{k+1}(AA) = G^*_{k+1}(AA)N^*_{k+1}$ with* *similar results for the other genotypes. Thus, from (1), valid for random mating,*

$$\mathscr{N}_{k+1}^{\bullet}(AA) = [P_k(A)]^2 N_{k+1}^{\bullet},$$
$$\mathscr{N}_{k+1}^{\bullet}(A\alpha) = 2P_k(A)P_k(\alpha)N_{k+1}^{\bullet}, \qquad (2)$$
$$\mathscr{N}_{k+1}^{\bullet}(\alpha\alpha) = [P_k(\alpha)]^2 N_{k+1}^{\bullet}.$$

From assumption (b), equal survival, a fraction r of each genotype of fertilized egg survives to the end of the $(k+1)$th generation, giving $\mathscr{N}_{k+1}(AA)$, $\mathscr{N}_{k+1}(A\alpha)$ and $\mathscr{N}_{k+1}(\alpha\alpha)$. Thus

$$\mathscr{N}_{k+1}(AA) = r\mathscr{N}_{k+1}^{\bullet}(AA) = r[P_k(A)]^2 N_{k+1}^{\bullet},$$
$$\mathscr{N}_{k+1}(A\alpha) = r\mathscr{N}_{k+1}^{\bullet}(A\alpha) = 2rP_k(A)P_k(\alpha)N_{k+1}^{\bullet}, \qquad (3)$$
$$\mathscr{N}_{k+1}(\alpha\alpha) = r\mathscr{N}_{k+1}^{\bullet}(\alpha\alpha) = r[P_k(\alpha)]^2 N_{k+1}^{\bullet}.$$

STEP 2: *The right-hand side of (2) contains the allele proportions $P_k(A)$ and $P_k(\alpha)$ (measured at the end of the generation). To obtain a difference equation we need to obtain the allele proportions at the end of the $(k+1)$th generation. Now*

$$P_{k+1}(\alpha) = \frac{\{\text{number of } \alpha\text{-alleles}\}}{\{\text{total number of alleles in gene pool}\}}$$

$$= \frac{2\mathscr{N}_{k+1}(\alpha\alpha) + \mathscr{N}_{k+1}(A\alpha)}{2\mathscr{N}_{k+1}(AA) + 2\mathscr{N}_{k+1}(A\alpha) + 2\mathscr{N}_{k+1}(\alpha\alpha)}$$

Substituting from (3), we thus obtain,

$$P_{k+1}(\alpha) = \frac{[P_k(\alpha)]^2 + P_k(A)P_k(\alpha)}{[P_k(A)]^2 + 2P_k(A)P_k(\alpha) + [P_k(\alpha)]^2}.$$

To eliminate $P_k(A)$ we use the identity

$$P_k(A) + P_k(\alpha) = 1$$

obtaining, after some straightforward algebra, the rather simple difference equation

$$P_{k+1}(\alpha) = P_k(\alpha). \qquad (4)$$

Thus we have obtained a difference equation for $P_k(\alpha)$ as required.

This simple difference equation is a remarkable result. It states that the α-allele proportion in the next generation is equal to the α-allele proportion in the current generation and hence that the α-allele proportion remains the same throughout all generations.

On reflection this result makes perfect sense in the case where each couple has just two offspring, all of whom survive to the end of the generation, since the gene pool always contains exactly the same genes. In the more general case it is the relative numbers of genes which remain the same since there is no mechanism built into the assumptions to change this. Note that the result is independent of the survival fraction r.

Having established that under random mating and equal survival the allele proportions do not change it is natural to ask whether the genotype proportions change over time. The following example shows how to determine this.

Example 3. *Suppose that an initial generation of flowering plants consists of 50% white flowers (all of genotype AA) and 50% yellow flowers (all of genotype αα). Calculate the expected genotype proportions in each subsequent generation, assuming that Model I is valid for the flowering plants.*

Solution. *First we obtain the initial conditions for the difference equation. Let the number of plants in the initial generation (for which k = 0) be N_0. The numbers of the various genotypes in this generation are then given by*

$$\mathcal{N}_0(AA) = \frac{1}{2}N_0, \quad \mathcal{N}_0(A\alpha) = 0, \quad \mathcal{N}_0(\alpha\alpha) = \frac{1}{2}N_0.$$

Hence, by (3a) of Section 10.1,

$$P_0(\alpha) = \frac{1}{2}. \tag{5}$$

The solutions of the difference equation (4) are all constant functions of k. From the initial condition (5) it therefore follows that

$$P_k(\alpha) = P_0(\alpha) = \frac{1}{2} \quad (k = 0, 1, 2, \ldots).$$

Since $P_k(A) + P_k(\alpha) = 1$, it follows that also

$$P_k(A) = \frac{1}{2} \quad (k = 0, 1, 2, \ldots).$$

Equations (1) now give the expected genotype proportions as constant function of k:

$$G^{\bullet}_{k+1}(AA) = \frac{1}{4}, \quad G^{\bullet}_{k+1}(A\alpha) = \frac{1}{2}, \quad G^{\bullet}_{k+1}(\alpha\alpha) = \frac{1}{4} \quad (k = 0, 1, 2, \ldots).$$

From assumption (b), equal survival, the genotype proportions cannot change during a generation, so it follows that

$$G_{k+1}(AA) = \frac{1}{4}, \quad G_{k+1}(A\alpha) = \frac{1}{2}, \quad G_{k+1}(\alpha\alpha) = \frac{1}{4} \quad (k = 0, 1, 2, \ldots)$$

Note that $G_{k+1}(AA) + G_{k+1}(A\alpha) + G_{k+1}(\alpha\alpha) = 1$.

In the above example, the expected genotype proportions remain the same from the first generation onwards. This can be proved to occur always with the assumptions of Model I and is known as the *Hardy-Weinberg law*.

There is an interesting historical anecdote associated with the discovery of this law. At a meeting in the early 1900s to discuss developments in genetics, R.C. Punnet (a famous geneticist who invented the Punnet square) was presenting a talk on Mendel's theory of heredity. During question time Punnet was asked to explain why, although the allele responsible for brown eyes is dominant to the allele responsible for blue eyes, there are still so many blue-eyed people in the population. Punnet was not able to provide his audience with a convincing answer to this question, but he did realize it had a mathematical answer. Each weekend Punnet played in the same Cambridge University cricket team as the mathematician G.H. Hardy, and so he related the problem during a game. Hardy subsequently solved the problem and the result now bears his name together with that of the German geneticist W. Weinberg, who independently solved the problem at about the same time.

Exercises 10.2

1. In each of the following cases state whether the assumption is consistent with the assumptions in the equal survival model (Model I). If the answer is '*no*', state which assumption of Model I would be violated.

(a) *Between birth and parenthood 10% of each of the genotypes AA, Aα, αα die.*

(b) *Between birth and parenthood 5% of the genotype AA die, 10% of Aα and 15% of αα die.*

(c) *The genotypes AA mate only with the genotypes AA or Aα.*

(d) *The genotypes AA have, on average, twice as many offspring as the genotypes αα.*

2. Suppose that a population satisfies the assumptions of Model I. Derive a difference equation for the proportion of *A*-alleles.

3. Let g_0 be a number between 0 and 1 and suppose that in an initial generation of some population

$$G_0(AA) = (g_0)^2, \quad G_0(A\alpha) = 2g_0(1 - g_0), \quad G_0(\alpha\alpha) = (1 - g_0)^2.$$

(a) Calculate $P_0(A)$ and $P_0(\alpha)$.

(b) Assuming Model I is applicable, write down the allele proportions $P_k(A)$ and $P_k(\alpha)$ for each $k \geq 0$.

(c) Use your answer to part (b), and the Punnet square, to calculate

$$G_k(AA), \quad G_k(A\alpha), \quad G_k(\alpha\alpha) \quad (k = 1, 2, 3, \ldots).$$

(d) The genotype proportions are said to have reached equilibrium when there is no further change in their values with the passage of time. In which generation do they reach equilibrium in the current exercise?

4. Suppose, as seems likely, that the simple random mating model is applicable to members of the MN blood group in humans. Suppose, furthermore, that the genotype proportions have already reached equilibrium values, as in Exercise 3. Given that 36% of the members of this blood group are of genotype MM, calculate the expected percentages of genotypes MN and NN.

10.3 Lethal recessives, selection and mutation

In this section we look at two possible ways for the allele proportions to change over time. The first is where the equal survival assumption is violated. We will consider a lethal combination of genes, where each individual with that combination dies before reaching maturity, and we will also consider where different genotypes have different relative chances of surviving to pass on their genes, an important situation from the point of view of evolution. Finally we look at a second way in which the allele proportions can change. Here the changes can occur because the alleles can mutate from one form to another.

A lethal recessive gene — model II

An obvious example where the equal survival assumption (b), from Section 10.2, is not valid is where the recessive gene α is such that all of the genotype $\alpha\alpha$ die before they reach maturity. Such a gene is called a *lethal recessive*. An example of this is the genetic disorder, cystic fibrosis. Equivalently, a lethal recessive gene could result in all of the embryos being aborted.

Another situation, equivalent to the existence of a lethal recessive gene, is where animal breeders deliberately slaughter animals where the recessives manifest a certain undesirable characteristic, so as to improve their breed. This practice is known as culling.

Because the $A\alpha$ genotype still carries the lethal gene it will not be immediately lost from the gene pool. Thus we are interested in determining how quickly the proportion of the lethal recessives $P_k(\alpha)$ diminishes.

Model II for population genetics consists of two of the assumptions of the equal survival model (Model I), random mating (a), and also equal fertility (c), but with equal survival (b) replaced by (b″) below.

(b″) *A fraction r of the AA and Aα newborn genotypes survive to the end of the generation but* none *of the αα genotypes survive.*

Modelling this situation only requires a small modification to Step 1 of the procedure used in Section 10.2. The following example explains how to do this.

Example 1. *Find a difference equation for $P_k(\alpha)$ if $\alpha\alpha$ is a lethal recessive and a fraction r of the newborn genotypes AA and Aα survive to the end of the generation.*

Solution. *We adopt the same notation as in Example 1 of Section 10.2.*

STEP 1: *Assuming random mating we can use equations (1) from Section 10.2.*
Thus

$$\mathcal{N}^*_{k+1}(AA) = [P_k(A)]^2 N^*_{k+1},$$
$$\mathcal{N}^*_{k+1}(A\alpha) = 2P_k(A)P_k(\alpha)N^*_{k+1}, \qquad (1)$$
$$\mathcal{N}^*_{k+1}(\alpha\alpha) = [P_k(\alpha)]^2 N^*_{k+1}.$$

Given a fraction r of newborn genotypes AA and Aα survive to the end of the (k + 1)th generation, and none of the $\alpha\alpha$ genotype survives, then

$$\mathcal{N}_{k+1}(AA) = r\mathcal{N}^*_{k+1}(AA) = r[P_k(A)]^2 N^*_{k+1},$$
$$\mathcal{N}_{k+1}(A\alpha) = r\mathcal{N}^*_{k+1}(A\alpha) = 2rP_k(A)P_k(\alpha)N^*_{k+1}, \qquad (2)$$
$$\mathcal{N}_{k+1}(\alpha\alpha) = 0.$$

STEP 2: *Now*

$$P_{k+1}(\alpha) = \frac{\{\text{number of } \alpha\text{-alleles}\}}{\{\text{total number of alleles in gene pool}\}}$$

$$= \frac{2\mathcal{N}_{k+1}(\alpha\alpha) + \mathcal{N}_{k+1}(A\alpha)}{2\mathcal{N}_{k+1}(AA) + 2\mathcal{N}_{k+1}(A\alpha) + 2\mathcal{N}_{k+1}(\alpha\alpha)}.$$

Substituting from (2), we obtain

$$P_{k+1}(\alpha) = \frac{P_k(A)P_k(\alpha)}{[P_k(A)]^2 + 2P_k(A)P_k(\alpha)}.$$

Finally, to eliminate $P_k(A)$, the identity

$$P_k(A) + P_k(\alpha) = 1$$

is used and we obtain

$$P_{k+1}(\alpha) = \frac{P_k(\alpha)}{1 + P_k(\alpha)}. \qquad (3)$$

For clarity we introduce the notation $X_k = P_k(\alpha)$ and thus (3) can be written as

$$X_{k+1} = \frac{X_k}{1 + X_k}, \qquad (4)$$

which is the required difference equation.

The difference equation (4) is a non-linear difference equation so the standard techniques developed in Chapter 8 are not directly applicable to it. It is generally unlikely that a closed-form solution can be found for a non-linear difference equation. We are fortunate in the case of (4), however, that the solution can be easily guessed by iteration. The following example explains how to do this.

Example 2. *Solve the difference equation*

$$X_{k+1} = \frac{X_k}{1 + X_k}$$

given an initial number X_0.

Solution. *Iteration gives*

$$X_1 = \frac{X_0}{1 + X_0},$$

$$X_2 = \frac{X_1}{1 + X_1} = \frac{\frac{X_0}{1+X_0}}{1 + \frac{X_0}{1+X_0}} = \frac{X_0}{1 + 2X_0},$$

$$X_3 = \frac{X_2}{1 + X_2} = \frac{\frac{X_0}{1+2X_0}}{1 + \frac{X_0}{1+2X_0}} = \frac{X_0}{1 + 3X_0}.$$

This suggests that for every k

$$X_k = \frac{X_0}{1 + kX_0}. \tag{5}$$

To verify that this is the solution we check that it satisfies both the initial condition and the difference equation. The initial condition is satisfied since putting $k = 0$ in (5) gives X_0. To check the difference equation is satisfied note that when (5) is used in (4)

$$LHS = \frac{X_0}{1 + (k + 1)X_0},$$

$$RHS = \frac{X_k}{1 + X_k}$$

$$= \frac{\frac{X_0}{1+kX_0}}{1 + \frac{X_0}{1+kX_0}} = \frac{X_0}{1 + (k + 1)X_0}.$$

Thus the two sides are equal, as required.

The closed-form solution is thus

$$P_k(\alpha) = X_k = \frac{X_0}{1 + kX_0},$$

where X_0 is the initial proportion of α-alleles. Suppose that $P_0(A) = P_0(\alpha) = 0.5$ then a simple calculation shows that it takes eight generations for the recessive allele proportion to be 0.1, it takes 98 generations to be 0.01 and it takes 998 generations for it to be 0.001. The elimination of the lethal recessive from the population is a fairly slow process. This is because the lethal gene can be carried by the hybrid genotype $A\alpha$ without any ill effect.

Natural selection — model III

The lethal recessive model is an extreme example where some genotypes are favoured over others. In the theory of evolution by natural selection, individuals with certain characteristics have a higher chance of surviving infancy and thus mating and passing on their genes to the next generation. This also violates the equal survival assumption (b) of Model I from Section 10.2.

Model III for population genetics consists of two of the assumptions in Model I of Section 10.2, the random mating assumption (a) and also equal fertility (c), but with (b) replaced by (b″) below.

> (b″) *A fraction r_1 of the AA and Aα newborn genotypes survive to the end of the generation and a different fraction r_2 of the newborn αα genotypes survive to the end of the generation.*

Note that we assume that both male and female are equally likely to survive. To illustrate the new assumption we consider the frequently studied example of the Peppered Moth *Biston betularia*. The dominant forms, *AA* and *Aα*, are a coal black colour whereas the recessive form, αα, is a pale speckled colour. For moths living in industrial cities the pale speckled coloured moths are less well camouflaged and thus are more likely to be eaten by predators. We thus say that the pale speckled moths are at a selective disadvantage compared with the coal black moths in industrial cities.

The details of setting up the model and deriving a difference equation is left to the exercises, where it is shown that the proportion of α-alleles, $X_k = P_k(\alpha)$, satisfies the non-linear difference equation

$$X_{k+1} = \frac{(\beta - 1)X_k^2 + X_k}{1 + (\beta - 1)X_k^2}, \tag{6}$$

where β is given by $\beta = r_2/r_1$. The number β is called the *relative fitness* of the genotype αα and measures the fitness to survive of the recessive αα genotype relative to the *AA* and *Aα* genotypes. Note that the special case $\beta = 1$ corresponds to the equal survival example from Section 10.2 (Model I) and the case and $\beta = 0$ corresponds to the lethal recessive gene model (Model II) of this section.

A numerical iteration of equation (6) has been carried out for various values of the parameter β starting with an allele distribution of 90% recessive and 10% dominant. The results are shown in Figure 10.3.1. Note that as β decreases the recessive allele proportion tends to decrease

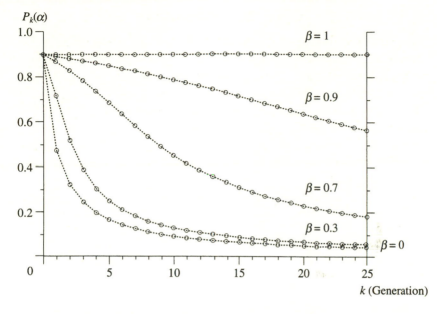

Fig. 10.3.1. Recessive allele proportions for Model III for various values of the parameter β starting from $P_k(\alpha) = 0.9$.

very rapidly within the first few generations. The decrease is more gradual for higher values of β.

It has been estimated that the relative fitness for the pale speckled Peppered Moth in Manchester, UK, compared with the coal black moth is approximately $\beta = 0.7$. Figure 10.3.1 shows that there is a rapid decrease from 90% to 40% in only 10 generations.

Another interesting application of these ideas is to the genetic disorder sickle cell anaemia in West Africa. Here a defective recessive gene causes a minor chemical change in the blood cells. Those who inherit the recessive gene from both parents have a low survival rate. However, the gene is not wholly recessive. The hybrid genotypes $A\alpha$ are slightly affected, but not enough to cause a fatal condition. In fact the hybrids have an enhanced resistance to malaria, which is prevalent in West Africa. Thus the hybrids, $A\alpha$, are the most likely to survive, followed by the pure dominants, AA, and then the recessives, $\alpha\alpha$. To model this situation it is necessary to postulate different survival fractions for *each genotype*. This case is considered, in detail, in the exercises.

Mutation — model IV

Another factor which can affect the distribution of genes in a population is *mutation*. This happens when external factors (for example, background radiation) cause an allele A to change into an allele α. The rate at which mutation of genes occurs is usually very small (typically a fraction 10^{-5} or 10^{-6} of the alleles per generation). Also, mutation more commonly acts to change a dominant gene into a recessive one.

Model IV for population genetics assumes random mating, equal survival of genotypes and also incorporates mutation of A-alleles into α-alleles. Incorporating mutation into the modelling of population genetics requires us to modify Step 2 of the procedure set out in Section 10.2. This is illustrated in the following example.

Example 3. *Find a difference equation for $P_k(\alpha)$ given equal survival and that the A-alleles mutate to the α-alleles at a rate of μ alleles per allele per generation.*

Solution. *Assume that, for each genotype, a fraction r survives from the beginning of the generation to the end of the generation, and assume that during a single generation a fraction μ of the A-alleles mutates into α-alleles.*

STEP 1: *This step is exactly the same as in Example 2 of Section 10.2 and we thus obtain*

$$\begin{aligned}
\mathcal{N}_{k+1}(AA) &= r[P_k(A)]^2 N^*_{k+1}, \\
\mathcal{N}_{k+1}(A\alpha) &= 2rP_k(A)P_k(\alpha)N^*_{k+1}, \\
\mathcal{N}_{k+1}(\alpha\alpha) &= r[P_k(\alpha)]^2 N^*_{k+1}.
\end{aligned} \tag{7}$$

STEP 2: *Now, with mutation present,*

$$\begin{aligned}
P_{k+1}(\alpha) &= \frac{\{\text{number of } \alpha\text{-alleles}\} + \mu\{\text{number of } A\text{-allelles}\}}{\{\text{total number of alleles in gene pool}\}} \\[2mm]
&= \frac{2\mathcal{N}_{k+1}(\alpha\alpha) + \mathcal{N}_{k+1}(A\alpha) + 2\mu\mathcal{N}_{k+1}(AA) + \mu\mathcal{N}_{k+1}(A\alpha)}{2\mathcal{N}_{k+1}(AA) + 2\mathcal{N}_{k+1}(A\alpha) + 2\mathcal{N}_{k+1}(\alpha\alpha)}.
\end{aligned} \tag{8}$$

Note that, even though some of the alleles have changed due to mutation, the total number of alleles remains the same so the denominator is still twice the sum of $\mathcal{N}_{k+1}(AA)$, $\mathcal{N}_{k+1}(A\alpha)$ and $\mathcal{N}_{k+1}(\alpha\alpha)$.

Substituting (7) into (8), and letting $X_k = P_k(\alpha)$, we obtain

$$X_{k+1} = (1 - \mu)X_k + \mu \tag{9}$$

as the required difference equation.

Note that the case $\mu = 0$ corresponds to the equal survival example (Model I) from the previous section, as expected. Equation (9) is a linear constant-coefficient difference equation. A closed-form solution can be found directly using the methods of Chapter 8. This solution is (see Exercises 10.3)

$$X_k = 1 - (1 - X_0)(1 - \mu)^k$$

where X_0 is the initial proportion of α-alleles. To get some feeling for how slow the process of mutation is let $X_0 = 0$ (no α-alleles initially) and take $\mu = 10^{-5}$. We calculate that it takes approximately 6932 generations for the α-allele proportion to reach 0.5.

Further discussions of mathematical models in population genetics may be found in Chapter 3 of Edelstein-Keshet (1988), Maynard-Smith (1968) and the article by Sandfur (1968). One of the pioneering articles in this field is the one by Haldane (1924).

Exercises 10.3

1. (a) For Model II (lethal recessives) show that the proportion of A-alleles satisfies the difference equation

$$Y_{k+1} = \frac{1}{2 - Y_k}$$

   where $Y_k = P_k(A)$.

   (b) Iterate and hence guess the solution of this difference equation. Prove that your guess is a solution.

2. Consider the difference equation from Example 2 (for Model II),

$$X_{k+1} = \frac{X_k}{1 + X_k}.$$

By making the transformation $Z_k = 1/X_k$ in the difference equation show that it reduces to a *linear* difference equation. Hence obtain the solution of the *original* difference equation.

3. Consider Model III (natural selection) where the recessive genotype $\alpha\alpha$ has a lower survival rate than the other genotypes.

   (a) Show that $X_k = P_k(\alpha)$ satisfies

$$X_{k+1} = \frac{(\beta - 1)X_k^2 + X_k}{1 + (\beta - 1)X_k^2}.$$

   where β is the fitness of the recessive relative to the other genotypes.

   (b) Find all the equilibrium solutions, and discuss their significance.

4. Find a difference equation for $P_k(A)$ for Model III.

5. For Model III show that the ratio of A-allele proportion to α-allele proportion U_k satisfies the difference equation

$$U_{k+1} = \frac{U_k(U_k + 1)}{\beta + U_k}.$$

Find solutions for the special cases $\beta = 1$ and $\beta = 0$.

6. Develop a model for sickle cell anaemia by assuming that a fraction r_1 of AA survives, a fraction r_2 of $A\alpha$ survives and a fraction r_3 of $\alpha\alpha$ survives to the end of each generation.

   You are given that the AA genotype has a fitness 0.81 relative to the hybrid $A\alpha$ and the $\alpha\alpha$ genotype has a fitness 0.2 relative to the hybrid $A\alpha$. Hence determine the α-allele proportions in the first, second and third generations. If you have access to a computer, determine the α-allele proportions in further generations.

7. Develop a difference equation for $P_k(A)$ for Model IV (mutation with equal survival).

8. Develop a model for mutation of the A-allele to an α-allele, where the α-allele is a lethal recessive, so that all $\alpha\alpha$ die before mating. Give difference equations for both the α and A-alleles.

9. Suppose that the A-allele mutates to the α-allele at a rate μ alleles per generation per allele and there is back mutation of the α-allele to the A-allele at a rate v alleles per generation per allele. Assume that random mating and equal survival applies.

   (a) Set up a difference equation for the recessive allele proportion $P_k(\alpha)$.

   (b) Solve the difference equation in (a) and use the solution to determine the number of generations for the recessive allele proportion to become 50% given that initially it is 10%. Take $\mu = 10^{-5}$ and $v = 10^{-6}$.

10. (a) Find all the steady-state solutions for the difference equations for $P_k(\alpha)$ for Models I–IV in the text.

   (b) *For those who have studied Section 9.4 of Chapter 9*: determine which of the steady-state solutions in (a) are attractors.

Part three
Models with Differential Equations

11

Continuous growth and decay models

In this chapter some problems of growth and decay will be studied for which differential equations, rather than difference equations, are the appropriate mathematical models. Such problems include:

- the growth of large populations in which breeding is not restricted to specific seasons,
- the absorption of drugs into the body tissues,
- the decay of radioactive substances.

The differential equations which arise from the above problems are all of the first order. The two methods of solution which we explain are sufficient to solve all the differential equations which arise in the next three chapters. The theoretical background for these two methods is contained in Chapter 5. The first of the two methods, which applies only to linear differential equations, is very similar to the method already given in Section 8.1 for solving linear difference equations. The continuous models used in this chapter are similar to the discrete models discussed in Chapter 9.

11.1 First-order differential equations

The two types of differential equations which you need to be able to solve in this chapter are called *linear with constant coefficients* and *variables separable* differential equations. The former arise from problems of unrestricted growth, while the latter appear when the growth is restricted. How to recognize and solve the two types of differential equations will now be explained.

Linear with constant coefficients

These differential equations, when of the first order, have the form

$$\dot{x} = ax + b \tag{1}$$

where a and b are constants. The differential equation is said to be *homogeneous* if $b = 0$. Replacing b by 0 in (1) gives the *homogenized equation* corresponding to (1). The linear differential equations occurring in this chapter will all be homogeneous, but in later chapters some inhomogeneous examples occur as well.

Homogeneous equations

The following example illustrates how to find all the solutions when the differential equation is homogeneous. It corresponds to the choice $a = 2$ and $b = 0$ in (1).

Example 1. *Find all the solutions of the first-order linear homogeneous constant-coefficient differential equation*

$$\dot{x} = 2x. \tag{2}$$

Solution. *To guess a solution, recall that the exponential function is its own derivative. Hence our first guess is $x = e^t$. This gives $\dot{x} = x$, however, which is out by a factor of 2. Hence our next guess is*

$$x = e^{2t} \tag{3}$$

Substitution of (3) into the differential equation (2) gives

$$LHS = \dot{x}$$
$$= 2e^{2t}$$
$$= 2x = RHS$$

Thus (3) is indeed a solution of (2).

Because the differential equation (2) is first-order linear homogeneous, it follows that every solution is some constant multiple of the particular solution (3) (by the superposition theorem for homogeneous equations in Section 5.3). Thus each solution has the form

$$x = Ce^{2t} \tag{4}$$

for some constant C.

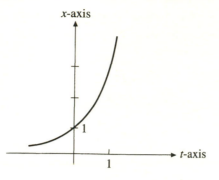

Fig. 11.1.1. Graph of a solution.

The graph of the solution (4) for which $C = 1$ is shown in Figure 11.1.1.

An alternative way to arrive at the particular solution (3) in the above example, which involves slightly less guesswork, is as follows: try for a solution of the form

$$x = e^{\lambda t} \tag{5}$$

where λ is a constant to be determined. Substitution of (5) in the differential equation (2) gives

$$\text{LHS} = \dot{x} = \lambda e^{\lambda t}$$
$$\text{RHS} = 2x = 2e^{\lambda t}.$$

Hence both sides will be equal if $\lambda = 2$. So once again we get the particular solution $x = e^{2t}$ for the differential equation (2). The rest of the solutions are then obtained as in Example 1.

Inhomogeneous equations

It is now easy to solve an inhomogeneous equation, such as

$$\dot{x} = 2x - 6. \tag{6}$$

First, try for a particular solution in which x is constant so that $\dot{x} = 0$. The equation (6) then becomes

$$0 = 2x - 6$$

and hence a particular solution is

$$x = 3$$

Second, add the solutions for the homogenized equation $\dot{x} = 2x$, which were obtained as equation (2) in Example 1, to get

$$x = 3 + Ce^{2t}$$

This formula gives all the solutions of (6) (by the superposition theorem for inhomogeneous equations in Section 5.3).

Variables separable

These differential equations have the form

$$\frac{dx}{dt} = f(x)g(t) \tag{7}$$

where f and g are known functions (which we assume to be smooth so that the existence–uniqueness theorem of Section 5.2 is applicable).

For example, the differential equation

$$\frac{dx}{dt} = x(1 - x)t \tag{8}$$

is of the variables separable type since it has the form (7) with $f(x) = x(1 - x)$ and $g(t) = t$. The reason for the term 'variables separable' in describing these equations will become clear later, after the procedure for solving them has been explained.

Constant solutions

These solutions, which are also called *equilibrium* or *steady-state* solutions, are important because they are easy to find and they provide a framework for the study of other solutions. To find these constant solutions, solve the equation $f(x) = 0$; in the example (8) this gives

$$x = 0 \text{ or } x = 1$$

It is easy to check that each of these is a solution of the differential equation since, when substituted in (8), each gives

$$\text{LHS} = \dot{x} = 0$$
$$\text{RHS} = x(1 - x)t = 0.t = 0$$

The graphs of these constant solutions $x = 0$ and $x = 1$ are horizontal lines, as shown in Figure 11.1.2.

If $x = \phi(t)$ is any other solution, its graph cannot intersect either of these lines (by the existence–uniqueness theorem of Section 5.2) and

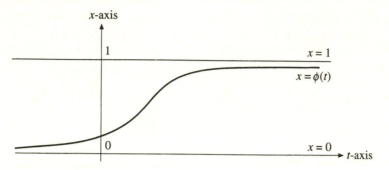

Fig. 11.1.2. Graphs of solutions cannot intersect.

hence its graph must lie entirely within one of the three horizontal strips determined by these lines.

Other solutions

The procedure for finding non-constant solutions of variables separable differential equations will now be illustrated. The idea behind the procedure is to try to 'separate' the variables so that x appears on one side, t on the other.

Example 2. *Find the solution $x = \phi(t)$ of the separable differential equation*

$$\frac{dx}{dt} = x(1-x)t$$

which satisfies the initial condition $x = \frac{1}{2}$ when $t = 0$.

Solution. *The constant solutions are $x = 0$ and $x = 1$. The initial condition places the solution $x = \phi(t)$ in the horizontal strip between the lines $x = 0$ and $x = 1$, as in Figure 11.1.2. Hence, for all time, this solution satisfies*

$$0 < x < 1. \tag{9}$$

STEP 1: *Divide both sides of the differential equation by $f(x) = x(1-x)$, which is non-zero. This gives*

$$\frac{1}{x(1-x)}\frac{dx}{dt} = t$$

STEP 2: *Integrate both sides, with respect to t, from the initial time 0 to the current time t. This gives*

$$\int_{t=0}^{t} \frac{1}{x(1-x)}\frac{dx}{dt}\,dt = \int_{t=0}^{t} t\,dt.$$

STEP 3: *Apply the rule for integrating by substitution to the LHS. This gives (since $\phi(0) = \frac{1}{2}$ and $\phi(t) = x$)*

$$\int_{x=\frac{1}{2}}^{x} \frac{1}{x(1-x)}\, dx = \int_{t=0}^{t} t\, dt.$$

STEP 4: *Work out the integrals and then solve for $x = \phi(t)$. By partial fractions*

$$\int_{x=\frac{1}{2}}^{x} \left(\frac{1}{1-x} + \frac{1}{x} \right) dx = \frac{1}{2} t^2$$

$$-\ln(1-x) + \ln(x)\big|_{x=\frac{1}{2}}^{x} = \frac{1}{2} t^2$$

Note that the arguments of log are both positive since $0 < x < 1$, by (9). Hence

$$
\begin{aligned}
\ln\left(\tfrac{x}{1-x}\right) - \ln(1) &= \tfrac{1}{2} t^2 \\
\tfrac{x}{1-x} &= e^{\frac{1}{2} t^2} \\
x &= 1 - \frac{1}{1+e^{\frac{1}{2} t^2}}
\end{aligned}
\tag{10}
$$

STEP 5: *Check the solution. Substituting (10) back into the differential equation gives*

$$LHS = \frac{dx}{dt} = \frac{t e^{\frac{1}{2} t^2}}{(1+e^{\frac{1}{2} t^2})^2},$$

$$RHS = x(1-x)t = \frac{e^{\frac{1}{2} t^2}}{1+e^{\frac{1}{2} t^2}} \frac{1}{1+e^{\frac{1}{2} t^2}} t.$$

Thus the differential equation is satisfied. Also, (10) shows that, when $t = 0$,

$$x = 1 - \frac{1}{1+e^0} = 1 - \frac{1}{2} = \frac{1}{2},$$

so that the initial condition is satisfied. Thus (10) does indeed give the required solution of the differential equation.

Note that, in Example 2, the initial conditions are used in evaluating the definite integrals. Hence the initial conditions are automatically included in the final solution. The need to evaluate an arbitrary constant C is thereby obviated.

Differential equations of the form

$$\frac{dx}{dt} = f(x)$$

are of the variables separable type since we can write $f(x)$ as $f(x)g(t)$ by choosing g to be the constant function 1. In particular, linear constant-coefficient differential equations

$$\frac{dx}{dt} = ax + b,$$

are of the variables separable type, although in practice it is quicker to ignore this and to stick to the method of Example 1. On the other hand, the variables separable method is the way to solve differential equations like

$$\frac{dx}{dt} = x^{3/2} \qquad \text{and} \qquad \frac{dx}{dt} = x(1-x),$$

which are non-linear.

Recognizing the variables

In the differential equations studied so far in this book the aim has been to solve for quantities denoted by x or y as functions of the time t. The use of the letters x and y was particularly appropriate in mechanics where the quantities they denoted were coordinates or displacements. In other branches of science there are other quantities to be expressed as functions of the time such as *temperature, pressure* and *concentration* and these are usually denoted by other letters such as u, p or c.

Thus, instead of a differential equation being written as

$$\frac{dx}{dt} = ax + b,$$

it may appear as

$$\frac{du}{dt} = au + b, \qquad \frac{dp}{dt} = ap + b \qquad \text{or} \qquad \frac{dc}{dt} = ac + b.$$

The quantity to be expressed as a function of the time is called the *dependent* variable (because it depends on the time) while, in this context, the time t is called the *independent variable*. Quantities other than the time may sometimes be used as independent variable.

Unless stated otherwise, all letters appearing in the differential equations other than the dependent and independent variables are to be regarded as constants. Thus, in the above equations, a and b are to be taken as constants (so all the equations are linear with constant coefficients).

Exercises 11.1

1. Copy and complete the following table to show the classification of the given first-order differential equations.

Differential equation	Linear homogeneous	Constant coefficient	Variables separable
$\dot{x} = tx$			
$\dot{x} + 2x = 0$			
$\dot{x} + 2x = 1$			
$\dot{x} = tx^{3/2}$			
$\dot{x} = t^2 + x^2$			

2. From the solutions found in the text for the differential equation $\dot{x} = 2x$,

    (a) find the solution which satisfies the initial condition $x = 3$ when $t = 0$,

    (b) show that the solution which satisfies the initial condition $x = x_0$ when $t = 0$ is

$$x = x_0 e^{2t}.$$

3.   (a) Use the method of Example 1 in the text to find all the solutions of the differential equation $\dot{x} = -3x$.

    (b) To which types of differential equations is the method used in (a) applicable?

    (c) Find the solution of the differential equation $\dot{x} = -3x + 6$ which satisfies the initial condition $x = x_0$ when $t = 0$.

4. Suppose that the graph of the solution of the differential equation $\dot{x} = \lambda x$ which satisfies the initial condition $x = 1$ when $t = 0$ is as in the following diagram. Sketch the graph of the solution of $\dot{x} = -\lambda x$ which satisfies the same initial condition.

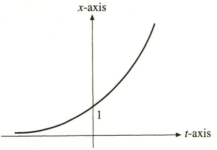

5. Consider the differential equation

$$\frac{dx}{dt} = xt$$

    (a) Although this differential equation is linear homogeneous, it is not of the type to which we apply the method of Example 1 in the text. Why?

(b) Which constant solution does the differential equation have? Find the solution which satisfies the initial condition $x = 1$ when $t = 0$ by separating the variables as in Example 2 in the text.

6. (a) Find the solution of the differential equation

$$\frac{dx}{dt} = x(1 - x)t$$

which satisfies the initial condition $x = 2$ when $t = 0$, by suitably changing the working for Example 2 in the text.

(b) Sketch the graph of the solution you have found.

7. Find the solution of the differential equation

$$\frac{dx}{dt} = x(1 - x)$$

which satisfies the initial condition $x = \frac{1}{2}$ when $t = 0$.

8. For each of the differential equations in the list below:

(a) identify the dependent variable, the independent variable and list the other symbols, which are constants (or parameters),

(b) indicate which of them are first-order linear, which are variables separable, and which are neither of these two types.

NOTE: You are NOT asked to solve these differential equations.

$$\text{(i)} \quad \frac{dm}{dt} = \lambda m^{2/3} \qquad\qquad \text{(ii)} \quad J = 2\pi r L \frac{du}{dr}$$

$$\text{(iii)} \quad \frac{du}{dt} = hA(u - u_s) \qquad \text{(iv)} \quad \frac{dh}{dx} = \alpha e^\gamma h - \beta h^2$$

$$\text{(v)} \quad \frac{dp}{dt} = atp + p^2 \qquad\quad \text{(vi)} \quad \frac{dc}{dt} = \alpha \sin(\beta t) - e^{\gamma t} c.$$

9. Radioactive iodine 131, produced by nuclear tests, settles on vegetation which is eaten by deer grazing on the vegetation. The iodine 131 accumulates in the thyroid gland of the deer. Let y be the amount of iodine 131 in the thyroid of a deer after t days and suppose I_0 is the initial amount deposited in the vegetation that will be eaten by a single animal. You are given that y satisfies

$$\frac{dy}{dt} = -\lambda_2 y + I_0 \lambda_1 e^{-\lambda_1 t}, \qquad \lambda_1 \neq \lambda_2$$

where λ_1 and λ_2 are positive constants.

(a) If there is no iodine 131 in the thyroid initially solve the differential equation and show that

$$y = \frac{I_0 \lambda_1}{\lambda_2 - \lambda_1} \left[e^{-\lambda_1 t} - e^{-\lambda_2 t} \right].$$

(b) After a nuclear test in Colorado in the USA in 1964 the following estimates of λ_1 and λ_2 were made for a Colorado deer population. They were

$$\lambda_1 = 0.126 \qquad \text{and} \qquad \lambda_2 = 0.107.$$

Use the model to estimate the maximum percentage increase in the amount of iodine 131 in the deers' thyroids.

11.2 Exponential growth

This section introduces the study of growth models in the context of population growth with no restrictions on the ultimate size of the population. The effect of imposing such restrictions will be examined in the next section. We begin with some remarks which are relevant to both sections.

Continuous models

The study of population models helps bring into focus the distinction between discrete and continuous models. The population models discussed in Chapter 9 were discrete, being appropriate for species of animals which breed during specific breeding seasons, equally spaced. The population models of interest in this chapter, however, are more appropriate for large populations which reproduce continuously, rather than at regular intervals. Human populations are naturally modelled in this way, as are certain types of microscopic organisms.

In the continuous models, the number of individuals in a population at time t will be modelled by a solution $N = \phi(t)$ of a differential equation. Thus both of the variables N and t will assume all the real values in some interval, fractional and irrational values included.

The justification for using N as a real variable in this way is that, when the population is sufficiently large, differences of one or two individuals are of little consequence. At the end of the problem, we simply round N to the nearest integer value. A similar justification applies to our use of t as a real variable: in the absence of specific breeding seasons, reproduction can occur at any time; for a sufficiently large population, it is then natural to think of reproduction as occurring continuously.

Microscopic organisms

Populations of microscopic organisms are attractive to model since they may be grown in the laboratory. This enables data concerning their growth to be collected easily and the environment to be controlled.

Examples of such micro-organisms for which data are available are *yeast cells* and *E. coli*. The former are involved in brewing and in the commercial production of certain vitamins; the latter are a species of bacteria which occur in the intestines of man and other animals.

Both of these examples are single-celled organisms which reproduce by dividing into two — a process known as binary fission. The cells absorb nutrients which are dissolved in the liquid in which they are immersed. Thus the amount of nutrients available to the cells can be controlled.

Exponential growth model

We consider a population consisting of yeast cells and we suppose that the environment in which the cells are multiplying does not alter with passage of time. In particular, we assume that nutrients are added to the liquid to replace those used by the cells and that the cells have enough room to multiply without overcrowding. On average then, each cell divides into two in a fixed time. Hence we could expect that

$$\left\{\begin{array}{c}\text{rate of}\\\text{increase of the}\\\text{number of cells}\end{array}\right\} \quad\text{is proportional to}\quad \left\{\begin{array}{c}\text{number of}\\\text{cells in the}\\\text{population}\end{array}\right\}. \qquad (1)$$

This leads us to propose as a mathematical model for the growth of the population the differential equation

$$\frac{dN}{dt} = aN \qquad (2)$$

where N (when rounded to the nearest integer) is the number of yeast cells in the population at time t and where a is a positive real constant. The constant a measures the average growth rate per unit time *per individual*.

Let us suppose that, at the time when observations are started ($t = 0$), the number of cells in the population is a positive number N_0. This gives the initial condition

$$N = N_0 \quad\text{when}\quad t = 0. \qquad (3)$$

Since the differential equation (2) is first-order linear homogeneous with constant coefficient, it can be solved as in Section 11.1. It is left as an exercise to show that the solution satisfying the initial condition (3) is

$$N = N_0 e^{at}. \qquad (4)$$

Thus the assumption (1) implies that the population of cells grows exponentially.

The formula (4) implies that the population grows indefinitely large at an ever-accelerating rate, as can be seen from the typical graph sketched in Figure 11.2.1.

Such unrestricted growth is impossible in practice since eventually the

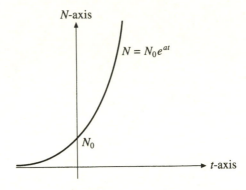

Fig. 11.2.1. Exponential growth.

population runs out of space and nutrients. The above 'J' curve is often observed, however, during the initial stages of a population's growth.

Finding the constants

In order to be able to use the formula (4) to predict the number of cells at any time $t > 0$, it is necessary to substitute numerical values for N_0 and a. Now N_0 is just the number of cells present when $t = 0$. If the number of cells is known at some later time, then the growth rate a can be found by solving a suitable equation. The following example shows how.

Example 1. *Suppose that, in a population of yeast cells which is growing exponentially, the initial number of cells is 1000 and ten minutes later it is 1500. Find the growth rate, a, for the population.*

Solution. *Substituting $N_0 = 1000$ in the formula (4) gives*

$$N = 1000e^{at}$$

But $N = 1500$ when $t = 10$ so that

$$1500 = 1000e^{10a}$$

Hence

$$e^{10a} = 1.5,$$
$$10a = \ln(1.5).$$

Thus, to two decimal places,

$$a = \ln(1.5)/10 = 0.04 \qquad (per\ minute\ per\ cell).$$

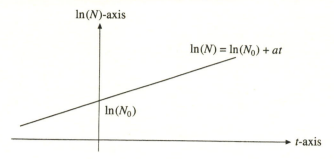

Fig. 11.2.2. Logarithm of exponential growth.

Testing the model

To test the accuracy of the above model for a particular population, one might determine the constant a as in the above example and then use the formula (4) to predict the size of the population at subsequent times. Comparison could then be made with the observed values.

A more convenient way is to first apply the function ln to both sides of the formula (4) for exponential growth. It is left as an exercise to show that the formula then becomes

$$\ln(N) = \ln(N_0) + at. \tag{5}$$

This shows that, while N is an exponential function of t, $\ln(N)$ is a *linear* function of t, as in Figure 11.2.2. The slope of this linear function, moreover, is just the growth rate a.

Thus to test the population for exponential growth, we simply plot the *logarithms* of the observed population values against the time. The extent to which the resulting points lie along a straight line is a measure of the accuracy of the model.

A nice illustration of a population which is growing exponentially was given by Monod (1949), whose results are also described in Rubinow (1975). He allowed a population of the bacteria E.coli to grow in a medium containing glucose as the nutrient and observed the density D of the population (dry weight of the cells per unit volume) at various times. In Figure 11.2.3 we have plotted the natural logarithms of the population densities which he observed. The points lie very neatly along a straight line, indicating the appropriateness of the exponential growth model — at least during the period of observation.

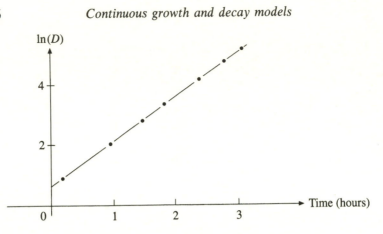

Fig. 11.2.3. Exponential growth of *E. coli* bacteria.

Human populations

In adapting the preceding growth model to human populations, we must take into account deaths as well as births. A plausible assumption is that births and deaths both occur at a rate which is proportional to the size N of the population at any time t. Hence we may write

$$\frac{dN}{dt} = \alpha N - \beta N = (\alpha - \beta)N \qquad (6)$$

where α and β are positive constants denoting the average rate of births and deaths, on average (per head of population per year). If $\alpha - \beta > 0$ then (6) has the same form as (2) with $a = \alpha - \beta$, and so the model predicts exponential growth, given by (4). It is left as an exercise to show that, if a population grows exponentially, then there is a fixed length of time that it takes to double.

An interesting historical account of theories of population growth is given in Hutchison (1978). The idea that populations grow exponentially can be found in the works published by Graunt (1662) and Malthus (1798), who made estimates of the times taken for various populations to double. All sorts of data concerning the growth of populations around the world can be found in the *Demographic Yearbook* published by the United Nations. For example, the following annual growth rates for the world's population, shown in Table 11.2.1, were obtained from this source.

An article in *The Age* newspaper (25 May 1989) quoted United Nations sources as stating that the world's population growth, after having slowed

Table 11.2.1. *Annual growth rate (per annum) for*
the population of the world.

Years	Growth rate
1965–70	1.9%
1970–75	1.9%
1975–80	1.8%
1980–85	1.7%

down in the 1970s, was speeding up again, and that the current population of about 5.25 billion people would double in 39 years, at present rates.

The above figures are in rough agreement with an exponential growth model for the world's population — for the period considered — with a growth rate a of about 0.02 new individuals per head of population per annum. The exponential growth model cannot be valid in the long term, however, as the population would run out of food and space. A more realistic model of population growth, taking such limitations into account, will be described in the next section.

Exercises 11.2

1. Let a be any real constant. Find all the solutions of the differential equation

$$\frac{dN}{dt} = aN$$

by using the appropriate method from Section 11.1. Hence show that the solution which satisfies the initial condition $N = N_0$ when $t = 0$ is

$$N = N_0 e^{at}.$$

2. Show that if N and N_0 are positive numbers then

$$N = N_0 e^{at} \quad \Rightarrow \quad \ln(N) = \ln(N_0) + at.$$

Prove the converse of this implication also.

3. Suppose that a population is growing exponentially, in accordance with the formula (4) in the text. Prove that, if the population doubles during the first T hours, then it doubles during every T hour period.

4. Find the growth rate a for the world's population, given that it doubles every 39 years. Is your answer consistent with the United Nations data quoted in the text?

5. Find approximately the value of the growth rate a for the population in Figure 11.2.3.

6. The population of Australia at the various census dates since Federation is shown in the following table. Decide whether the data is consistent with exponential growth. If so, find an approximate value for the growth rate a.

Date	Population
1901	3 773 801
1911	4 455 005
1921	5 435 734
1933	6 629 839
1947	7 579 358
1954	8 986 530
1961	10 508 186
1966	11 550 462
1971	12 755 638
1976	13 548 472
1981	14 576 330

Can you think of any historical reasons for any anomalies in the data?

11.3 Restricted growth

We now discuss a useful modification of the exponential growth model which takes into account the fact that, in practice, there is a limit to the size to which a population can grow. The ideas underlying the model are similar to those explained in Section 9.2, but now lead to a *differential* equation, called the logistic equation. Unlike the *discrete* logistic equation, however, its solutions can be expressed in closed form and there is no chaotic behaviour.

Logistic growth model

A population growing in a favourable environment can initially grow very rapidly, in accordance with the exponential growth law. Restrictions on space and food supply, however, eventually come into play and impose a limit on the maximum sustainable size for the population. This is called the *carrying capacity* of the environment, and is denoted by K.

If N is the size of the population at time t, we can think of the ratio

$$\frac{1}{N} \frac{dN}{dt}$$

as the *growth rate at this time*, giving the increase in population per unit time per head of population. In the exponential growth model this

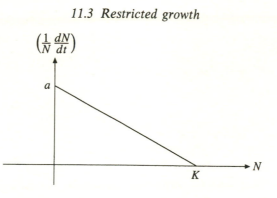

Fig. 11.3.1. Logistic decline in growth rate.

assumes a constant value a. In the logistic model, however, the growth rate is assumed to start at a when $N = 0$, and then decreases linearly until it reaches the value 0 when $N = K$.

Thus the growth rate is given by the formula in Figure 11.3.1. Hence the logistic model says that

$$\frac{1}{N}\frac{dN}{dt} = a\left(1 - \frac{N}{K}\right) \tag{1}$$

or

$$\frac{dN}{dt} = aN\left(1 - \frac{N}{K}\right). \tag{2}$$

This differential equation, for N as a function of t, is the *logistic equation*. It involves the parameters a and K. The parameter a, which is relevant to the initial phase of the population's growth before the restrictions on growth are significant, is the *unrestricted* growth rate of the population for uncrowded conditions.

The differential equation (2) is first order and of the variables separable type. It is left as an exercise to show that, if $0 < N_0 < K$, then the solution which satisfies the initial condition $N = N_0$ when $t = 0$ is

$$N = \frac{K}{(\frac{K}{N_0} - 1)e^{-at} + 1}. \tag{3}$$

Note that this formula gives the correct initial condition $N = N_0$ when $t = 0$. As $t \to \infty$, moreover, the term involving e^{-at} approaches 0 as a is positive. Hence

$$\lim_{t \to \infty} N = K.$$

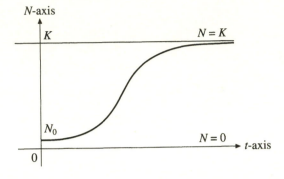

Fig. 11.3.2. Logistic growth curve.

It also follows from (2) that dN/dt is positive. Hence the graph of the solution (3) is the 'S'-shaped curve shown in Figure 11.3.2.

Thus the model predicts that the population increases steadily from the value N_0 and approaches the carrying capacity K as the time becomes arbitrarily large. The relevance of the model to various types of populations will now be discussed.

Microscopic organisms

The logistic model is reputed to give reasonably good predictions for the behaviour of populations of yeast cells, bacteria, and protozoans (the most primitive form of animal life), when grown under suitable laboratory conditions.

To test the relevance of the model to the growth of such populations, we shall refer to Table 11.3.1, which is based on actual laboratory measurements of Carlsen (1913).

When the points in Table 11.3.1 are plotted, as in Figure 11.3.3, they are seen to lie along an 'S'-shaped curve. In this respect, at least, they are in agreement with the predictions of the logistic model.
Graphs similar to that in Figure 11.3.3 for the growth of the population of yeast cells may be seen for example in Emlen (1984), page 43, Emmel (1976), page 103, Hutchinson (1971), page 24, and Kormondy (1976), page 78.

How can we further test the agreement with the logistic model? A simple geometric answer is provided by Figure 11.3.1: the logistic model

Table 11.3.1. *Growth of yeast cells.*

Time in hours t	Number of yeast cells N
0	10
2	29
4	71
6	175
8	351
10	513
12	584
14	641
16	651
18	662

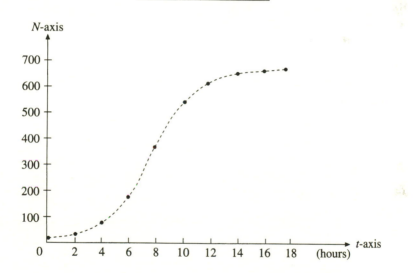

Fig. 11.3.3. Population of yeast cells showing logistic growth.

is characterized by a linear decline in the growth rate

$$\frac{1}{N}\frac{dN}{dt}$$

when considered as a function of N. Now, by using a ruler, we can approximate tangents to the graph of N as a function of t and hence find the approximate values of dN/dt at various points. This process, when applied to Figure 11.3.3, yields the results shown in Table 11.3.2.

The points on the graph of $(1/N)dN/dt$ as a function of N given by

Table 11.3.2.

Time in hours t	2	4	6	8	10	12	14
Number of yeast cells N	29	71	175	351	513	584	641
Slope of tangent dN/dt	15	31	75	117	57	29	14
Growth rate $(1/N)dN/dt$	0.52	0.44	0.43	0.33	0.10	0.05	0.02

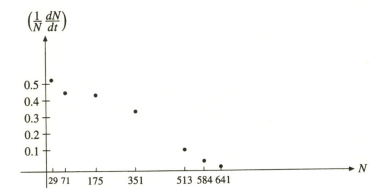

Fig. 11.3.4. Linear decline in growth rate.

this table have been plotted in Figure 11.3.4. It can be seen that the points do indeed lie approximately on a straight line, in accordance with the logistic model.

Fisheries management

Models for population growth find ready application in the fishing industry, which aims at maintaining a permanent supply of fish. Too much fishing in a particular year might so deplete the population that it would take a long time to recover or it might even become extinct. Too little fishing, on the other hand, might leave the population intact but result in a smaller harvest than necessary. Biologists employed by the fishing industry are therefore interested in determining the maximum rate, somewhere in between these two extremes, at which fish can be harvested without reducing the population in the long term. Models for population growth play an important part in determining this maximum rate, as we now show.

First, it is clear that in order to maintain the fish population at a constant level, only the *increase* in population should be harvested during any one season. Hence, to maximize the harvest,

the population should be kept at the size N for which its rate of increase dN/dt is a maximum.

The value of N which this condition determines will depend on which population model is being used. For the logistic model, we simply choose N to maximize the RHS of the logistic equation (2), which is the quadratic

$$aN\left(1 - \frac{N}{K}\right).$$

A little calculation shows that the desired choice is $N = K/2$. This means that the population should be maintained at half the carrying capacity. To get the maximum value of dN/dt we now substitute this value of N back into the quadratic to get

$$\left(\frac{dN}{dt}\right)_{\max} = \frac{aK}{4}.$$

This is *the maximum rate at which fish can be harvested*, if the population is to be kept at a constant size.

A discussion of how the answer for the maximum value depends on the particular population model is given in Ginzburg (1985), pages 130,131. Some pros and cons of the logistic model, when used in this context, are discussed in Walter (1981).

Human populations

The data in Section 11.2 suggest that the world's population is currently growing exponentially with a growth rate of about 0.02 per year. If this were to continue for the next three centuries, however, the population would increase by a factor of $(1.02)^{300}$, which is about 380. Hence the average density of the world's population (over the surface area of the inhabited countries) would increase from its 1985 value of 36 people per km^2 to about $13\,680$ per km^2. This latter figure is truly fantastic: *less than one square metre of land for each living person.* Long before this happened, of course, the food supply would have been exhausted and the population would have exceeded its maximum sustainable size.

As a more realistic alternative, the logistic model was proposed for the growth of human populations by Verhulst in the middle of the nineteenth

Table 11.3.3. *Actual and predicted values of the population
of the USA in millions.*

Year	Actual	Predicted
1790	3.929	3.929
1800	5.308	5.336
1810	7.240	7.228
1820	9.638	9.756
1830	12.866	13.109
1840	17.069	17.506
1850	23.192	23.191
1860	31.443	30.412
1870	38.558	39.371
1880	50.156	50.177
1890	62.948	62.769
1900	75.995	76.870
1910	91.972	91.972

century. He used the logistic model to estimate the maximum values for
the population of various countries.

The use of the logistic model to study human populations was revived
in 1920 by Pearl and Reed. They compared the census figures for the
population of the USA from the years 1790 to 1910 with the values which
could be predicted from the logistic model. The remarkable agreement
between the actual and predicted values is shown in Table 11.3.3.

To get the predicted values, Pearl and Reed assumed that the logistic
equation (2) was satisfied with N denoting the population of the USA at
a time t years after some initially chosen year; hence the population is
given as a function of the time by the solution (3) of the logistic equation.
They chose the parameters a and K in such a way as to make the formula
(3) give the actual values of the population in the years 1790, 1850 and
1910. As explained in the exercises, the values they obtained for these
parameters were

$$a = 0.03134 \quad \text{per year,}$$

$$K = 197\,273\,000 \quad \text{individuals.}$$

The formula (3) then gives the remaining predicted values in Table 11.3.3.

As figures from later censuses became available, the remarkable agreement between actual and predicted values for the population of the USA
continued, as can be seen from the first column of Table 11.3.4. After
1950, however, the predicted values consistently underestimated the ac-

Table 11.3.4. *Good then bad news for the logistic model.*

Year	Actual	Predicted
1920	105.711	107.395
1930	122.775	122.398
1940	131.670	136.318
1950	150.679	148.678
1960	179.323	159.231
1970	203.235	167.944
1980	226.546	174.942

tual size of the population, and by 1980 the actual population was well in excess of the previously estimated carrying capacity, K.

The failure of the Pearl and Reed model to give a realistic prediction of the maximum sustainable population for the USA, highlights the difficulties of making predictions about the growth of human population in the long term. An obvious difficulty is that human beings can change their environment in such a way as to invalidate the values of the parameters previously relevant to the model. Thus, for example, technological advances in agricultural production and distribution can improve the supply of food and thereby increase the carrying capacity. Advances in medical science can decrease the death rate and thereby increase the growth rate. It is easy to think of many other factors under the control of human beings which can affect the values of the parameters.

Human population models are discussed in Braun (1983), Section 1.5, Hutchinson (1978), pages 22–23, and in Keyfitz (1977), pages 213–220.

Exercises 11.3

1. This exercise is about the population of yeast cells from which the data in Table 11.3.1 were obtained.

(a) Estimate the growth rate a and the carrying capacity K for this population by comparing Figure 11.3.4 with Figure 11.3.1. What is the accuracy of your estimates?

(b) Hence compare the number N of yeast cells at time t predicted by the formula (3) in the text with the observed values given in the table, for $t = 0, 2, 4, \ldots, 18$ hours.

2. Let a, K, and N_0 be positive real numbers such that $0 < N_0 < K$. Use separation of variables to show that the solution of the logistic equation

$$\frac{dN}{dt} = aN\left(1 - \frac{N}{K}\right)$$

which satisfies the initial condition $N = N_0$ when $t = 0$ is

$$N = \frac{K}{\left(\frac{K}{N_0} - 1\right)e^{-at} + 1}.$$

3. Repeat Exercise 2, but this time suppose that $N_0 > K$. Sketch the graph of a typical solution.

 (a) What does your graph tell you about what happens in the long term if the population initially exceeds the carrying capacity?

4. Let N be given as a function of the time by the solution of the logistic equation in Exercise 2. Let N_0, N_1 and N_2 be the values of N when $t = 0$, $t = T$ and $t = 2T$ respectively.

 (a) Show that

$$\frac{N_1}{N_0}e^{-aT} = \frac{K - N_1}{K - N_0},$$
$$\frac{N_2}{N_0}e^{-2aT} = \frac{K - N_2}{K - N_0}.$$

 (b) Eliminate T and solve for K to show that

$$K = N_1\frac{N_2(N_1 - N_0) - N_0(N_2 - N_1)}{N_1^2 - N_0N_2}.$$

5. (a) Show that if the population N reaches half its carrying capacity when the time $t = t_1$ then the solution of the logistic equation in Exercise 2 may be written

$$N = \frac{K}{e^{-a(t-t_1)} + 1}.$$

 (b) On page 32 of Braun (1975) it is stated that Pearl and Reed calculated that the population of the USA reached half its carrying capacity in April 1913. It follows from the above formula that in year t

$$N = \frac{197\,273\,000}{e^{-0.03134(t-1913.25)} + 1}.$$

Does this formula give the values predicted in Table 11.3.2? Comment.

6. Growth of a tumour. A tumour grown in a laboratory with a plentiful supply of nutrients provides an example of restricted growth which does not follow the logistic model. The number of cells in the tumour is proportional to its volume V at time t. Its growth rate is then

$$\frac{1}{V}\frac{dV}{dt}$$

The growth rate, instead of decreasing linearly with V (as in the logistic model), decreases exponentially with t, its value at time t being found empirically to be

$$\alpha e^{-\lambda t}$$

where α and λ are positive constants.

(a) Express this empirical law as a differential equation.

(b) Show, by separation of variables, that the solution which satisfies the initial condition $V = V_0$ when $t = 0$ is

$$V = V_0 \exp \left(\frac{\alpha}{\lambda}(1 - e^{-\lambda t}) \right)$$

(c) Sketch the graph of a typical solution. What happens to the volume as $t \to \infty$?

[The law of growth for the tumour is called the *Gompertz growth law*. Further discussion may be found, for example in Braun (1983), and Rubinow (1975), page 43.]

11.4 Exponential decay

A variety of processes involving the decay of some substance can be usefully modelled by the assumption that the rate of decrease of the substance at any time is proportional to the amount of the substance present.

Radioactive decay

A typical example of such a process is the decay of a radioactive element, such as radium. Since the decrease in mass is caused by the emission of alpha particles, the decay is really a discrete process. The mass of an alpha particle, however, is very small compared with the mass of the sample and hence it is appropriate to regard the mass as a quantity which can change continuously. On average, the larger the sample, the greater will be the number of alpha particles emitted per unit time. The simplest way to model this is to assume that, at any time,

$$\left\{ \begin{array}{l} \text{rate of decrease} \\ \text{of mass of sample} \end{array} \right\} \text{ is proportional to } \left\{ \begin{array}{l} \text{mass of sample} \\ \text{still present} \end{array} \right\}.$$

To express this as a differential equation, let m denote the mass of the sample still present at time t and so obtain

$$\frac{dm}{dt} = -km \qquad (1)$$

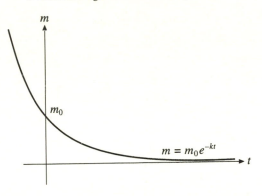

Fig. 11.4.1. Exponential decay.

where k is a positive constant. The minus sign ensures that the derivative of m with respect to t is negative; hence the mass of the sample *decreases* as time goes on.

Since this differential equation is linear homogeneous with constant coefficient, it can be solved by the method of Section 11.1. The solution satisfying the initial condition $m = m_0$ when $t = 0$ is found in this way to be

$$m = m_0 e^{-kt}. \tag{2}$$

The graph of a typical solution is sketched in Figure 11.4.1 and the process just modelled is said to be an example of *exponential decay* (in contrast to the exponential growth of populations in Section 11.2).

When a substance decays exponentially, it takes a fixed time T for the amount to decrease by a factor of $\frac{1}{2}$ (just as with exponential growth, it takes a fixed time for a population to double its size). The time T is called the *half-life* of the decaying substance and is related to the *decay constant* k by the equation

$$T = \frac{\ln(2)}{k}. \tag{3}$$

Radium has a half-life of 1600 years while the extremely dangerous radioactive element plutonium (used in atomic weapons and nuclear power stations) has a half-life of 24100 years. A table of the half-lives of radioactive elements is given in Giancolo (1985). Some uses and dangers of radioactive elements are discussed in Marion (1976). An interesting application of the exponential decay model to carbon-14 dating of archaeological finds is explained later in an exercise.

Drug absorption

Another process which also leads to an exponential decay model is the absorption of drugs from the bloodstream into the body tissues. When a drug is administered by an injection, it mixes with the blood. As time goes on, the amount of the drug in the bloodstream diminishes, being absorbed by the body tissues or excreted from the body. When medical staff administer a drug it is important for them to know how much to give in the next injection — too little and the drug is ineffective, too much and undesirable side effects could result.

The significant quantity to monitor is the *concentration* of the drug in the bloodstream, which is defined as *the amount of drug per unit volume of blood*, and is usually measured in mg/litre. For most drugs the rate of absorption from the bloodstream increases with higher concentration. As with radioactive decay, the simplest model consistent with this is the assumption that, at any time,

$$\left\{ \begin{array}{l} \text{rate of decrease} \\ \text{of concentration} \end{array} \right\} \quad \text{is proportional to} \quad \{\text{concentration}\}.$$

With c denoting the concentration of the drug in the bloodstream at time t this may be written as a differential equation

$$\frac{dc}{dt} = -\mu c \tag{4}$$

where μ is a positive constant. This differential equation has an obvious analogy with the differential equation (1) for radioactive decay, and can be solved in a similar way.

Exercises 11.4

1. In the differential equation (1) in the text:

   (a) What quantities do the symbols m, t, and dm/dt stand for?

   (b) Why is there a minus sign?

   (c) Assume mass is measured in grams and time in years. What are the SI units for the decay constant k? What are its dimensions?

2. Obtain the solution of the differential equation (1) in the text which satisfies the initial condition $m = m_0$ when $t = 0$. What makes the solution decrease more rapidly: large k or small k?

3. Show from the solution (2) in the text that, if the mass of a radioactive sample decays from m_0 to $m_0/2$ in time T, then $T = \ln(2)/k$.

4. Given that the half-life of radium is 1600 years, what is the value of its decay constant k? How long does it take for the mass of a given sample to decrease to $\frac{1}{3}$ of its value? $\frac{1}{4}$ of its value? $\frac{1}{n}$ of its value?

5. Carbon-14 dating. While a plant or animal is living, the ratio of ^{14}C to ^{12}C in its tissues is a small constant, the same for all living tissue. When a plant or animal dies, however, this ratio decays exponentially with a half-life of 5730 years.

  A sample of charcoal was found at the cave at Lascaux in France containing the famous prehistoric painting, for which the ratio of ^{14}C to ^{12}C had decayed to 14.5% of its original value. How many years ago did the wood grow?
(Further information about carbon-14 dating is given in Braun (1983), Section 1.3.)

6. (a) Match each symbol occurring in the differential equation (1) in the text with the symbol which plays a similar role in the differential equation (4).

   (b) Why is there a minus sign in the differential equation (4)?

   (c) Suppose the concentration is measured in mg/litre and time is measured in hours. What are the units for μ?

   (d) Use the solution given in the text for (1) to write down the solution of (4) which satisfies the initial condition $c = c_0$ when $t = 0$.

   (e) How long does it take for the concentration to reduce to half its initial value? Express your answer in terms of μ.

7. The concentration of drug in a patient's bloodstream reduces to half its initial value in 30 minutes. What is the concentration after 2 hours?

8. Suppose that after 4 hours an *additional* injection is given to the patient in Exercise 7.

   (a) Find the concentration 5 hours after the initial injection was given. See if you can identify the constituents from the first and second injections.

   (b) Sketch a graph of the concentration against the time.

9. When the drug Theophylline is administered for asthma, a concentration in the blood below 5 mg/litre of blood has little effect while undesirable side effects appear if the concentration exceeds 20 mg/litre. Suppose a dose corresponding to 14 mg/litre of blood is administered initially. The concentration satisfies the differential equation

$$\frac{dc}{dt} = -\frac{c}{6}$$

where the time t is measured in hours.

   (a) Find the concentration at time t.

   (b) Show that a second injection will need to be given after about 6 hours to prevent the concentration becoming ineffective.

   (c) Given that the second injection also increases the concentration by 14 mg/litre, how long is it before another injection is necessary?

(d) What is the shortest safe time that a second injection may be given so that side effects do not occur?

10. One method of administering a drug is to feed it continuously into the blood stream by a process called intravenous infusion. This may be modelled by the linear differential equation

$$\frac{dc}{dt} = -\mu c + D$$

where c is the concentration in the blood at time t, μ is a positive constant, and D is also a positive constant which is the rate at which the drug is administered.

(a) Find the constant (or equilibrium) solution of the differential equation.

(b) Given $c = c_0$ when $t = 0$, find the concentration at time t. What limit does the concentration approach as $t \to \infty$? Compare with your answer to part (a).

(c) Sketch the graph of a typical solution.

12

Modelling heat flow

Some typical processes from everyday life which involve the flow of heat from one region to another are the heating of beverages, food and living areas, and the cooling of foodstuffs in refrigerators. The flow of heat involved in such processes is best described by mathematical models. This chapter introduces some simple mathematical models which are based on Newton's law of cooling and Fourier's law of heat conduction. These laws lead to very simple differential equations of the type studied in Chapter 11. At the end of this chapter these ideas are used to model the loss of heat from an insulated water pipe. The model makes some unexpected predictions.

The only concept from physics which is assumed initially is that of temperature — which indicates the hotness of a body, and is measured with a thermometer.

12.1 Newton's model of heating and cooling

A hot cup of coffee, when left standing for a while, cools as heat is lost to the surrounding air. The temperature of the coffee drops and, if the coffee is left standing for long enough, its temperature eventually reaches that of its surroundings. This example is typical of many processes involving cooling, and heating, which occur in a wide variety of situations. Fortunately there is a very simple mathematical model for such problems, due to Newton., which is both reliable and versatile.

In this section the simplest version of Newton's model is described, which uses only the concept of temperature. The model will be refined in the next section so as to include the effects of such factors as the size of the heated body and the material of which it is composed. This refinement, however, will involve us in a discussion of the concept of the

232

amount of heat in a body, which is a little more sophisticated than that of temperature.

The model

Although the model for cooling applies to any heated object, we shall stay with the cup of coffee as an illustration. We aim at predicting how the temperature of the coffee changes with time.

The intuitive starting point for modelling this problem is the idea that the greater the difference between the temperature of the coffee and that of the surrounding room, the greater will be the *rate of cooling* of the coffee. The simplest mathematical model consistent with this requirement is to have

{rate of cooling} is proportional to {temperature difference}. (1)

This is known as *Newton's law of cooling*. In proposing this law, Newton assumed that the coffee was fanned by a continuous stream of air at the temperature of the surroundings.

To express the model in terms of mathematical symbols we let

$$u = \left\{ \begin{array}{c} \text{temperature of the coffee} \\ \text{at a time } t \\ \text{after being placed in the room} \end{array} \right\},$$

$$u_s = \left\{ \begin{array}{c} \text{temperature of} \\ \text{the surrounding room} \end{array} \right\},$$

and we assume u_s to be constant. Note that u is a function of the time t. The law (1) may then be written as a differential equation for $u = \phi(t)$,

$$\frac{du}{dt} = -\lambda(u - u_s) \tag{2}$$

where $\lambda > 0$ is the constant of proportionality. Note that the minus sign is required so that for $u > u_s$ (that is, the coffee is hotter than the surroundings) we obtain $du/dt < 0$ (that is, temperature is a decreasing function of time).

Equation (2) is also applicable to situations in which a cold object is placed in a hot room. Here $u - u_s < 0$ so that $du/dt > 0$ (that is, the temperature of the object increases with time).

Solving the differential equation

Newton's law of cooling thus provides us with the differential equation (2) for the temperature u as a function of the time t. By solving this differential equation, with a given initial condition, we can find how the temperature varies with the time. The value of parameter u_s, being the temperature of the surrounding room, is easily determined, while ways of determining the value of λ will be discussed later.

Example 1. *A cup of coffee is initially at boiling point, 100 °C. The temperature of the room is 20 °C. Find the temperature of the coffee as a function of the time.*

Solution. *Let u be the temperature of the coffee after time t. Since $u_s = 20$, the differential equation (2) is now*

$$\frac{du}{dt} = -\lambda(u - 20) \tag{3}$$

with the initial condition $u = 100$ when $t = 0$. The differential equation, being linear constant coefficient, may be solved by the method of Section 11.1.

First, we find all solutions of the homogenized equation

$$\frac{du}{dt} = -\lambda u. \tag{4}$$

Try $u = e^{mt}$ where m is a constant to be determined. By substituting in the homogenized equation (4) we find that $m = -\lambda$. Hence one solution of (4) is $u = e^{-\lambda t}$. As (4) is homogeneous, every solution therefore has the form

$$u = Ce^{-\lambda t} \tag{5}$$

for some real constant C.

Second, we guess the original equation (3) has a particular solution in which u is a constant. This means that $du/dt = 0$ and hence by (3) that $u - 20 = 0$. Thus a particular solution of (3) is the constant solution

$$u = 20. \tag{6}$$

Finally, all the solutions of the original equation (3) are obtained by adding the solutions (5) and (6) to get

$$u = Ce^{-\lambda t} + 20. \tag{7}$$

The initial condition when used in (7) gives

$$100 = Ce^0 + 20 = C + 20$$

and hence $C = 80$. Thus the required solution is

$$u = 80e^{-\lambda t} + 20. \tag{8}$$

It is left as an exercise to check that (8) satisfies both the differential equation (3) and the initial condition. This formula gives the temperature u of the coffee as a function of the time t.

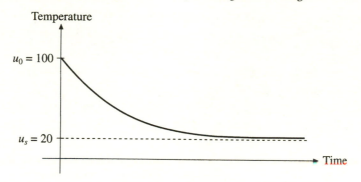

Fig. 12.1.1. Graph of temperature of coffee against time.

Behaviour of solutions

The constant solution (6) obtained in the course of the above working has the interesting physical interpretation that if the coffee is initially at room temperature 20 °C, then it will stay at this temperature indefinitely. It is therefore called the *steady-state temperature* for the coffee.

Note that the formula (8) does not provide a complete answer to Exercise 1 since the value of the parameter λ has not yet been specified. In spite of this, however, it is possible to indicate the general shape of a typical graph of temperature against time, as in Figure 12.1.1.

The graph was obtained from (8) by observing that

$$u = 100 \qquad \text{when} \quad t = 0, \tag{9}$$

$$\frac{du}{dt} < 0 \qquad \text{for} \quad t \geq 0 \tag{10}$$

and

$$\lim_{t \to \infty} u = 20. \tag{11}$$

The property (11) means that, as the time approaches ∞, the temperature of the coffee approaches the steady-state temperature, equal to that of the surroundings.

Determining the parameter λ

To determine the value of the temperature of the coffee at any time, it is necessary to know the value of the parameter λ. This requires information additional to that given in Example 1. One way is to give the temperature

at some other time, besides the initial one. A second way, which will
be explored in the next section, is to reformulate the model, taking into
account the physics of heat transfer. This will show how the parameter λ
depends on such factors as the mass of the heated object, the material of
which it is composed, and its surface area. Such information will make
the model more versatile and will play an essential role later in our study
of the effect of insulating a hot water pipe.

Exercises 12.1

1. An object is at a temperature u which is colder than the temperature u_s of
its surroundings. Let $\lambda > 0$. Which of the following differential equations predict
that the temperature of the object will increase with time?

(a) $\dot{u} = \lambda u$ (b) $\dot{u} = \lambda(u_s - u)^3$

(c) $\dot{u} = -\lambda(u - u_s)^2$ (d) $\dot{u} = -\lambda|u_s - u|$

2. Let λ, u_s and u_0 be constants. Show that the solution of the differential
equation

$$\frac{du}{dt} = -\lambda(u - u_s)$$

which satisfies the initial condition $u = u_0$ when $t = 0$ is

$$u = (u_0 - u_s)e^{-\lambda t} + u_s$$

[Hint: Use the method of Example 1 in the text, which utilizes the fact that the
differential equation is linear constant coefficient.]

3. Repeat Exercise 2, but use the method of separation of variables.

4. A cold beer is at a temperature of $10\,°C$. After 10 minutes the beer is at
a temperature of $15\,°C$. Find how long it takes for the beer to warm to $20\,°C$,
given that the temperature of the room is $30\,°C$.

5. Sketch the general shape of the graph of temperature against time in Exercise
2, assuming $\lambda > 0$ and $u_0 < u_s$.

6. From the expression in Exercise 2 for temperature as a function of time, say
what happens when $u_0 = u_s$. Explain this physically.

7. Suppose that, instead of Newton's law of cooling (2) in the text, the law of
cooling is

$$\frac{du}{dt} = f(u - u_s)$$

for some function f. Explain the physical interpretations of each of the following
conditions on the function f and state whether they seem realistic.

(a) $f(0) = 0$. (b) $f(x) > 0$ for $x > 0$

(c) $f(x) < 0$ for $x < 0$ (d) $f(x) = -f(-x)$.

8. A student was seen to enter a tutor's office at 4.00 p.m. The tutor was later at the bar at 4.30 p.m. drinking heavily. At 6 p.m. the cleaners discovered the student's body in the tutor's office and called the police. The police *first* measured the temperature of the corpse at exactly 6.30 p.m. as 30 °C and later at 8.30 p.m. as 27 °C. The temperature of the office remained at a constant 25 °C. [Hint: Assume Newton's Law of cooling, and use the solution obtained in Exercise 2.]

(a) What is u_s?

(b) Why is it a good idea to set $t = 0$ to correspond to 6.30 p.m.? What is the initial temperature?

(c) Write down an expression for the temperature at time t and hence determine λ from the information given in the question.

(d) Hence determine the time of death, assuming that the temperature of the student just before the murder was 37 °C (normal body temperature).

9. As mentioned in the text, Newton's law of cooling assumes that air at room temperature is blown past the cooling body ('forced cooling'). For cooling in still air ('natural cooling') a better model is to assume that the rate of temperature decrease of the cooling body is directly proportional to the 5/4th power of the difference between the temperature of the body and the temperature of the surrounding air.

(a) Introduce appropriate notation and thus write the law for natural cooling as a differential equation. Is it linear?

(b) Show that the temperature u of the cooling body at time t is given by the formula

$$u = u_s + \left((u_0 - u_s)^{-\frac{1}{4}} + \lambda t/4 \right)^{-4}$$

where $u = u_0$ when $t = 0$ and where u_s is the temperature of the surrounding air and λ is a constant.

12.2 More physics in the model

The model for cooling in the previous section involved the constant of proportionality λ. In this section we make the model more versatile by explaining how the parameter λ depends on physical aspects of the cooling body, such as its mass and size. We do this by reformulating the model, taking into account more of the underlying physics. The idea is to express the model in terms of loss of heat, rather than loss of temperature.

Heat and temperature

Recall that temperature is measured with a thermometer in °C and indicates how hot a substance is. Heat, on the other hand, is a quantity

which can flow from a hotter substance to a colder one, thereby raising
its temperature. The microscopic origin of heat is the motion of atoms
and molecules which compose the substance. Heat is a form of energy,
which in the SI system is measured in joules (J).

The change in the heat of a given substance depends on both the mass
of the substance and the change in temperature, in a manner which we
now describe.

As to the dependence on mass suppose, for example, that 1 joule of
heat flows into 1 kg of the substance, raising its temperature by 1 °C. It
then seems reasonable to suppose that 2 joules of heat will be required
to raise the temperature of 2 kg by the same amount. More generally,
we assume that when the temperature of a given substance is raised by
a fixed amount

{change in heat}   is proportional to   {mass of substance}.   (1)

If, furthermore, we keep the original mass of the substance fixed but
wish to raise its temperature by 2 °C we might expect that it would
take twice as much heat energy and, in general, for a fixed mass of the
substance,

{change in heat}   is proportional to   {change in temperature}.   (2)

Measurements show, however, that (2) is only approximately true. It is
reasonably accurate provided that the temperature stays close enough
to some initially fixed temperature of say 20 °C. To simplify our model,
however, we shall simply assume that (2) always holds.

We now combine the assumptions (1) and (2) in a single formula.
Firstly we define some notation by letting

$m = $ {mass of a given substance},

$H = $ {amount of heat in the sample of mass m},

$u = $ {temperature of the sample.}

Here H and u vary with time and m is a constant. Now if δu denotes the
change in temperature due to a change in the amount of heat δH then
(1) and (2) combine to give

$$\delta H = cm\,\delta u \qquad\qquad (3)$$

where c is a positive constant of proportionality.

For our purposes it will be more useful to recast (3) in a form which
refers to the *rate* of change of heat. Suppose, therefore, that the changes

Table 12.2.1. *Specific heat c for some common substances, taken at 20 °C*
except where otherwise indicated.

Substance	$c\,(\text{J}\,\text{kg}^{-1}\,^{\circ}\text{C}^{-1})$	Substance	$c\,(\text{J}\,\text{kg}^{-1}\,^{\circ}\text{C}^{-1})$
Aluminium	896	Asbestos	841
Copper	383	Brick	840
Iron	452	Concrete	837
Stainless steel	461	Glass	800
Water (at 0 °C)	4226	Butter fat	2300
Water (at 20 °C)	4182	Lamb	3430
Water (at 100 °C)	4211	Potatoes	3520

in (3) occur during a time interval of length δt. Dividing both sides of
(3) by δt and then letting δt approach 0 gives in the limit

$$\frac{dH}{dt} = cm\frac{du}{dt}. \tag{4}$$

This gives the desired relationship between rate of change of heat and
rate of change of temperature at any given instant.

The constant c depends on the type of substance being heated (aluminium, brick, glass, etc.) and is called the *specific heat* of that substance.
Specific heats of some common substances are shown in Table 12.2.1. In
the table the specific heats are said to be 'taken at 20 °C' to indicate that
the formula (3) is valid provided the temperatures stay close to 20 °C.
Thus we can see from Table 12.2.1 and equation (3) that metals such
as aluminium, copper and iron have lower values of c and thus require
much *less* heat energy to raise their temperature than does water and
food products such as butter fat, lamb and potatoes (which contain a
substantial proportion of water).

Newton's law of cooling revisited

A useful formula for the parameter λ occurring in Newton's law of
cooling will now be derived. The derivation will be based on the idea
that a hot object, placed in cold surroundings, cools by giving up heat
to the surroundings. Similarly, a cold object heats up by gaining heat
from its surroundings. This, of course, means that the temperature of
the surroundings will change as it gains or loses heat. Since the heat is
spread over such a large region by convection, however, the change in
temperature is usually neglected.

Table 12.2.2. *Some heat transfer coefficients.*

Type of convection at surface	$h\,(\mathrm{W\,m^{-2}\,^{\circ}C^{-1}})$
Plate in still air	4.5
Airflow at 2 m/s over plate	12
Airflow at 35 m/s over plate	75
Cylinder, 5 cm diameter, in still air	6.5
Cylinder, 2 cm diameter, in still water	890

We now model how the heat is lost to (or gained from) the surroundings. The key quantity to consider is the *rate of change of heat energy contained within the object*, which we have denoted by dH/dt.

First, since the heat loss occurs at the surface of the object, it seems reasonable to suppose that

$$\left\{\begin{matrix}\text{rate of change}\\ \text{of heat}\end{matrix}\right\} \quad \text{is proportional to} \quad \left\{\begin{matrix}\text{surface area}\\ \text{of object}\end{matrix}\right\}. \qquad (5)$$

Second, when cooling is expressed as a loss of heat, Newton's law of cooling says that

$$\left\{\begin{matrix}\text{rate of change}\\ \text{of heat}\end{matrix}\right\} \quad \text{is proportional to} \quad \left\{\begin{matrix}\text{temperature}\\ \text{difference}\end{matrix}\right\}. \qquad (6)$$

Denoting A as the surface area of the object, (5) and (6) may be combined to give

$$\frac{dH}{dt} = -hA(u - u_s) \qquad (7)$$

where h is a positive constant of proportionality.

The constant h is known by several different names — the *convective heat transfer coefficient, surface conductance*, and the *surface convection coefficient*. Experimentally determined values of this coefficient are shown in Table 12.2.2 under various circumstances. The unit for h is watt metre^{-2} $^{\circ}$C^{-1} (where the watt is the unit of power, or 1 Joule s^{-1}).

We can now use the relationship (4) between heat and temperature to write (7) as the following differential equation for the temperature.

$$\frac{du}{dt} = -\frac{hA}{mc}(u - u_s).$$

Comparison of this differential equation with Newton's law of cooling

(2) in the previous section shows that

$$\lambda = \frac{hA}{cm}.\qquad(8)$$

Thus, by including some of the physics of heat transfer in our model of cooling, we have been able to derive a formula for the constant of proportionality λ. The formula involves the heat transfer coefficient h, the surface area A of the heated object, the specific heat c of the substance of which the object is composed, and the mass m of the object.

Exercises 12.2

1. Discuss the significance of the minus sign multiplying the RHS of the differential equation (7) in the text.

2. Find the parameter λ in Newton's law of cooling for an iron plate whose total surface area is $2\,m^2$ which is cooling in a stream of air flowing over the plate at $35\,m/s$. Assume the mass of the plate is $2\,kg$.

3. Consider a $3\,kg$ plate of iron with total surface area $2\,m^2$, initially at a temperature $150\,°C$. In each of the following cases, find how long it takes to cool to a temperature $100\,°C$ if the temperature of the surroundings is $20\,°C$ and

(a) the plate is in still air,

(b) air flows over the plate at a speed of $35\,m/s$.

Useful data are given in various tables in the text.

12.3 Conduction and insulation

This section is about how heat flows through a sample of a given material, this process being known as *conduction* of heat. In some situations it is desirable to maximize the conduction of heat through the material so as to ensure efficient heating; in others the aim is to minimize conduction so as to reduce heat loss. A material which does not conduct heat readily is called an *insulator*.

Steady-state conduction

We will consider the conduction of heat through insulating material between the inner and outer walls of a house, as in Figure 12.3.1 below. Suppose that the inner wall is at a temperature of $20\,°C$ and the outer wall is at a temperature of $10\,°C$. Thus heat flows from the inner wall

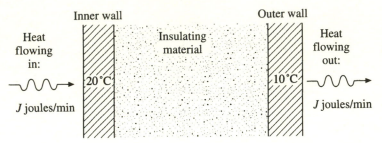

Fig. 12.3.1. Heat flow through a wall.

to the outer wall. The temperature inside the insulating material varies continuously between the temperatures of the inner and outer walls.

As heat flows from the inner to the outer wall, some of the heat goes into raising the temperature of the insulation. We suppose that eventually the temperature at each point inside the insulation reaches a steady-state; this steady-state temperature is independent of the time but varies continuously with respect to distance, from inside to outside. When the temperatures inside the insulation have reached this steady-state, the rate of flow of heat going into the insulation must equal the rate of flow of heat coming out.

Fourier's law for heat flow

A model for heat conduction will now be described which can be used to predict the rate at which heat flows through the insulation, once the steady-state has been reached.

First we need to introduce some notation. Imagine a cross-section through the insulating material which is perpendicular to the direction of heat flow, as in Figure 12.3.2. Suppose that this cross-section is at a distance x from the inside wall, and has area A. Now put

$$J = \left\{ \begin{array}{l} \text{rate at which heat is flowing through a} \\ \text{cross-section of area } A \text{ in the } x\text{-direction} \end{array} \right\}. \tag{1}$$

Intuition suggests that a cross-section of double the area will have double the heat flowing through it, and, more generally, that

$$J \text{ is proportional to } \left\{ \begin{array}{c} \text{the area } A \text{ through which} \\ \text{the heat flows} \end{array} \right\}. \tag{2}$$

Intuition also suggests that the rate of heat flow, along the x-direction,

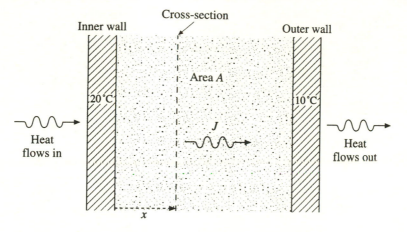

Fig. 12.3.2. Rate of heat flow at distance x.

will depend on the drop in temperature per unit length in this direction. A larger drop per unit length will produce a larger rate of heat flow. The simplest model consistent with this idea is to assume that

$$J \quad \text{is proportional to} \quad \{\text{temperature gradient}\}. \tag{3}$$

This law of heat conduction is named after the famous mathematical physicist *Joseph Fourier*, who proposed it in 1822.

Once the temperature in the insulation has reached a steady-state, the temperature will be a function of x only, say $u = \psi(x)$, and so the temperature gradient at any point is just the derivative du/dx at that point. Thus, combining (2) and (3) into the one formula gives

$$J = -kA\frac{du}{dx} \tag{4}$$

where k is a positive constant of proportionality. Reference to Figure 12.3.2 indicates why the minus sign is necessary: if temperature is decreasing along the x-direction then J is positive (as heat flows from hot to cold) while du/dx is negative.

The constant k which occurs in (4) is called the *conductivity* of the material in the insulation. In Table 12.3.1 are listed conductivities of some common materials. Note that a large value of k indicates a good conductor, while a low value indicates a good insulator.

How good a model is Fourier's law? If we want (4) to hold exactly, then for most materials we must allow k to vary with temperature. For small ranges of temperature, however, our assumption that the

Table 12.3.1. *Thermal conductivities of some common materials. These are measured at 20 °C except where otherwise indicated.*

Substance	$k\,(\mathrm{W\,m^{-1}\,^\circ C^{-1}})$	Substance	$k\,(\mathrm{W\,m^{-1}\,^\circ C^{-1}})$
Aluminium	204	Asbestos	0.113
Copper	386	Brick	0.38–0.52
Iron	73	Concrete	0.128
Stainless Steel	14	Glass	0.81
Water (at 0 °C)	0.57	Wood	0.15
Lamb (at 5 °C)	0.42	Rock wool	0.04
Butter (at 5 °C)	0.20	Polystyrene	0.157

thermal conductivity is constant is a good approximation to what actually happens.

If in (1) we assume that the area of the cross-section stays constant as x varies, then the rate of heat flow J will also stay constant with respect to x, once a steady-state has been reached. Fourier's law (4) then gives a particularly simple differential equation,

$$\frac{du}{dx} = -\frac{J}{kA} = \text{constant},$$

which can be solved by antidifferentiation for the temperature u as a function of the distance x. This in turn enables us to determine the heat flow J, as in the following example.

Example 1. *In a furnace the temperature of an inner wall of area* 3 *metre²* is 500 °C. *The temperature of the outer wall is* 100 °C. *There is* 1 *metre of asbestos insulation between the walls (the furnace having been built before the carcinogenic property of asbestos was realized). How much heat escapes in one minute?*

Solution. *Let J joules per minute be the rate at which heat flows across a cross-section of area* 3 m² *parallel to the walls at a distance x metre from the inner wall (as in Figure 12.3.2). We assume steady-state temperatures, hence J is a constant, which is to be determined.*

Let u be the temperature at distance x. Hence by (4) the differential equation

$$\frac{du}{dx} = -\frac{J}{kA}$$

is satisfied where $A = 3$ m². The initial condition is $u = 500$ °C when $x = 0$. Solving the differential equation by antidifferentiation gives

$$u = -\frac{J}{kA}x + 500$$

Now $u = 100\,°C$ *when* $x = 1\,m$. *Hence*

$$J = 400kA$$
$$= 1200k \quad (as \quad A = 3)$$
$$= 135.6\,J \quad (as \quad k = 0.113).$$

Thus 136 joules of heat escape per minute.

Exercises 12.3

1. Complete the solution to Example 1 in the text by verifying the claims made concerning the solution of the differential equation satisfying the given initial condition.

2. Give reasons why the following would not be suitable to use in place of Fourier's Law.

(a) $J = -A\dfrac{du^2}{dx}$.

(b) $J = -A^2\dfrac{du}{dx}$.

[Hint: Consider dimensions.]

Exercises 3,4,5,6,7,8 refer to the rectangular slab of material shown below. It may be regarded as a wall of a heated room or of a furnace.

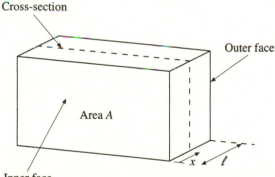

Cross-section

Outer face

Area A

Inner face

    The slab, of thickness ℓ, has an inner face and an outer face, each of area A. The inner face is at a uniform temperature u_a, and the outer face is at a cooler temperature u_b. Heat is assumed to flow straight through the slab from inner to outer face. All points on a cross-section at distance x from the inner face have the same temperature u. Heat flows through this cross-section at the rate J. Assume the temperature has reached the steady-state. In Exercise 7(c) below, you will express J in terms of the other parameters.

3. In terms of the notation introduced above, what is the value of the temperature

(a) when $x = 0$,

(b) when $x = \ell$?

4. In our model we have assumed that the heat flows straight through the slab from inner to outer face, none of it escaping out the other sides. Is this assumption more appropriate when ℓ is large or when ℓ is small?

5. Given that the temperature has reached the steady-state, what follows about the value of J as x increases from 0 to ℓ?

6. On the basis of physical intuition, decide the effect on the value of J of each of the following separate changes.

(a) Increasing u_a.

(b) Increasing u_b.

(c) Decreasing ℓ.

(d) Increasing A.

(e) Replacing the material by one with greater thermal conductivity k.

7. Recall that Fourier's law

$$J = -kA\frac{du}{dx}$$

implies that, once the steady-state has been reached,

$$\frac{du}{dx} = \text{constant}.$$

(a) What does this tell you about the shape of the graph of u against x? Hence state why this graph is as shown in the diagram below.

(b) Use the diagram below and the interpretation of the derivative as a slope to obtain du/dx in terms of u_a, u_b, and ℓ.

(c) Deduce that

$$J = kA\,\frac{u_a - u_b}{\ell}$$

and hence verify your answers to Exercise 6.

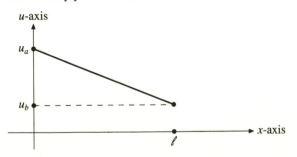

8. (a) Write down Fourier's law as a differential equation for the temperature u as a function of the distance x. The differential equation will involve the parameters k, A, and J. Why can the differential equation be solved by antidifferentiation?

(b) Find the solution of the differential equation which satisfies the initial condition $u = u_a$ when $x = 0$.

(c) Now use the value of the temperature u at the outer face to find J in terms of the parameters k, A, u_a, u_b, ℓ.

(d) Does your answer agree with that found in Exercise 7(c)?

9. Suppose a stone slab has a surface area of $10\,\mathrm{m}^2$ and thickness $2.7\,\mathrm{m}$. The inner and outer faces are at steady temperatures of $20\,^{\circ}\mathrm{C}$ and $0\,^{\circ}\mathrm{C}$ respectively. Given that the thermal conductivity of stone is $2.7\,\mathrm{W\,m^{-1}\,^{\circ}C^{-1}}$, calculate J from your answer to Exercise 7(c).

Exercises 10,11,12,13,14,15 below refer to the figure below (which gives an end-on view of the slab in the diagram above).

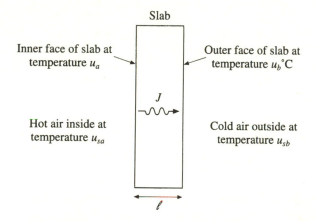

Slab

Inner face of slab at temperature u_a

Outer face of slab at temperature $u_b\,^{\circ}\mathrm{C}$

J

Hot air inside at temperature u_{sa}

Cold air outside at temperature u_{sb}

ℓ

Besides the assumptions already listed for Exercises 3,4,5,6,7,8, the following additional assumptions also apply. The outer face of the slab, at a temperature of u_b, is cooled by a stream of cold air at a temperature of u_{sb}.

Newton's law of cooling applies to the loss of heat from the outer face of the slab to the cold air. From (7) of Section 12.2, this means that

$$\frac{dH_b}{dt} = -h_b A(u_b - u_{sb})$$

where $-dH_b/dt$ is positive and denotes the rate at which heat is being lost to the cold air at the outer face. h_b denotes the heat transfer coefficient between the outer face and the cold air. Similarly, the inner face of the slab, at a temperature of u_a, is heated by a stream of hot air at a temperature of u_{sa}. Newton's law of cooling applies to this face also. In Exercise 15 you will express the rate of heat flow J in terms of the temperatures of the hot and cold air (rather than the temperatures of the inner and outer faces, over which we have no direct control).

10. (a) Arrange the temperatures u_a, u_b, u_{sa}, u_{sb} in increasing order.

(b) What are the signs of (i) $u_b - u_{sb}$ and (ii) $u_a - u_{sa}$?

(c) Extend the diagram shown in Exercise 6 to show also the temperature of the hot air (corresponding to points with $x < 0$) and to show the temperature of the cold air (corresponding to points with $x > \ell$).

(d) At which points is there a discontinuity in the graph you have drawn in (c)?

11. Check, from the formula given above for dH_b/dt and your answer to Exercise 10(b), that $-dH_b/dt$ is positive.

12. (a) The steady-state temperatures having been attained, what is the relationship between the rate J at which heat is arriving at the outer face and the rate $-dH_b/dt$ at which heat is being lost from the outer face? Deduce that

$$J = h_b A(u_b - u_{sb}).$$

  (b) Hence find the temperature u_b at the outer face in terms of the temperature u_{sb} of the cold air (and the parameters J, h_b, A).

13. Newton's law of cooling at the inner face may be written, with a suitable choice of notation, as

$$\frac{dH_a}{dt} = -h_a A(u_a - u_{sa}).$$

What is the sign of dH_a/dt? What is the physical significance of this quantity?

14. (a) The steady-state temperature having been attained, what is the relationship between the rate J at which heat is entering into the slab from the inner face and the rate dH_a/dt at which heat is being transferred to the inner face from the hot air? Deduce that

$$J = h_a A(u_{sa} - u_a).$$

  (b) Hence find the temperature u_a of the inner face in terms of the temperature u_{sa} of the hot air (and the parameters J, h_a, A).

15. (a) From your answers to Exercises 7(c), 12(b) and 14(b), show that

$$J = \frac{A(u_{sa} - u_{sb})}{h_a^{-1} + h_b^{-1} + \ell k^{-1}}.$$

  (b) What features of this answer agree with your physical intuition?

Exercise 16 refers to the diagram below, which shows two slabs of material joined together.

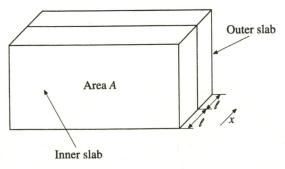

Outer slab

Area A

Inner slab

The notation to be used for each slab is similar to that used in the previous exercises. The inner and outer faces of the combination will be assumed to be at the respective temperatures u_a and u_b, where $u_a > u_b$. The thermal conductivities of the respective slabs are denoted by k_1 and k_2. Assume the temperatures have reached the steady-state; hence J is the same for each slab of material. For the first slab

$$J = -k_1 A \frac{du}{dx} \qquad (0 \le x \le \ell).$$

16. (a) Write out in words the meaning of the above differential equation for the temperature in the first slab.

    (b) Write down a similar differential equation for the temperature in the second slab.

    (c) Solve these two differential equations, using the fact that $u = u_a$ when $x = 0$ and $u = u_b$ when $x = 2\ell$.

    (d) Hence show that

$$J = \frac{A(u_a - u_b)}{\ell(k_1^{-1} + k_2^{-1})}.$$

12.4 Insulating a pipe

In a normal Australian house, narrow pipes are used to convey hot water from a supply to the taps. In cold weather it is desirable to reduce heat loss from such pipes so as to minimize heating costs. For cold water pipes, moreover, sufficient heat may be lost to freeze the water in the pipe. The usual answer to these problems is to insulate the pipes.

We shall formulate a mathematical model of the heat flow through a layer of insulation wrapped around a hot water pipe. The model will be used to investigate how thick the insulation should be. The presentation in this section concentrates on the formulation of the model and the analysis of the results, leaving the detailed calculations for the exercises.

The problem stated

We consider a length L of a typical hot water pipe, as in Figure 12.4.1. The pipe has an outer radius a and is surrounded by a layer of insulation so that the radius of the exposed surface is b, where $b > a$. Thus the thickness of the insulation is $b - a$.

The pipe is assumed to be at the same temperature, say u_w, as the hot water. The surrounding air is at a cooler temperature u_s. Heat is lost by

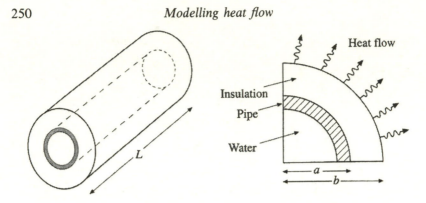

Fig. 12.4.1. An insulated pipe. On the right is a cross-section of the pipe.

flowing through the insulation and then escaping to the surrounding air. The problem is to determine the extent to which the insulation reduces loss of heat from the pipes.

The model

Our model will be based on Fourier's law of heat conduction (to describe the flow of heat outwards through the insulation) and Newton's law of cooling (to describe the loss of heat from the outer surface of the insulation to the surrounding air). Instead of imagining the heat as flowing across plane faces as in Section 12.3, however, it will now be regarded as flowing across cylindrical surfaces, which we now describe.

For each $r > 0$, the points which are at the same distance r from the axis of the pipe form a cylinder of radius r, as shown in Figure 12.4.2. (The cylinder is a *surface, not a solid*.) We assume the cylinder has the same length L as the pipe. The cylinder coincides with the outer surface of the pipe when $r = a$ and with the outer surface of the insulation when $r = b$. If r lies between a and b, then the cylinder lies inside the insulation.

Note that, since a circle of radius r has circumference $2\pi r$, the surface area of this cylinder is given by

$$A(r) = 2\pi r L. \tag{1}$$

Heat is assumed to flow radially outwards through these cylinders with the temperature having the same value at each point of a given cylinder.

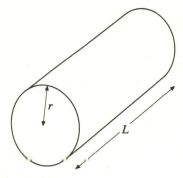

Fig. 12.4.2. Coordinate system for the pipe.

Let

$$u = \left\{ \begin{array}{l} \text{temperature at each point} \\ \text{of the cylinder of radius } r, \end{array} \right\}$$

$$J = \left\{ \begin{array}{l} \text{rate of heat flow radially} \\ \text{outwards through the cylinder} \end{array} \right\}.$$

We assume that the steady-state temperatures have been attained. It follows that J is constant with respect to r, and that u is a function of r, say $u = \phi(r)$. The temperature gradient in the direction of the heat flow can hence be written as du/dr.

Fourier's law of heat conduction, introduced in Section 12.3, will be used to model heat flow through the insulation. Recall that, according to this law, the rate of heat flow is jointly proportional to the temperature gradient and to the area through which heat flows. Thus

$$J = -kA(r)\frac{du}{dr} \tag{2}$$

where the positive constant k denotes the thermal conductivity of the insulation. By (1) this can be written as

$$\frac{du}{dr} = -\left(\frac{J}{2\pi kL}\right)\frac{1}{r}. \tag{3}$$

Because the steady-state has been attained, J and $-J/(2\pi kL)$ are constants; hence (3) is a very simple differential equation for u as function of r.

The inner boundary of the insulation is assumed to be at the common temperature of the pipe and the water. Hence the initial condition for the solution of (3) is

$$u = u_w \qquad \text{when} \quad r = a. \tag{4}$$

This solution of (3) will contain the parameter J, whose value is to be determined. An extra equation for determining J comes from information about the temperature of the insulation at the outside boundary. Thus Newton's law of cooling, in the version given in Section 12.2, gives

$$\frac{dH}{dt} = -hA(b)(u_b - u_s) \tag{5}$$

where $-dH/dt$ is the rate of heat loss from the insulation to the surrounding air, h is the heat transfer coefficient from the insulation to the air, and u_b is the temperature of the insulation at the boundary. In the steady-state,

$$-\frac{dH}{dt} = J$$

(since in the steady-state the rate at which heat is being lost to the air must equal the rate at which heat is flowing through the insulation). The last two equations give

$$J = hA(b)(u_b - u_s) \tag{6}$$

The unknown u_b, furthermore, is the value of u when $r = b$. Hence u_b can be found by solving the differential equation (3) with the initial condition (4). This gives (as will be shown in an exercise)

$$u_b = -\frac{J}{2\pi k L} \ln\left(\frac{b}{a}\right) + u_w. \tag{7}$$

When this value for u_b is substituted into (6), we get an equation for J in terms of known quantities. The solution (as will be shown in an exercise) is

$$J = 2\pi(u_w - u_s)hL \left(\frac{b}{1 + \frac{h}{k}b\ln(\frac{b}{a})}\right). \tag{8}$$

Predictions from the model

Our aim is to find from (8) how J varies with respect to b while all the other parameters are held constant. This tells us how the rate of loss of heat varies as we increase the amount of insulation wrapped around the pipe. To illustrate the possibilities contained in the equations, we now substitute some typical values for the parameters.

Example 1. *A hot water pipe has outside diameter* 15 mm *and* 5 mm *of insulation. The temperature of the water is* 60 °C *and that of the surrounding air is* 15 °C. *The insulation is made of fibreglass with conductivity* 0.05 W m^{-1} °C^{-1} *and the surface heat transfer coefficient is* 10 W m^{-2} °C^{-1}. *Compare the rate of loss of heat per metre length of this pipe with that of a pipe when there is no insulation and the surface heat transfer coefficient is* 8 W m^{-2} °C^{-1}.

Solution. *We convert all quantities to SI units. Thus for the first pipe*

$$a = 0.015 \, \text{m} \qquad u_s = 15 \, °\text{C} \qquad k = 0.05 \, \text{W m}^{-1} \, °\text{C}$$
$$b = 0.020 \, \text{m} \qquad u_w = 60 \, °\text{C} \qquad h = 10 \, \text{W m}^{-2} \, °\text{C}.$$

Substituting into (8) gives

$$J = 26.3 \, \text{W}.$$

In the case of no insulation, $b = 0.015$ *and* $h = 8$, *giving*

$$J = 33.93 \, \text{W}.$$

Thus addition of 5 mm *of fibreglass insulation has led to a* 22% *decrease in heat loss.*

Example 2. *Repeat Example 1 but now assume the pipe has outside diameter* 5 mm *and has* 2 mm *of asbestos insulation. For asbestos take* $k = 0.11$ W m^{-1}°C^{-1} *and* $h = 8$ W m^{-2}°C^{-1}.

Solution. *We obtain* $J = 13.5$ W *with the insulation and* $J = 11.3$ W *without it.*

Note that, in this second example, adding insulation causes an *increase* in heat lost. This is quite a surprising result and it takes a little thought to work out why.

The expected effect of adding insulation is to increase the resistance to heat flow. However there is another important effect for heat flow in cylinders. Adding insulation increases the surface area and heat is lost to the surroundings at a rate proportional to the surface area. There are competing effects; sometimes the first effect wins and heat loss is reduced as more insulation is added (as in Example 1) but sometimes the second effect wins and heat loss is increased as more insulation is added (as in Example 2). Clearly it is important to decide what will happen before deciding to insulate a pipe.

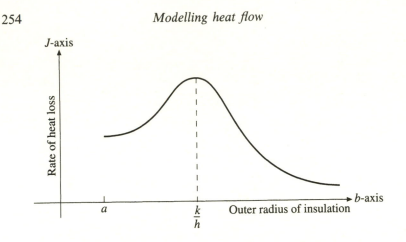

Fig. 12.4.3. Graph showing rate of heat loss verses outer radius of insulation.

Practical considerations

To obtain a rule of thumb as to whether a pipe should be insulated or not we graph the rate of heat loss J against the outer radius b, as given in equation (8). The details have been left to the exercises and the graph is presented in Figure 12.4.3. The turning point is at

$$b = \frac{k}{h}. \tag{9}$$

If b is below this value then the rate of heat loss initially *increases* as b increases. If b is above this value, however, then adding more insulation *decreases* the rate of heat loss.

One way to guarantee that adding insulation *always* decreases the rate of heat loss is to make sure that the outer radius of the pipe is *greater than the critical value* given in equation (9). Thus in Example 2, this condition requires the pipe to have outside diameter $a > k/h = 0.11/8$ m or 13.75 mm. Alternatively, for a given pipe size we should choose the type of insulating material to give a critical value (9) less than the radius of the given pipe.

This chapter has been based on a model introduced in the Open University module: 'Modelling Heat' (1975).

Exercises 12.4

1. (a) State briefly the meaning of each of the symbols occurring in equation

(2) in the text,

$$J = -kA(r)\frac{du}{dr}.$$

(b) Derive the differential equation (3) in the text,

$$\frac{du}{dr} = -\left(\frac{J}{2\pi kL}\right)\frac{1}{r}.$$

(c) Steady-state temperatures having been attained, which symbols in the above differential equation are constants? What are the dependent and independent variables? Solve the differential equation subject to the initial condition $u = u_w$ when $r = a$.

(d) Hence derive the formula (7) in the text,

$$u_b = -\frac{J}{2\pi kL}\ln\left(\frac{b}{a}\right) + u_w.$$

2. (a) Explain the meaning of the symbols occurring in Newton's law of cooling (5) in the text,

$$\frac{dH}{dt} = -hA(b)(u_b - u_s)$$

(b) Deduce, under suitable conditions which should be stated, that

$$J = hA(b)(u_b - u_s).$$

(c) Deduce from the results of part (b) and Exercise 1(d) that the rate at which heat flows through the insulation is given by (8) in the text,

$$J = 2\pi(u_w - u_s)hL\left(\frac{b}{1 + \frac{h}{k}b\ln(\frac{b}{a})}\right).$$

(d) The water being at a higher temperature than the surrounding air, what is the sign of $u_w - u_s$? Given the geometrical interpretations of a and b in the text, what is the sign of $\ln(b/a)$ in the above formula?

3. (a) What do you expect to be the effect on the rate of heat loss from the heated pipe of each of the following separate changes?

(i) The temperature of the water is increased.

(ii) The temperature of the surroundings is increased.

(iii) The length of the pipe is increased.

(iv) The thermal conductivity of the material insulating the pipe is decreased.

(b) Check your answers to part (a) of this exercise by using the formula from Exercise 2(c).

4. (a) Suppose that f is a differentiable function which does not assume the value 0 at any point of its domain. Show that $(c/f)'(x) = 0$ precisely for those x such that $f'(x) = 0$, where c is any non-zero constant.

(b) Let J be given as a function of b by the formula in Exercise 2(c). Show that

$$\frac{dJ}{db} = 0 \qquad \text{precisely when} \quad b = \frac{k}{h}$$

as claimed in the text. [Hint: you can simplify your calculations by using the result from part (a).]

13

Compartment models of mixing

Compartment modelling is a means of constructing a differential equation for a complicated process by considering just the inputs and outputs of the process, during a small time interval. The basic ideas are developed in the context of a model describing the mixing of a dye and water. Compartment models are then formulated for the pollution in a lake and the temperature of a domestic hot water system. The latter model uses ideas about the flow of heat from Chapter 12. The differential equations obtained are mainly of the first-order linear constant-coefficient type.

13.1 A mixing problem

One of the aims of modelling is to isolate the most important factors in a problem and ignore those which may not be important. Even very complicated processes can initially be analysed using very simple mathematical models which may later be extended to more complex and realistic models by incorporating more features. In problems involving the mixing of two or more substances, simple models may be formulated by considering the input and output to a compartment containing the quantity of interest.

The following problem will be used to illustrate these ideas. The problem is illustrated in Figure 13.1.1.

Statement of problem

In a dye factory a large vat is used to mix dye and water. The water flows in at a rate of 6 litres/minute and the dye flows in at a rate of 2 litres/minute. The mixture is drawn off at a rate of 8 litres/minute. Ini-

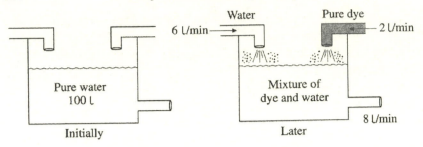

Fig. 13.1.1. Mixing dye and water.

tially the vat contains 100 *litres of pure water. How does the strength of dye in the water change with time?*

We shall set up a compartment model for the mixing of dye and water within the vat to produce a mixture of the two liquids. In this problem, since the total flow rate of ingredients into the vat equals the flow rate of mixture out of the vat, the volume of mixture in the vat is constant. However, the amount of dye in the tank changes with time.

We now introduce some relevant terminology for problems involving mixing of liquids. We then look at some introductory examples of the small time-interval technique which we use to formulate differential equations for this type of problem.

Concentration

In mixing problems in general we refer to the mixture as the *solution*, the substance we are introducing as the *solute* and the liquid which dissolves the solute as the *solvent*. In our problem the dye–water mixture is the solution, the dye is the solute, and the pure water is the solvent.

To obtain a measure of the strength of dye in the dye–water mixture we introduce the *concentration* of the mixture, which is the ratio of the amount of solute to the amount of solution. In our problem it is convenient to define the concentration as a ratio of volumes:

$$\{\text{concentration}\} = \frac{\{\text{volume of solute}\}}{\{\text{volume of solution}\}}. \tag{1}$$

This definition assumes that the solute is homogeneously mixed in the solution. Note that the maximum concentration, which is unity, occurs when the whole of the mixture is pure dye; the minimum concentration, which is zero, corresponds to pure water.

In other problems it may be more useful to define concentration as a ratio of mass of solute to volume of solution. For example, many drug prescriptions have concentrations measured in milligrams per litre. In chemistry, concentration is often defined as the ratio moles of solute per unit volume of solution. (The mole is defined so that the mass of one mole of Carbon-12 atoms is exactly 12 grams.)

Formulating a differential equation

In this model we are concerned only with the input and output of dye to the vat. We are not concerned with any fine detail about the distribution of dye within the vat. This is an appropriate assumption to make when the contents of the vat are *well stirred*. The next phase of the model building process involves formulating a *differential equation* for the amount of dye in the tank, at time t. This is done using a *small interval approximation*, which we now illustrate.

To formulate a differential equation we need to account for the volume of dye entering and leaving the vat. We need to determine how the flow rate of mixture leaving the vat affects the amount of dye leaving the vat in a given time interval. Let us first consider an illustrative example in which the concentration of dye in the mixture remains constant.

Example 1. *Mixture flows out of a tank at a rate of 8 litres/min. Suppose there are 6 litres of dye in the mixture of 100 litres. What volume of dye leaves the tank in a time δt minutes?*

Solution. *Firstly,*

$$\left\{ \begin{array}{c} \text{volume of mixture} \\ \text{flowing out} \end{array} \right\} = 8\delta t \ (litres).$$

Of the mixture, a constant fraction 6/100 is dye. Thus

$$\left\{ \begin{array}{c} \text{volume of dye} \\ \text{leaving tank} \end{array} \right\} = 8\delta t \times \frac{6}{100} = \frac{12}{25}\delta t \ litres.$$

Note that the fraction $\frac{6}{100}$ is the concentration *of the dye.*

More realistically, we would expect the volume of dye in the vat, and hence also the concentration of dye in the mixture, to change with time. For very small time intervals, however, the volume of dye and the concentration do not change significantly. Thus we can approximate the volume of dye and the concentration of dye by constants *over this small time interval*.

Now let us obtain an expression for the change in the volume of dye in the mixture over some small time interval δt. Let

$$x = \phi(t) = \left\{ \begin{array}{c} \text{volume of dye in vat} \\ \text{at time } t \end{array} \right\}$$

and let δx denote the change in volume in a small time interval δt.

The starting point for formulating the differential equation is to write down an equation relating the change in volume of dye to the input and output of dye form the system. Thus

$$\delta x = \left\{ \begin{array}{c} \text{volume of dye} \\ \text{entering vat} \end{array} \right\} - \left\{ \begin{array}{c} \text{volume of dye} \\ \text{leaving vat} \end{array} \right\}. \tag{2}$$

Note that the *volume* of dye is the appropriate measure of the amount of dye here since the concentrations in the problem are given as the ratio of *volume* of dye in the mixture per unit volume of the mixture. In the following example we obtain *approximately* the volume of dye which leaves the vat in a small time interval δt.

Example 2. *Find the approximate change in volume of dye δx in the vat shown in Figure 13.1.1, during a small time interval δt.*

Solution. *Pure dye flows into the vat at a rate 6 litres/minute. So*

$$\left\{ \begin{array}{c} \text{volume of dye} \\ \text{entering vat} \\ \text{in time } \delta t \end{array} \right\} = 6 \, \delta t \quad \text{litres.} \tag{3}$$

The mixture flows out at a rate 8 litres/minute. So

$$\left\{ \begin{array}{c} \text{volume of mixture} \\ \text{leaving vat} \\ \text{in time } \delta t \end{array} \right\} = 8 \, \delta t \quad \text{litres.} \tag{4}$$

Because the mixture is homogeneous (i.e. well stirred) a fraction $x/100$ of the mixture flowing out is dye — at the beginning of the time interval. We assume this does not vary significantly over the small time interval δt. Thus, of the $8 \, \delta t$ litres leaving the vat in a time δt, we obtain

$$\left\{ \begin{array}{c} \text{volume of dye} \\ \text{leaving vat} \\ \text{in time } \delta t \end{array} \right\} \simeq 8 \, \delta t \, \frac{x}{100} \quad \text{litres,} \tag{5}$$

with the approximation becoming more accurate as δt becomes smaller. Hence, using (2), (3) and (5) we obtain

$$\delta x \simeq 6 \, \delta t - \frac{8x}{100} \delta t \quad \text{litres} \tag{6}$$

as the approximate change in the volume of dye in the vat in a small time interval δt.

Having accounted for the input and output of dye to the vat we are finally ready to derive a differential equation for the volume of dye in the vat at time t. First we divide (6) by δt and then we let $\delta t \to 0$. Since

$$\frac{dx}{dt} = \lim_{\delta t \to 0} \frac{\delta x}{\delta t}$$

we obtain

$$\frac{dx}{dt} = 6 - \frac{8x}{100}. \tag{7}$$

But the problem was posed in terms of finding how the strength, or concentration, of the dye in the mixture varied with time. To answer this, a differential equation for the concentration of dye in the vat at time t is more relevant.

Equation for the concentration

Let

$$c = \left\{ \begin{array}{c} \text{concentration of dye in the vat} \\ \text{at time } t \end{array} \right\}. \tag{8}$$

Since the volume of mixture in the tank remains constant at 100 litres, then from (1)

$$c = \frac{x}{100} \quad \text{or} \quad x = 100c. \tag{9}$$

Substituting (9) into (7) we obtain

$$\frac{d}{dt}(100c) = 6 - \frac{8 \times 100c}{100}$$

which simplifies to

$$\frac{dc}{dt} = 0.06 - 0.08\,c. \tag{10}$$

The initial condition for this differential equation is

$$c = 0 \quad \text{at} \quad t = 0 \tag{11}$$

since there was no dye in the vat initially. It is to be verified in Exercise 13.1.4 below that the solution of the differential equation (10) subject to the initial condition (11) is

$$c = 0.75(1 - e^{-0.08t}). \tag{12}$$

A sketch of the solution (12) is given in Figure 13.1.2 below. The value 0.75 corresponds to the steady-state solution, obtained from the

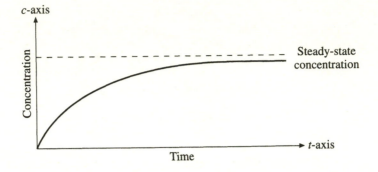

Fig. 13.1.2. Graph of how the concentration of dye varies with time, from equation (12).

differential equation by setting $\dfrac{dc}{dt} = 0$. From (10) we see that, if $c < 0.75$, then $\dfrac{dc}{dt} > 0$. Thus all concentrations with $c_0 < 0.75$ increase with time. They all tend to the steady-state value 0.75.

Summary

A summary of the procedure for formulating a differential equation for mixing problems is now given.

STEP 1: *Draw a diagram showing the rates of flow and concentration of the liquids in and out of the mixture. Check to see if the total volume of the mixture remains constant.*

STEP 2: *Write down word equations relating the change in the amount of solute in the system to the amount input and output, in a small time interval δt. Convert to symbols using data given in the problem.*

STEP 3: *Divide by δt and take the limit as δt → 0, obtaining a differential equation for the amount of solute present as a function of time.*

STEP 4: *If necessary, use a substitution to obtain a differential equation for another quantity of interest (e.g. concentration).*

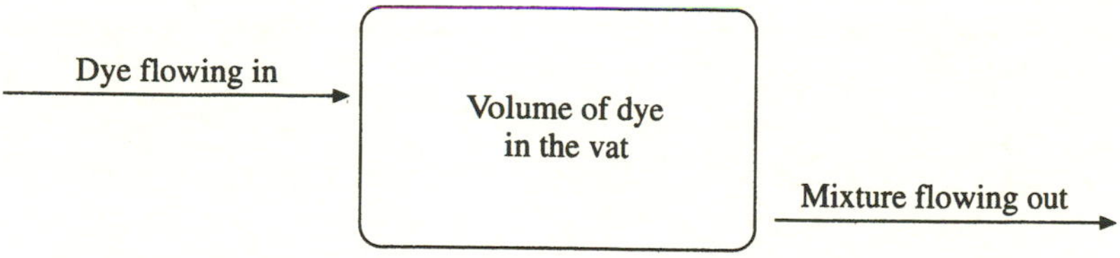

Fig. 13.1.3. Compartment diagram for the dye mixing problem.

This type of approach can be applied to a variety of problems other than the mixing of liquids. For example, in the next section we model pollution in a lake. Here the quantity flowing out of the lake is the *mass* of pollutant. In Section 13.3 we model the heat loss from a hot water system where the quantity going into and out of the hot water system is the *heat energy*.

The approach used to analyse such problems is called *compartment modelling*. This involves drawing a compartment diagram. For the dye mixing problem in this section, the compartment diagram consists of a box showing the amount of dye contained in the vat together with an arrow showing the input through dye entering the vat and another arrow showing the output through dye leaving the vat as the mixture is drawn off. A compartment diagram for this problem is shown in Figure 13.1.3. Note that the input of water is not relevant to the compartment diagram here since the amount of dye is the quantity of interest.

Compartment diagrams, while apparently trivial for the dye mixing problem of this section, are useful aids when dealing with more complicated models where there is more than one compartment (and sometimes many compartments) and there is an interaction between compartments. In Chapters 18 and 19 we consider some examples of two-compartment models including interaction between species of animal, insulin and glucose interaction in diabetics and models of combat between two armies.

Exercises 13.1

1. In which of the following situations is the volume of the mixture constant?

   (a) Dye flows into a container at a rate 3 litres/min and water flows in at a rate 11 litres/min. The mixture flows out at a rate 12 litres/min.

   (b) Pure water flows in at a rate 5 litres/s, a chemical flows in at a rate 2 litres/s and the mixture flows out at a rate 7 litres/s.

2. In each case determine the *amount* of the substance entering the compartment in a time interval δt. Give the appropriate units.

   (a) Pure dye flows in at the rate 9 litres/min. Give the amount of dye.

   (b) Dye, at a concentration 5%, flows into a vat at a rate of 2 litres/min. Give the amount of dye.

   (c) Water containing 0.5 kg of salt per litre enters a tank at a rate of 3 litres/min. Give the amount of salt.

3. In each case below assume that there is an amount x of the substance present at time t and then determine the amount δx of the substance entering and the amount leaving the compartment in the small time interval from t to $t + \delta t$. Indicate units, and also indicate where your answer is exact, and where it is approximate for very small δt. Hence determine the differential equation for the amount of substance and then for the concentration in each case.

   (a) Pure dye flows into a vat at a rate of 4 litres per minute and pure water flows into the vat at a rate of 8 litres per minute. Initially there are 50 litres of pure water in the vat. The well stirred mixture is drawn off at a rate of 12 litres per minute.

   (b) Dye of concentration 40% by volume flows into a vat at a flow rate of 2 litres per minute. Initially there are 100 litres of pure water in the vat. The well-stirred mixture flows out at the same rate.

   (c) A salt-water mixture containing 15 grams of salt per litre flows into a lake of volume 2000 litres at a flow rate of 10 litres per minute. The flow rate out of the lake is 16 litres per minute and pure water flow into the lake at a flow rate of 6 litres per minute. Assume the mixture is well stirred.

4. Solve the differential equation

$$\frac{dc}{dt} = -\alpha c + \beta$$

where α and β are constants, using the method explained in Section 11.1.

5. Dye at 50% concentration enters a tank at a rate of 1 litre/min. Fresh water enters the tank at a rate 2 litres/min. The well-stirred mixture leaves at a rate 3 litres/min. The tank contains initially 50 litres of a 50% concentration of dye and water.

   (a) Formulate a differential equation for the concentration.

   (b) What is the concentration after 15 minutes?

   (c) What is the steady-state concentration?

6. At 6.00 pm on a Friday night a public bar opens and is rapidly filled with clients of whom the majority are smokers. The bar is equipped with ventilators which exchange the smoke–air mixture with fresh air. Unfortunately cigarette smoke contains 4% carbon monoxide, CO, and a prolonged exposure to a concentration of more than 0.012% of CO can be fatal. The bar has dimensions of 20 m by 15 m by 4 m and it is estimated that smoke enters the room at a constant rate 0.006 m^3/minute. The ventilators remove the mixture of smoke and air at 10 times the rate that smoke is produced. *The problem is to find the time when the concentration of* CO *reaches* 0.012%.

(a) Formulate a differential equation for the concentration of CO at time t.

(b) By solving the differential equation find at what time the lethal concentration will be reached.

7. Using the data of the previous exercise determine the rate at which the air-conditioners should operate if the concentration is *never* to reach the lethal level.

8. A dam contains 10^6 litres of water. Fresh water enters the dam at a rate of 10^4 litres per day and the same amount flows out each day. Suppose someone spills a barrel of pesticide into the dam.

(a) Set up a differential equation for the concentration of pesticide in the dam.

(b) Suppose that at a given time the concentration is 5 times the safe level for use by stock. How long before the stock can use the dam?

9. A 100 litre tank originally contains 50 litres of fresh water. Beginning at time $t = 0$, water containing 50% of pollutant flows into the tank at a rate 2 litres per minute and the well-stirred mixture leaves at a rate of 1 litre per minute. This exercise involves a situation where the volume of the mixture is not constant.

(a) Find the volume of mixture in the tank as a function of time.

(b) Formulate a differential equation for the volume of pollutant in the tank.

(c) Hence show that the concentration of pollutant at the time the tank overflows is approximately 48%.

13.2 Modelling pollution in a lake

Surrounding the Great Lakes system, on the eastern United States–Canada border, is an area of extensive industrial activity. It has been common practice to dump waste products into the Lakes. As a result of this the Lakes have been seriously polluted. In this section we adopt the compartment modelling approach to estimate how long it takes to reduce the pollution levels in a lake through the flow of water into and out of the lakes given that no more waste products are put into the lakes.

In particular, we look at Lake Superior, the largest of the Great Lakes (see Figure 13.2.1). The volume of Lake Superior is approximately 1.2×10^{13} litres and it is estimated that 6.5×10^{10} litres of fresh water enters the lake each year. We shall consider the following problem.

Suppose the authorities suddenly stop all pollution flowing into the lake. Find how long it takes for the concentration of the pollutant to decrease to 10% of its original value.

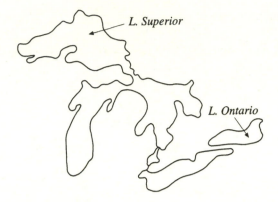

Fig. 13.2.1. The Great Lakes system on the eastern USA and Canadian border.

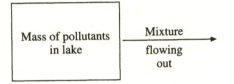
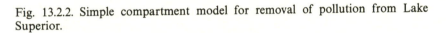

Fig. 13.2.2. Simple compartment model for removal of pollution from Lake Superior.

The model

The mechanism for clearing the lake of pollution is the inflow of pure water which dilutes the water–pollutant mixture. This process is what we wish to model. We start with the following assumptions:

- the lake is well stirred so that pollutant is uniformly mixed throughout the lake,
- the flow rate of mixture out of the lake is equal to the flow rate of pure water into the lake (i.e. we ignore the net effect of rainfall and evaporation).

From the second of these two assumptions we see that the volume of the lake remains constant.

The quantity of interest is the amount of pollutant in the lake at any given time, measured in tonnes. In this problem we choose initial time to be when pollution input to the lake stops. A compartment diagram for this situation is shown in Figure 13.2.2.

We now define some notation. Let us write

$$m = \phi(t) = \left\{ \begin{array}{l} \text{mass of pollutant} \\ \text{in lake at time } t \end{array} \right\}.$$

and define the constants

$$V = \{\text{volume of lake}\} = 1.2 \times 10^{13} \text{ litres} \qquad (1)$$

and

$$F = \left\{ \begin{array}{l} \text{flow rate of mixture} \\ \text{out of lake} \end{array} \right\} = 6.5 \times 10^{10} \text{ litres/year.} \qquad (2)$$

We also define the concentration of pollutant in the lake c by

$$c = \frac{m}{V}$$

We would expect the concentration to be a decreasing function of time in this problem.

Formulating the differential equation

We now write down the fundamental equation which relates the change in the mass of pollutants to the input and output of pollutant. Let δm denote the change of pollutant in the lake in a small time interval from t to $t + \delta t$. For $t \geq 0$ there is no flow of pollutant into the lake and so

$$\delta m = 0 - \left\{ \begin{array}{l} \text{mass of pollutant} \\ \text{flowing out} \end{array} \right\}. \qquad (3)$$

Now,

$$\left\{ \begin{array}{l} \text{volume of mixture} \\ \text{flowing out of lake} \\ \text{in time } \delta t \end{array} \right\} = F\delta t \text{ litres} \qquad (4)$$

since F is the flow rate of mixture in litres per unit time. Thus

$$\left\{ \begin{array}{l} \text{mass of pollutant} \\ \text{flowing out of lake} \\ \text{in time } \delta t \end{array} \right\} = \left\{ \begin{array}{l} \text{volume of mixture} \\ \text{flowing out of lake} \\ \text{in time } \delta t \end{array} \right\} \times \left\{ \begin{array}{l} \text{fraction of} \\ \text{pollutant} \\ \text{in mixture} \end{array} \right\}$$

$$\simeq F\delta t \times \frac{m}{V} \text{ tonnes} \qquad (5)$$

since m/V gives the mass of pollutant per unit volume of the mixture.
Hence by (3) and (5) we obtain

$$\delta m \simeq -\frac{F}{V} m \, \delta t. \qquad (6)$$

Now, dividing by δt and letting $\delta t \to 0$, we obtain

$$\frac{dm}{dt} = -\left(\frac{F}{V}\right)m \tag{7}$$

as the differential equation for the mass of pollutant in the lake at time t.

We define the concentration

$$c = \left\{\begin{array}{c}\text{concentration of}\\\text{pollutant in lake}\\\text{at time } t\end{array}\right\} = \frac{\left\{\begin{array}{c}\text{mass of}\\\text{pollutant}\\\text{in lake}\end{array}\right\}}{\left\{\begin{array}{c}\text{volume}\\\text{of lake}\end{array}\right\}} = \frac{m}{V}. \tag{8}$$

Substituting into (7) we obtain the differential equation

$$\frac{dc}{dt} = -\left(\frac{F}{V}\right)c \tag{9}$$

for the concentration of pollutant in the lake as a function of time.

Now we need to find an initial condition. We have not been given the initial concentration or initial mass of pollutant in the lake, so we introduce the symbol c_0 for the initial concentration. The solution of (9) with the initial condition $c = c_0$ when $t = 0$ is

$$c = c_0 e^{-(F/V)t}. \tag{10}$$

Thus to find the time $t = T$ when the concentration reaches 10% of its initial value, we obtain the equation

$$\frac{1}{10}c_0 = c_0 e^{-(F/V)T}. \tag{11}$$

Solving for T (omitting the algebra) yields

$$T = \frac{V}{F}\ln(10). \tag{12}$$

Substituting the values for F and V for Lake Superior from (1) and (2) we find that our model predicts that *it takes approximately 425 years for the concentration of pollutant to be reduced to 10% of its initial value, even if no more pollutants are put into the lake.*

Limitations of the model

The model developed in this section ignores many factors which might influence the clearance of pollution from a lake. Some of these are as follows:

- The flow of water through the lake depends on the existence of stagnant regions cased by eddy currents, thermal layers and wind. Thus the assumption of a well-stirred mixture is not always true.
- Some pollutants may settle on the bottom of the lake.
- Bacterial action can affect the concentrations.
- The volume of the lake may not be constant over a whole year.

Despite these limitations, the model does provide a starting point for further investigations. A detailed discussion of this model and its applicibility is given in the article by Rainey (1967). In the exercises we also look at simple extensions. These include a model where pollutants are fed into one lake from an outflow from another lake and a model where bacteria consume some of the pollution.

Exercises 13.2

1. For Lake Erie the flow rate of water into and out of the lake is 1.75×10^{11} litres/year. The volume of the lake is 4.6×10^{11} litres. How long does it take for the concentration of the lake to reduce to one quarter of its initial value?

2. For Lake Ontario, about 84% of its inflow comes from Lake Erie. Using the data in Exercise 1 together with the volume of Lake Ontario as 1.6×10^{11} litres, find an expression for the concentration of pollutants at time t. [Hint: You will neeed to account for a variable concentration in the input of pollutant to Lake Ontario.]

The next two exercises involve a slightly different type of problem which nevertheless uses the same technique of a small interval analysis to formulate the governing differential equation.

3. In a cylindrical container, water drains through a small circular hole at the bottom of the container. The radius of the circular hole is a, the radius of the cylinder is b and the height of the cylinder is H. The water emerges with velocity v.

(a) Using a small interval analysis show that the height of the water in the tank at time t is given by

$$\frac{dh}{dt} = -\left(\frac{a^2}{b^2}\right)v.$$

(b) Toricelli's law states that $v^2 = 2gh$ where g is acceleration due to gravity. By solving the differential equation show that the time taken for the cylinder to empty, given that it was initially full, is

$$T = \frac{b^2}{a^2}\sqrt{\frac{2H}{g}}.$$

(c) Design your own experiment to test the validity of this model.

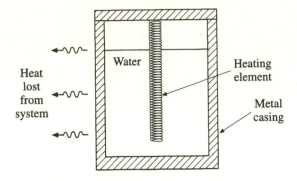

Fig. 13.3.1. Hot water system to be modelled.

4. It is well known that a stream of water emerging from a cylindrical hole contracts so that its cross-section area is 0.6 times the area of the hole. How would this change the model of the previous exercise? Does this help the model agree better with your experimental observations?

13.3 Modelling heat loss from a hot water tank

In this section we will formulate a mathematical model which describes heat loss from a typical domestic hot water tank. The model is formulated using the compartment approach adopted in this chapter. It also uses some of the ideas developed in Chapter 12.

Suppose we have a cylindrical tank, which is partially full of water (see Figure 13.3.1). The water is heated by a heating element which is immersed in the water. Heat is supplied at a constant rate 3000 watts. Some heat is lost to the surroundings, at temperature 15 °C, from the surface of the tank. *Our problem is to find how long it takes to heat the water to a comfortable* 60 °C, given that the water is initially at the same temperature as the surroundings, 15 °C.

The model

The main idea of the model is to account for the flow of heat into and out of the water. We assume that it only requires a negligible amount of heat to raise the temperature of the metal casing to that of the water. A compartment diagram of the heat flow is shown in Figure 13.3.2.

To arrive at a differential equation we examine the heat energy input

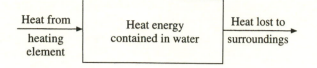

Fig. 13.3.2. Compartment diagram for the model.

and output to the system in a very small time interval δt. First, we define

$$u = \phi(t) = \left\{ \begin{array}{c} \text{temperature} \\ \text{at time } t \end{array} \right\}$$

$$H = \psi(t) = \left\{ \begin{array}{c} \text{heat contained} \\ \text{in water} \\ \text{at time } t \end{array} \right\} \tag{1}$$

We also define symbols for the following constants:

$$q = \left\{ \begin{array}{c} \text{rate at which heat} \\ \text{is supplied to water} \end{array} \right\} = 3000 \text{ watts,}$$

$$u_s = \left\{ \begin{array}{c} \text{temperature of} \\ \text{surroundings} \end{array} \right\} = 15\,^\circ\text{C} \tag{2}$$

$$u_0 = \left\{ \begin{array}{c} \text{initial temperature} \\ \text{of water} \end{array} \right\} = 15\,^\circ\text{C}.$$

The change in heat energy in the water, δH, in the time interval δt is given by

$$\delta H = \left\{ \begin{array}{c} \text{heat entering water} \\ \text{from heater} \end{array} \right\} - \left\{ \begin{array}{c} \text{heat lost} \\ \text{to surroundings} \end{array} \right\}. \tag{3}$$

Since q is the *constant rate* of heat supplied to the water, then

$$\left\{ \begin{array}{c} \text{heat entering water} \\ \text{from heater} \end{array} \right\} = q\delta t. \tag{4}$$

The heat lost to the surroundings is modelled using Newton's law of cooling. From Section 12.2, the rate of heat loss is $hA(u - u_s)$ where A is the surface area of the tank and h is the heat transfer coefficient. Thus

$$\left\{ \begin{array}{c} \text{heat lost} \\ \text{to surroundings} \end{array} \right\} \simeq hA(u - u_s)\delta t. \tag{5}$$

Note that this is positive for $u > u_s$. Note also that this is only an approximate expression since the temperature u changes with time. Over a small time interval, however, we can neglect the variation in the temperature u.

Substituting (4) and (5) into (3) we obtain

$$\delta H = q\delta t - hA(u - u_s)\delta t. \tag{6}$$

Dividing by δt and then letting $\delta t \to 0$ we obtain

$$\frac{dH}{dt} = q - hA(u - u_s). \tag{7}$$

Our objective is to obtain a differential equation for the temperature. To do this we must relate heat H to temperature u. We saw how to do this in Section 12.2 where we argued $\delta H = cm\,\delta u$, where c is the specific heat and here m the mass of water in the tank. Dividing by δt and letting $\delta t \to 0$ we thus obtain the relation

$$\frac{dH}{dt} = cm\frac{du}{dt} \tag{8}$$

which we then substitute into (7). Hence

$$\frac{du}{dt} = \frac{q}{cm} - \frac{hA}{cm}(u - u_s) \tag{9}$$

is the differential equation for the temperature of the water as a function of the time. From (2), the initial condition for (9) is

$$u = u_0 \quad \text{at} \quad t = 0. \tag{10}$$

Time to heat the water

To find the time to heat the water to $60\,°C$ we must first solve the differential equation. We can see the form of the differential equation easier by lumping the constants together. Thus we write (9) and (10) in the simpler form

$$\frac{du}{dt} + \alpha u = \beta, \qquad u = u_0 \text{ at } t = 0 \tag{11}$$

with the 'lumped' constants α and β given by

$$\alpha = \frac{hA}{cm} \quad \text{and} \quad \beta = \frac{q}{cm} + \frac{hA}{cm}u_s.$$

The solution is calculated (Exercises 13.3.1) as

$$u = u_0 e^{-\alpha t} + \frac{\beta}{\alpha}(1 - e^{-\alpha t}).$$

To find the time, $t = T$, when the water reaches $60\,°C$, we need to solve the equation

$$60 = u_0 e^{-\alpha T} + \frac{\beta}{\alpha}[1 - e^{-\alpha T}]$$

for T. After some simple algebra, we obtain the expression

$$T = \frac{1}{\alpha} \ln \left(\frac{\beta/\alpha - u_0}{\beta/\alpha - 60} \right).$$

(12)

Note that (12) is a well-defined expression since it can be shown that the term inside the brackets is always positive and thus we never take the log of zero or a negative number.

We now substitute the appropriate values into (12). Using (2) together with the typical values

$m = \{\text{mass of water}\} = 50\,\text{kg}$

$c = \{\text{specific heat of water}\} = 4200\,\text{J}\,\text{kg}^{-1}\,{}^\circ\text{C}^{-1}$

$A = \{\text{surface area of tank}\} = 1\,\text{m}^2$

$h = \{\text{heat transfer coefficient}\} = 10\,\text{W}\,\text{m}^{-2}\,{}^\circ\text{C}^{-1},$

we obtain the values $\alpha = 4.76 \times 10^{-5}$ and $\beta = 1.5 \times 10^{-2}$ and hence

$$T \simeq 3413 \text{ seconds} \simeq 57 \text{ minutes.}$$

(Note that the calculation gives the time in seconds since all quantities in the problem have been converted to SI units.) Thus our model predicts that it takes approximately 57 minutes for the water to be heated from 15°C to 60°C.

In this problem we have introduced symbols for all the constants. Not only does this make the algebra simpler but it can also be easier to perform dimensional and physical checks on the answer. Also, now that we have a general expression we can easily look at times for heating with different values of the various constants. One can then easily determine how these times change as one of the parameters (for example, the surface area of the tank) changes.

Discussion

Is this a good model? To answer this we should check the prediction of the model with a real water heater. One possible problem with this model is that we assumed that the temperature was the same at all points in the water. If the water was well stirred then there would not be a problem. This will not be true in practice, however, since the hot water will rise to the top. Nevertheless, the model does incorporate most of the essential physics. It should be able to be used, albeit cautiously, to gain some *insight* into the following questions:

- How much better is it to have the heater inside the house rather than outside?
- What advantages are there in insulating the system?
- How much saving is made by using off peak heating?

To use the model to help answer some of these questions you should attempt some of the exercises. Further discussion may be found in Open University module 'Modelling Heat' (1975).

Exercises 13.3

1. Obtain the solution, given in the text, of the differential equation

$$\frac{du}{dt} + \alpha u = \beta \qquad \text{with} \quad u = u_0 \quad \text{when} \quad t = 0,$$

where α and β are constants.

2. What is the steady-state temperature for the problem in the previous exercise? Hence argue that the expression

$$\ln\left(\frac{\beta/\alpha - u_0}{\beta/\alpha - 60}\right)$$

is a well-defined expression.

3. Imagine that the water heater is inside the house (at temperature $20\,°C$). Recalculate the time taken to heat the water.

4. Plot a graph showing the time taken to heat the water as the surface area of the tank is varied. For all other data use the values in the text.

5. Plot a graph showing the time taken to heat the water as the heat transfer coefficient h is varied. For all other data use the values in the text.

6. Plot a graph showing the time taken to heat the water as the outside temperature u_a is varied. Assume that the initial temperature is the same as the outside temperature. For all other data use the values in the text.

7. Suppose the water heater is *perfectly* insulated so that no heat is lost to the surroundings. Modify the expression obtained in the text for the time to heat the water to $60\,°C$. Calculate this time using the data given in the text.

8. Modify the model used in the text to take account of heat lost to the metal casing. You are given that the specific heat of the metal is $400\,J\,kg^{-1}\,°C$ and the mass of the metal casing is $5\,kg$. How much difference does this make to the result for the time to heat the water? Express your answer as a percentage.

9. The cost of heating is proportional to the time it takes to heat the water. Is it cheaper to switch the heater off all night (for eight hours) and have it turn on in the morning or is it better to use a thermostat which switches the heater on whenever the temperature falls below $60\,°C$? Assume the temperature of the surroundings is $5\,°C$. (You will have to find what temperature the water cools to in eight hours overnight and how long it takes to reheat to $50\,°C$, amongst other things.)

Part four
Further Mechanics

14

Motion in a fluid medium

The refinement of some of the simple mechanics models obtained in Part A is to be undertaken in this chapter and the next. Thus, for example, in Chapter 2 the problem of free fall under gravity was considered. The medium through which the object moves was completely ignored and so was the size and shape of the falling object. In this chapter these features will be included and it will be seen that two new forces become relevant — the drag force and the buoyant force. These forces provide the mechanism for some phenomena not present in the model of Chapter 2: the decrease in acceleration of all free falling objects, and the ability of some objects such as balloons to rise rather than fall.

The differential equations obtained in this chapter are first-order linear with constant coefficients or first-order separable. Knowledge of Section 11.1 of Chapter 11 is therefore required.

14.1 Some basic fluid mechanics

As an object moves through a fluid, a force is exerted by the fluid on the object which is in the opposite direction to the motion of the object. This force is called the *drag force*. To gain an understanding of the quantities influencing the drag force in a fluid, it is necessary first to discuss two fundamental properties of a fluid: the viscosity and Reynolds' number.

Viscosity

Gases and liquids are collectively known as fluids since they can both be made to flow if a force is applied. While all fluids will flow, we know that pouring water out of a bottle is a faster process than pouring, for example, cream or honey. One obvious distinction between these three

Table 14.1.1. *Coefficients of viscosity of selected liquids and gases.*

Liquid	Viscosity η (poise)
Water (at 0 °C)	1.792×10^{-2}
Water (at 20 °C)	1.005×10^{-2}
Water (at 40 °C)	0.656×10^{-2}
Ethyl alcohol (at 20 °C)	1.20×10^{-2}
Castor oil (at 20 °C)	9.86×10^{-2}
Mercury (at 20 °C)	1.55×10^{-2}
Gases	Viscosity η (poise)
Air (at 0 °C)	1.71×10^{-4}
Air (at 20 °C)	1.81×10^{-4}
Air (at 40 °C)	1.90×10^{-4}
Hydrogen (at 20 °C)	0.93×10^{-4}
Ammonia (at 20 °C)	0.97×10^{-4}
Carbon dioxide (at 20 °C)	1.46×10^{-4}

fluids is their 'thickness', which is an intuitive measure of how close the fluid is to a rigid object. Here we are using the word 'thickness' as in the phrase 'thickened cream' or in the saying 'blood is thicker than water'. A precise measure of the thickness of a fluid is found in fluid mechanics and it is called the *coefficient of viscosity* which is sometimes denoted by the symbol η.

The dimensions of η are $ML^{-1}T^{-1}$ (mass $\times$ length^{-1} $\times$ time^{-1}). The SI unit of viscosity is 1 kg m^{-1} s^{-1}, but it is common practice to use a unit known as the *poise*, which is one-tenth of the SI unit. A value of 10^{-4} poise is typical of a gas whereas liquids typically have a value of η of about 10^{-2} poise (see Table 14.1.1). It is common experience that in hot weather honey flows rapidly in comparison with its behaviour in colder weather. This is saying that the coefficient of viscosity is temperature dependent, as is evident from an inspection of Table 14.1.1.

Reynolds' number

When an object falls through a fluid medium, a disturbance is caused as the fluid is pushed away from the path of the object. The fluid dynamicist Osborne Reynolds (1883) demonstrated that the fluid flow can change character as the speed of the object relative to the fluid increases. This is illustrated in Figure 14.1.1 where the curves, known as streamlines,

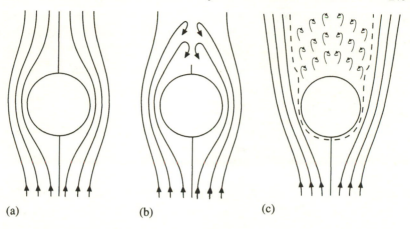

(a)   (b)   (c)

Fig. 14.1.1. Three types of flow of a fluid around a sphere. Here the frame of reference is such that the spheres are held fixed.

indicate the motion of the fluid around the sphere. For low speeds the flow looks the same upstream and downstream as in Figure 14.1.1(a). As the speed increases a wake is formed formed behind the object and the fluid recirculates inside the wake, as in Figure 14.1.1(b). The type of flow in Figure 14.1.1(a) and Figure 14.1.1(b) is called laminar. For very high speeds the flow in the wake region is no longer smooth but turbulent as in Figure 14.1.1(c). It might be expected that the modelling of the drag will be different in each of these three typical situations.

Of course Reynolds realized that not only the velocity of the object but also the properties of the fluid would affect the type of flow. To take account of this he introduced a dimensionless parameter (now known as the *Reynolds' number* of the motion). The Reynolds' number is defined by the formula

$$R = \frac{\rho_f |\dot{x}| d}{\eta}$$

where ρ_f denotes the density (mass per unit volume) of the fluid, $|\dot{x}|$ denotes the speed of the object, η is the coefficient of viscosity of the fluid, and d denotes a characteristic length of the object (for example, if the object is a sphere the number d could denote the diameter of the sphere). The formula shows that the Reynolds' number is proportional to the speed of the object, but inversely proportional to the viscosity of the fluid.

For a smooth, falling sphere, a Reynolds' number in the range $R <$

1 typically gives a flow of the type in Figure 14.1.1(a), a Reynolds'
number in the range $1 < R < 3000$ typically gives a flow of the type
in Figure 14.1.1(b) and for a Reynolds' number $R > 3000$ the flow is
turbulent, as in Figure 14.1.1(c). The critical Reynolds' number for when
the flow becomes turbulent is generally smaller for spheres with rough
surfaces (e.g. a baseball or cricket ball).

Example 1. *Calculate the Reynolds' number for the flow past a raindrop of radius*
1 mm falling in air at 5 m/s. (Data: $\rho_f = 1.23\,\text{kg/m}^3$, $\eta = 1.8 \times 10^{-5}\,\text{kg/m s}$).

Solution. *Choosing the characteristic length as the diameter of the raindrop gives*

$$d = 0.002\,\text{m}.$$

Substituting this value and the given data in the formula for the Reynolds' number
gives

$$R = 670.$$

Since $0 < R < 3000$, we might expect that the flow around the raindrop will be like
that in Figure 14.1.1(b).

It is to be expected that modelling of the drag will be different for
each of the three flows shown in Figure 14.1.1 corresponding to low,
intermediate and high Reynolds' numbers. Thus it is to be expected that
the drag will depend on the Reynolds' number (and hence on the density
and viscosity of the fluid, and the velocity and size of the object) and
also on the shape of the object. Our concern will be solely with spherical
objects.

Stokes' law

For a sphere of radius r moving at velocity $\dot{x}$, Stokes in 1845 derived the
following expression for the magnitude of the drag force:

$$|F_D| = 6\pi r\eta|\dot{x}|.$$

The formula is valid only for very low Reynolds' numbers ($R \ll 1$).
This requirement puts a severe limitation on the applicability of Stokes'
law. Example 1 shows that, even for a sphere as small as a raindrop,
the Reynolds' number is much larger than 1. Stokes' law is applicable,
however, to dust-like particles, such as smoke and other pollutants in
the air, and silt in lakes and streams. In Section 14.3, an application of
Stokes' law to the calculation of the charge of the electron, will be given.

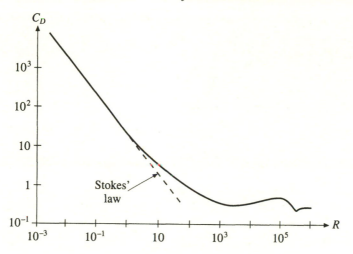

Fig. 14.1.2. Coefficient of drag against Reynolds' number for a sphere.

The velocity-squared drag law

For intermediate and large values of the Reynolds' number (typically, $R > 1$) no theoretical calculations of the drag force are available and we have to use experimental measurements of the dimensionless *drag coefficient* C_D obtained from the equation

$$F_D = \frac{1}{2}\rho_f C_D A \dot{x}^2.$$

Here A denotes the cross-sectional area of the object which presents itself to the fluid, ρ_f is the density of the fluid and $\dot{x}$ is the velocity of the object. For a sphere, A is the surface area of a disk and is thus given in terms of the radius r by $A = \pi r^2$.

A plot of C_D against the Reynolds' number for a sphere is given in Figure 14.1.2. Note that the plot in Figure 14.1.2 has been done on logarithmic graph paper. The value of C_D for R between 10^3 and 2.5×10^5 lies in the range 0.4 to 0.5. As an approximation, the value of C_D, for a sphere, will be taken as the constant 0.45, and furthermore this will be assumed to be the value for all values of R from 1 up to 2.5×10^5. The magnitude of the drag force is then given by

$$|F_D| = 0.225\pi \rho_f r^2 \dot{x}^2.$$

Exercises 14.1

1. (a) From the dimensions stated in the text for η, show that the Reynolds' number R is dimensionless.

   (b) Verify that Stokes' law is dimensionally correct.

   (c) Show that the drag coefficient C_D is dimensionless.

2. For each of the following motions a typical speed $|\dot{x}|$ and characteristic length d are given. Compute the corresponding Reynolds' number. For motion in air use $\rho_f = 1.22 \, \text{kg/m}^3$ and $\eta = 1.8 \times 10^{-5} \, \text{kg m}^{-1} \, \text{s}^{-1}$, while for motions in water use $\eta = 10^{-3} \, \text{kg m}^{-1} \, \text{s}^{-1}$ and $\rho_f = 1000 \, \text{kg/m}^3$. Indicate the cases for which Stokes' law should be applicable.

   (a) A peregrine falcon in a hunting dive: $|\dot{x}| = 70 \, \text{m/s}$; $d = 0.15 \, \text{m}$.

   (b) A minnow swimming in a quiet stream: $|\dot{x}| = 1 \, \text{m/s}$; $d = 0.03 \, \text{m}$.

   (c) Airborne dust particles settling on a calm day : $|\dot{x}| = 2 \times 10^{-4} \, \text{m/s}$; $d = 4 \times 10^{-6} \, \text{m}$.

   (d) A cruising yacht: $|\dot{x}| = 10 \, \text{m/s}$; $d = 10 \, \text{m}$.

   (e) Silt particles settling in a still lake: $|\dot{x}| = 1.6 \times 10^{-3} \, \text{m/s}$; $d = 4 \times 10^{-5} \, \text{m}$.

3. From the graph in Figure 14.1.2, for low Reynolds' numbers,

$$\log_{10} C_D \simeq K - \log_{10} R$$

where K is a constant.

   (a) Deduce a formula for C_D in terms of R.

   (b) Substitute the formula defining R into the RHS of the formula you found in (a).

   (c) Deduce from (b) and the equation $F_D = \frac{1}{2} \rho_f C_D A \dot{x}^2$ that

$$|F_D| = H r \eta |\dot{x}|$$

   where H is another constant. Compare this with Stokes' formula.

14.2 Archimedes' Principle

We all know that not all objects released in a fluid medium will fall. Air bubbles in a carbonated drink and helium-filled balloons are typical examples. The reason for this behaviour is the presence of a *buoyant force* acting in the opposite direction to the weight force. Archimedes, in 250 BC, formulated what is now known as *Archimedes' principle*:

The buoyant force acting on a body either fully or partially immersed in a fluid is equal in magnitude but opposite in direction to the weight of the fluid displaced by the body.

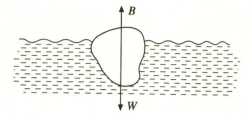

Fig. 14.2.1. Force diagram for a floating chunk of ice. *B* denotes the buoyant force and *W* the weight force.

The story goes that Archimedes discovered this principle when observing that the level in the municipal bath increased upon his entry. In his excitement he ran naked through the streets shouting 'Eureka' (which is Greek for 'I have found it').

It is not too difficult to understand the nature of Archimedes' principle. Consider any undisturbed region of the fluid. There are two types of force acting on this region: a gravitational force equal to the weight of the particles, and the pressure force exerted at the boundary of the region by the fluid particles outside this region. If the region of fluid is to remain stationary, the pressure must exactly balance the weight of the fluid particles. If a solid object displaces the fluid particles, the same pressure forces act on the object's surface as once acted on the now displaced fluid. Hence the object experiences a buoyant force equal in magnitude, but opposite in direction, to the weight of fluid it displaced.

Example 1. *Use Archimedes' principle to calculate the fraction of a chunk of ice which is below water level in a fresh water lake (data: the densities of water and ice are given by $\rho_{\text{water}} = 1000 \, \text{kg/m}^3$ and $\rho_{\text{ice}} = 920 \, \text{kg/m}^3$ respectively).*

Solution. *The calculation can be divided into steps:*

STEP 1: Draw a diagram indicating the forces acting on the chunk of ice (see Figure 14.2.1).

STEP 2: *Use Archimedes' principle to calculate B:*
    If the downwards direction is taken as positive then

$$B = -m_f g,$$

where m_f is the mass of the fluid displaced.

STEP 3: *Apply Newton's second law to obtain a relation between W and B:*
    If an object is floating there is no acceleration so the two forces must add to zero. Thus

$$W = -B$$

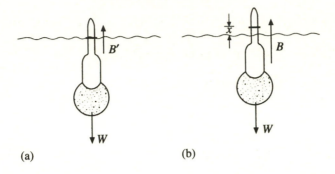

(a)   (b)

Fig. 14.2.2. A hydrometer in (a) water, and (b) a liquid more dense than water. Here $|B| > |B'|$ and $x > 0$.

which implies

$$m = m_f,$$

where m denotes the mass of the chunk of ice.

STEP 4: *Express the masses in terms of volumes by using the given densities:*
    In general

$$\{\text{mass}\} = \{\text{volume}\} \times \{\text{density}\}.$$

Thus

$$m = 920\,V, \qquad m_f = 1{,}000\,V_s,$$

where V is the total volume, and V_s is the submerged volume of the ice, and so

$$V_s/V = 0.92.$$

Hence 92% of the chunk of ice is below water level.

The hydrometer

A study of the previous example shows that the fraction of any particular floating object above the surface level depends only on the density of the fluid. This is the principle behind the hydrometer (see Figure 14.2.2) which is used to measure the density of a liquid.

It is first necessary to mark on the stem of the hydrometer the surface level of distilled water (density $1000\,\text{kg/m}^3$), and to calculate the submerged volume, $V_s^{(0)}$ say. The hydrometer is then floated in the liquid whose density is to be determined, and the displacement x of the surface level, which is measured as the distance below the mark for distilled water, is noted. If the stem of the hydrometer has cross-sectional area A, then it is straightforward to show (see Exercise 14.2.4 below) that

$$\rho_f = \frac{1000\, V_s^{(0)}}{V_s^{(0)} - xA},$$

where ρ_f denotes the density of the fluid.

Exercises 14.2

1. There's a well known saying: 'That's just the tip of the iceberg'. Given that the density of ice is $920\,\mathrm{kg/m^3}$ and the density of seawater is $1020\,\mathrm{kg/m^3}$, use Archimedes' principle to calculate the fraction of the total volume of an iceberg which is submerged under water.

2. A block of wood with dimensions $2 \times 5 \times 30\,\mathrm{cm}$ and density $400\,\mathrm{kg/m^3}$ is held beneath the surface of a bucket of water by a straight string. Calculate the tension in the string.

3. A 70 kg pig is marooned on a wooden board which is floating down a flooded river. If the board is 15 cm thick and 2 m long and has a density of $600\,\mathrm{kg/m^3}$, what is the width of the board if the top surface is level with the water?

4. Use Archimedes' principle to show that, for the hydrometer of Figure 14.2.2,

$$\rho_f = \frac{1000 V_s^{(0)}}{V_s^{(0)} - xA},$$

where the symbols are defined in the text.

5. A 'striped' cocktail is a drink made of three different liqueurs, which lie in layers one on top of the other. The colours and densities of these liqueurs are shown in the diagram below. The densities are in gram/ml, and each layer is 1 cm thick.

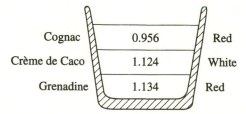

Cognac	0.956	Red
Crème de Caco	1.124	White
Grenadine	1.134	Red

A small plastic cube of volume $1\,\mathrm{cm^3}$ and density $1.130\,\mathrm{g/ml}$ is carefully lowered into the drink.

(a) Between which layers does it float?

(b) Show that 3/5 of its volume is in one layer and 2/5 in another.

6. The molecular weight of helium is $4.00\,\mathrm{gram/mole}$. The molecular weight of air is $28.9\,\mathrm{gram/mole}$. A helium filled balloon rises because it displaces air in excess of its own weight. The molar volume is approximately the same for all gases: $0.0224\,\mathrm{mole/m^3}$ at a temperature of $0\,^\circ\mathrm{C}$ and a pressure of 1 atm. Estimate the volume of helium required to lift a mass of 100 kg. What is the required radius of the balloon?

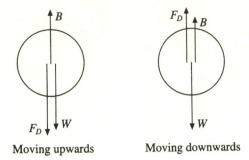

Fig. 14.3.1. Forces on a moving sphere in a fluid medium.

14.3 Falling sphere with Stokes' resistance

The above sections tell us that a sphere falling through a fluid medium is subject to three distinct forces:

- a weight force (W),
- a drag force (F_D) and
- a buoyant force (B).

As indicated in Figure 14.3.1, the direction of F_D depends on whether the sphere is moving upwards or downwards.

Furthermore, as discussed in Section 14.1, there are two different forms of the drag force F_D to consider (depending on the size of the Reynolds' number). The first form of F_D will be considered in this section while the second form is left until the next section. Thus Stokes' law will be considered, so that the magnitude of the resistive force is

$$|F_D| = 6\pi\eta r|\dot{x}|.$$

To determine the sign of F_D, both the cases of an upward moving sphere and a downward moving sphere must be considered separately. Suppose the displacement x is measured upward from the ground. When the sphere moves upward $\dot{x}$ is positive and, from Figure 14.3.1, F_D is negative. Thus

$$F_D = -6\pi\eta\dot{x}.$$

When the sphere is moving towards the ground, $\dot{x}$ is negative and, from Figure 14.3.1, F_D is positive. Thus

$$F_D = 6\pi\eta|\dot{x}| = -6\pi\eta\dot{x}$$

since $\dot{x} < 0$, and so here the formula for F_D is in fact independent of whether the sphere is moving upwards or downwards.

Newton's second law gives the equation of motion

$$m\ddot{x} = W + B + F_D.$$

Since upwards has been chosen as the positive direction, Figure 14.3.1 shows that the weight force W is negative while the buoyant force B is positive. Inserting the formulae for all the forces in the equation of motion then gives

$$m\ddot{x} = -(m - m_f)g - 6\pi\eta r\dot{x}$$

where m is the mass of the sphere and m_f the mass of the displaced fluid. The quantities of interest can now be found from the differential equation by applying the theory from Chapter 11.

Example 1. *Obtain a first-order equation for the velocity. Calculate the terminal velocity.*

Solution. *Writing*

$$v = \dot{x} \qquad (\text{and thus } \dot{v} = \ddot{x})$$

in the differential equation, and then slightly rearranging the result gives

$$\dot{v} = -\frac{6\pi\eta r}{m}v - \frac{(m - m_f)}{m}g.$$

This equation is a first-order linear inhomogeneous differential equation with constant coefficients of the type studied in Chapter 11. The terminal velocity v_t occurs when $\dot{v} = 0$. The differential equation is then simply solved for y to give

$$v_t = -\frac{(m - m_f)g}{6\pi\eta r}$$

as the required terminal velocity.

The general solution of the differential equation, to be obtained in the exercises, is

$$\dot{x} = v_0 e^{-t/\tau} + v_t(1 - e^{-t/\tau})$$

where v_0 is the initial velocity and

$$\tau = \frac{m}{6\pi\eta r}$$

is called the *characteristic time*.

From the general solution, it follows that if $v_0 = 0$, then $t = \tau$ corresponds to the time it takes the sphere to reach a fraction $1 - e^{-1} \simeq$ 67% of its terminal velocity. For times $t > 5\tau$, the sphere has reached more than 99.9% of its terminal velocity. Highly accurate calculations can then be carried out by assuming that the velocity is precisely the terminal velocity. This is especially relevant when the characteristic time is small in comparison with the duration of the motion.

The Millikan oil drop experiment

In a famous experiment performed in 1909, R. Millikan used a modification of the expression for the terminal velocity v_t of a sphere falling under Stokes' law to determine the charge on the electron. The experiment consisted of observing v_t for small oil droplets under the influence of an electric field created by two parallel plates.

If there is no electric field, the unknown radius r of the oil droplet can be expressed in terms of the terminal velocity $v_t^{(0)}$ say. This follows from manipulation of the formula for the terminal velocity, obtained in Example 1, giving

$$r = \left(\frac{9|v_t^{(0)}|\eta}{2(\rho - \rho_f)g} \right)^{\frac{1}{2}}$$

where ρ denotes the density of the oil. As the characteristic time is very small, the oil droplet can be assumed to be falling at the constant velocity $v_t^{(0)}$. Thus, by measuring the time t it takes the droplet to fall a distance ℓ, the terminal velocity is determined by the simple formula

$$|v_t^{(0)}| = \ell/t.$$

Suppose that now the electric field is switched on. Since the oil droplets have acquired a charge of a few excess electrons when they were formed by being sprayed from an atomiser, they feel an electric force F_E. If the top plate is kept at potential V, the bottom plate at zero potential (i.e. earthed) and the oil droplet has excess charge $q = ne$, the laws of electrostatics give

$$F_E = n|e|\frac{V}{d},$$

where d is the plate separation and the upwards direction has been taken as positive (see Figure 14.3.2).

When the oil drop is moving, a drag force obeying Stokes' law will act in addition to the forces in Figure 14.3.2. The equation of motion is therefore

$$n|e|\frac{V}{d} - (m - m_f)g - 6\pi\eta r\dot{x} = m\ddot{x}.$$

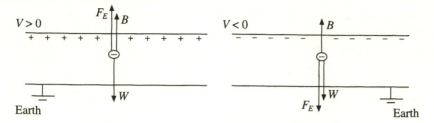

Fig. 14.3.2. Forces on a charged oil drop at rest in an electric field created by the Millikan apparatus.

Putting $\ddot{x} = 0$ and solving for $\dot{x}$ gives the terminal velocity v_t, say, as

$$v_t = \frac{n|e|V}{6\pi\eta rd} + v_t^{(0)}.$$

By measuring v_t for a positive and negative voltage of the same magnitude, and subtracting, this formula gives a value

$$\frac{n|e|V}{3\pi\eta rd}.$$

It is possible to vary the charge $n|e|$ on the oil droplet by using radiation. The above formula predicts that the resulting differences in upward and downward terminal velocities will always be an integer multiple of some constant. This was found by Millikan, and the value of the constant accurately gave the magnitude of the electronic charge. Further details on the Millikan oil drop experiment can be found in Melissinos (1968).

Exercises 14.3

1. It was shown in the text that the equation of motion for a falling sphere subject to Stokes' resistance is

$$\dot{y} = -\frac{6\pi\eta r}{m}y - \frac{(m - m_f)g}{m}.$$

  (a) Identify the symbols in this equation.

  (b) Find the general solution of the homogenized form of this equation.

  (c) Find a particular solution and hence determine the general solution of the original equation.

2. The differential equation for the velocity of a sphere subject to Stokes resistance is given in Exercise 1. If the sphere is released from rest, it follows, as a special case of the solution to Exercise 1(c), that

$$\dot{x} = v_t(1 - e^{-t/\tau}).$$

Use antidifferentiation to calculate the displacement x as a function of time, where x is measured downwards from the point of release.

3. The turbidity (muddiness) of lakes and streams — often a serious environmental problem — depends on the concentration of solid particles suspended in them. To understand the persistence of small particles in lakes and streams, calculate the time it takes a silt-sized particle ($r \simeq 2 \times 10^{-5}$ m) to fall one metre at terminal speed. (Data: $\eta = 10^{-3}$ kg/m s, $\rho = 2800$ kg/m^3, $\rho_f = 10^3$ kg/m^3).

4. A typical oil droplet in the Millikan experiment has radius 10^{-6} m. If $\eta = 1.8 \times 10^{-5}$ kg m^{-1} s^{-1}, $\rho = 883$ kg/m^3 and $\rho_f = 1.29$ kg/m^3, calculate the characteristic time for the fall of a droplet with the electric field turned off. Use your answer to explain why accurate calculations can be performed by assuming that the sphere falls at precisely its terminal velocity.

5. For the Millikan oil drop experiment, suppose that an experimenter measures the time it takes a droplet to

(a) fall a distance ℓ without the electric field turned off (time $t^{(0)}$)

(b) fall a distance ℓ with the electric field on (time $t^{(V)}$)

(c) rise a distance ℓ with the electric field reversed (time $t^{(-V)}$)

The results are tabulated in the following table. Different rows of the table correspond to the charge of the droplet having been altered by radiation.

$t^{(0)}$(s)	$t^{(V)}$ (s)	$t^{(-V)}$ (s)
19.38	2.46	3.18
20.3	1.7	2.0
20.18	4.34	7.74
19.52	4.82	7.96

(a) Give some possible reasons why $t^{(0)}$ is not precisely the same in each case. What is the average value of $t^{(0)}$?

(b) Use the average value of $t^{(0)}$ to calculate the radius of the drop r. (Data: $\eta = 1.60 \times 10^{-5}$ kg/m^3, $\ell = 6.1 \times 10^{-3}$ m, $\rho = 883$ kg/m^3, $\rho_f = 1.29$ kg/m^3.)

(c) Use an appropriate formula from the text to calculate $n|e|$ for each line in the above table. (Data: $V = 500$ volts.) You should find that, to a good approximation, each of the values is an integer multiple of the smallest such value.

(d) From your answer to part (c) calculate the approximate value of the magnitude of the charge of the electron $|e|$. The unit of charge in the MKS system is the Coulomb.

14.4 Falling sphere with velocity-squared drag

In Section 14.1 the magnitude of the velocity-squared drag was given as

$$|F_D| = 0.225\pi\rho_f r^2 \dot{x}^2$$

Fig. 14.4.1. Choice of a coordinate for a sphere falling over a cliff.

for a sphere of radius r. The right hand side of this expression is always positive, irrespective of the sign of the velocity $\dot{x}$, whereas the direction of the drag force F_D must always be opposite the direction of motion (recall Figure 14.3.1). The equation of motion can be written down correctly if this is kept in mind. The solution of the equation of motion requires the technique of separation of variables, discussed in Section 11.1.

Example 1. *A wooden sphere of radius 5 cm is released from rest at time $t = 0$ at the top of a cliff. Find the subsequent velocity of the sphere assuming the $\dot{x}^2$ resistive law. (Data: $\rho = 600\,kg/m^3$, $\rho_f = 1.2\,kg/m^3$, $g = 9.8\,m/s^2$.)*

Solution. *The method of solution can be broken into a number of distinct steps.*

STEP 1: Choose a coordinate system and indicate this on a diagram. *In Figure 14.4.1, the displacement of the sphere has been measured from the point of release. The downwards direction is therefore positive.*

STEP 2: Draw a force diagram for the sphere, and use Newton's second law to obtain the equation of motion. *The force diagram is given in the second diagram of Figure 14.3.1. By Newton's second law, the equation of motion is*

$$m\ddot{x} = W + B + F_D$$

Since the downwards direction is positive, the second diagram of Figure 14.3.1 shows that the weight force W is positive while the buoyant force B and drag force F_D are negative. Substituting the formulae for the forces in the equation of motion gives

$$m\ddot{x} = (m - m_f)g - 0.225\pi\rho_f r^2 \dot{x}^2 \tag{1}$$

where m is the mass of the sphere, and m_f is the mass of the air displaced by the sphere.

STEP 3: Write m and m_f in terms of the density of the fluid and the density of the sphere respectively, and rewrite the differential equation with $y = \dot{x}$. *Since the volume V of a sphere is given in terms of its radius r by*

$$V = \frac{4}{3}\pi r^3$$

and

$$\{mass\} = \{density\} \times \{volume\},$$

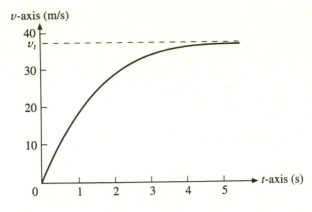

Fig. 14.4.2. The velocity–time graph for the falling sphere of Example 1.

we see

$$m = \frac{4}{3}\pi r^3 \rho \qquad and \qquad m_f = \frac{4}{3}\pi r^3 \rho_f.$$

Now divide both sides of the differential equation (1) by m and make the above substitutions for m and m_f, and then put $v = \dot{x}$ and thus $\dot{v} = \ddot{x}$. The differential equation (1) then becomes

$$\dot{b} = -A^2 b^2 + B^2$$

where

$$A = 0.411 \left(\frac{\rho_f}{r\rho}\right)^{\frac{1}{2}} \qquad and \qquad B = \left[\left(1 - \frac{\rho_f}{\rho}\right)g\right]^{\frac{1}{2}}.$$

STEP 4: Solve the differential equation using the technique of separation of variables. *It is to be shown in the exercises that this method gives for the solution*

$$\dot{x} = \frac{B}{A}\left(\frac{e^{2ABt} - 1}{e^{2ABt} + 1}\right).$$

A plot of this solution is given in Figure 14.4.2.

Terminal velocity and characteristic time

In Example 1, it was shown that the differential equation for the velocity of a falling sphere could be written in the form

$$\dot{v} = -A^2 v^2 + B^2.$$

The terminal velocity v_t occurs when $\dot{y} = 0$. Thus

$$v_t = B/A,$$

which is in agreement with the behaviour seen in Figure 14.4.2 with appropriate values of A and B.

The exact solution

$$\dot{x} = \frac{B}{A}\left(\frac{e^{2ABt}-1}{e^{2ABt}+1}\right)$$

mentioned in Example 1 suggests the choice

$$\tau = \frac{1}{2AB}$$

for a characteristic time. In this model, if $t = \tau$, the sphere has reached approximately 46% of its terminal velocity. For times $t > 8\tau$, the sphere has reached more than 99.9% of its terminal velocity.

Exercises 14.4

1. Obtain the Reynolds' number for the wooden sphere of Example 1 falling at terminal velocity. Does the value obtained lie in the range appropriate for the choice of drag constant $C_D = 0.45$ (recall Figure 14.1.2)?

2. Consider the differential equation

$$\dot{v} = -A^2v^2 + B^2,$$

with initial condition $v = 0$ at $t = 0$.

(a) State why $0 < v < B/A$ for $t > 0$.

(b) What are the dimensions of A and B, given that $[v] = LT^{-1}$?

(c) Obtain the partial fractions expansion

$$\frac{1}{B^2 - A^2v^2} = \frac{1}{2B}\left(\frac{1}{B-Av} + \frac{1}{B+Av}\right).$$

(d) Show that the antiderivative of $\dfrac{1}{B-Av}$ with respect to v is

$$-\frac{1}{A}\ln(B - Av)$$

and obtain a similar result for $\dfrac{1}{B+Av}$.

(e) Use the technique of separation of variables and the results of (b) and (c) to show that

$$\ln\frac{B+Av}{B-Av} = 2ABt.$$

Check that both sides of this equation are dimensionless.

(f) From part (e), obtain the result

$$v = \frac{B}{A}\left(\frac{e^{2ABt}-1}{e^{2ABt}+1}\right).$$

3. Suppose a sphere of radius r and mass m is thrown vertically upwards with velocity v_0.

   (a) If the coordinate x is measured upwards from the point of release, and the sphere is subject to the velocity-squared resistance law, show that the equation of motion is

   $$-(m - m_f)g - 0.225\pi\rho_f r^2 \dot{x}^2 = m\ddot{x}.$$

   Identify all the symbols not previously defined in the question.

   (b) If $m > m_f$, show that the above equation of motion can be written in the form

   $$\dot{v} = -A^2 y^2 - B^2$$

   where A and B are defined in Example 1.

   (c) Verify that the antiderivative of $\dfrac{1}{A^2 v^2 + B^2}$ with respect to v is

   $$\frac{1}{AB} \arctan\left(\frac{Av}{B}\right).$$

   Use this result and the method of separation of variables to show that for the differential equation in part (b)

   $$\arctan\left(\frac{Av}{B}\right) - C = -ABt$$

   where $C = \arctan Av_0/B$.

   (d) From (c) obtain the result

   $$v = \frac{B}{A}\tan(C - ABt).$$

   (e) How long is the ball moving upwards? For which values of t is the solution in (d) valid?

4. For a lacrosse ball, $r = 3\,\text{cm}$ and $\rho = 600\,\text{kg/m}^3$. If the ball is propelled vertically upward with velocity $15\,\text{m/s}$, use the answer to Exercise 3(d) to calculate the time it takes before the ball begins to fall. Do you expect the time it takes to fall to be less than, equal to, or greater than the time it takes to reach its maximum height?

5. Draw the velocity–time graph of an object projected vertically upwards with speed $10\,\text{m/s}$ ignoring the effect of air-resistance. On the same axis sketch the velocity–time graph of the same object *with* air-resistance. The precise form of the drag force is unimportant here; the relevant feature is that it opposes the direction of motion. In each case compare the speed of the object at a certain point on its rise and the same point on its descent.

15

Damped and forced oscillations

In Chapter 6, the equation of motion for a particle on a spring was set up and solved. This model will now be extended to include damping and driving forces. As expected, the effect of a damping force is to decrease the amplitude of the oscillation of the spring. On the other hand, forced motion of a spring can lead to the phenomenon of resonance, when the amplitude of the oscillations becomes very large.

The differential equations which occur in this chapter are of the linear second-order, constant-coefficient type. Both homogeneous and inhomogeneous equations occur. Their solution relies on the general theory of Chapter 5, and as a computational tool complex numbers are utilized. We assume basic familiarity with the complex number system, including real and imaginary parts and the polar form of a complex number.

15.1 Constant-coefficient differential equations

In this section we briefly explain a technique for solving linear constant-coefficient second-order differential equations. We first show how to find the solution of the homogenized equation. We then show how to find particular solutions of differential equations with trigonometric forcing terms.

The homogeneous equation

The differential equation

$$a\ddot{x} + b\dot{x} + cx = 0 \qquad (1)$$

where a, b and c are constants is second-order linear homogeneous and has constant coefficients. A method of solution of such equations is to look for solutions of the form

$$x = e^{\lambda t}.$$

Substitution into (1) gives

$$(a\lambda^2 + b\lambda + c)e^{\lambda t} = 0.$$

Since $e^{\lambda t}$ is never zero, this implies λ must satisfy the quadratic equation

$$a\lambda^2 + b\lambda + c = 0. \tag{2}$$

This is called the *characteristic equation* of (1).

Solution of the characteristic equation allows the general solution of the differential equation to be obtained. There are three distinct cases:

(i) the roots are real and distinct,
(ii) the roots are complex conjugate,
(iii) the roots are equal to each other.

Complex variables. To deal with case (ii) above where λ is complex it is necessary to understand the meaning of the exponential of a complex number. For a complex number $a + ib$, $a, b \in \mathbb{R}$, we define

$$e^{a+ib} = e^a(\cos a + i \sin b).$$

For example,

$$e^{\pi i} = -1, \quad e^{\pi i/2} = i, \quad \text{and} \quad e^{(1+i)\pi/4} = \frac{e^{\pi/4}}{\sqrt{2}}(1 + i).$$

It is easy to verify from this definition that the exponential of a sum is the product of exponentials (just as for exponentials of real numbers). Now let $\lambda = a + ib$ be complex and let $t \in \mathbb{R}$. From the above definition

$$e^{\lambda t} = e^{at} \cos(bt) + ie^{at} \sin(bt).$$

This is an example of a function of the form

$$f(t) = f_1(t) + if_2(t), \qquad t \in \mathbb{R}$$

where f_1 and f_2 are real-valued functions. We assign a meaning to differentiating complex-valued functions by differentiating the real part $f_1(t)$ and the imaginary part $f_2(t)$ separately, so that

$$f'(t) = f_1'(t) + if_2'(t).$$

It is left as an exercise to show, using these definitions, that

$$\frac{d}{dt}\left(e^{\lambda t}\right) = \lambda e^{\lambda t}$$

holds even when λ is complex.

We now discuss how to obtain the general solution of the homogeneous constant-coefficient differential equation (1) depending on the nature of the solutions of the characteristic equation (2). We consider each of the three cases listed above: (i) real and distinct roots, (ii) complex roots, and (iii) equal real roots.

(i) Real distinct roots. This case is particularly straightforward, as is illustrated in the following example.

Example 1. *Find the general solution of the differential equation*

$$\ddot{x} + 3\dot{x} + 2x = 0.$$

Solution. *Looking for solutions of the form*

$$x = e^{\lambda t}$$

gives the equation

$$\lambda^2 e^{\lambda t} + 3\lambda e^{\lambda t} + 2e^{\lambda t}$$

from which we obtain the characteristic equation

$$\lambda^2 + 3\lambda + 2 = 0.$$

This has solutions

$$\lambda = -2 \qquad and \qquad \lambda = -1$$

and hence two linearly independent solutions of the differential equation are

$$x = e^{-2t} \qquad and \qquad x = e^{-t}.$$

Since the differential equation is second-order linear and homogeneous, the super-position theorem says that the general solution is a linear combination of the two particular solutions. Thus

$$x = c_1 e^{-2t} + c_2 e^{-t}, \qquad where \quad c_1, c_2 \in \mathbb{R},$$

is the required general solution.

(ii) Complex conjugate roots. When the roots are complex we first get complex solutions, using the complex exponential; *real* solutions can then be obtained from the complex ones. The following example shows how.

Example 2. *Find the general solution of the differential equation*

$$\ddot{x} + 2\dot{x} + 2x = 0.$$

Solution. *Looking for solutions of the form $e^{\lambda t}$ gives*

$$\lambda^2 e^{\lambda t} + 2\lambda e^{\lambda t} + 2e^{\lambda t} = 0$$

from which we obtain the characteristic equation,

$$\lambda^2 + 2\lambda + 2 = 0.$$

The solutions of this quadratic equation are

$$\lambda = -1 + i \text{ and } \lambda = -1 - i,$$

and thus two complex solutions of the differential equation are

$$\phi_1(t) = e^{(-1+i)t} \quad \text{and} \quad \phi_2(t) = e^{(-1-i)t}.$$

From the definition of the exponential of a complex number,

$$e^{a+ib} = e^a \cos(b) + ie^a \sin(b),$$

the two solutions can be written as

$$\phi_1(t) = e^{-t} \cos(t) + ie^{-t} \sin(t) \quad \text{and} \quad \phi_2(t) = e^{-t} \cos(t) - ie^{-t} \sin(t).$$

Since the equation is second order, homogeneous and linear, the superposition theorem says the general solution is a linear combination of two independent particular solutions. To get real solutions we choose the combinations

$$\frac{1}{2}(\phi_1(t) + \phi_2(t)) \quad \text{and} \quad \frac{1}{2i}(\phi_1(t) - \phi_2(t)).$$

Thus

$$x = c_1 e^{-t} \cos t + c_2 e^{-t} \sin t, \quad c_1, c_2 \in \mathbb{R}.$$

is the required general *real solution.*

(iii) Equal roots. When the roots of the characteristic equation are the same then we only obtain one solution of the form

$$\phi_1(t) = e^{\lambda t}$$

for the equation (1). It can be verified that another solution, linearly independent from the first, can be obtained by multiplying the first solution $\phi_1(t)$ by t:

$$\phi_2(t) = te^{\lambda t}.$$

The general solution is thus

$$x = c_1 e^{\lambda t} + c_2 te^{\lambda t}, \quad c_1, c_2 \in \mathbb{R}.$$

Inhomogeneous second-order equations

The equations which will be obtained in our study of the harmonic oscillator are of the form

$$\ddot{x} + b\dot{x} + cx = F \cos \omega t \tag{3a}$$

and

$$\ddot{x} + b\dot{x} + cx = F \sin \omega t. \tag{3b}$$

According to the superposition theorem of Chapter 5, the general solution of these equations consists of the solution of the homogenized equation plus a particular solution. The homogenized equation can be solved by looking for a solution of the form $x = e^{\lambda t}$ and solving the characteristic equation, as we have just seen. Our technique of finding a particular solution uses the complex exponential function. The following steps should be performed.

STEP 1: *In the differential equation, replace* $\cos(\omega t)$ *or* $\sin(\omega t)$ *by* $e^{i\omega t}$. (We do this because $e^{i\omega t} = \cos(\omega t) + i\sin(\omega t)$ so (3a) corresponds to taking the real part and (3b) corresponds to taking the imaginary part of the resulting complex equation.)

STEP 2: *Look for a particular solution of the new (complex) equation of the form* $\phi_p(t) = ae^{i\omega t}$ *where a is a complex number, which is to be determined.*

STEP 3: *Take the real or imaginary parts of the particular solution of the complex equation to obtain the particular solution to the original equation.*

These steps are illustrated by the following example.

Example 3. *Find a particular solution of the equation*

$$\ddot{x} - \dot{x} + 6x = \sin 2t. \tag{4}$$

Solution.

STEP 1: *Consider the complex equation, with* $\sin 2t$ *replaced by* e^{2it},

$$\ddot{x} - \dot{x} + 6x = e^{2it}. \tag{5}$$

STEP 2: *Look for a particular solution of the form*

$$\phi_p(t) = ae^{2it}.$$

When substituted into (5), this gives the equation

$$a(-2^2 - 2i + 6)e^{2it} = e^{2it}.$$

Solving for a and then writing a in polar form gives

$$a = \frac{1}{2-2i} = \frac{1}{4}(1+i) = \frac{1}{2\sqrt{2}}e^{i\frac{\pi}{4}}.$$

Hence a particular complex solution to (5) is

$$\phi_p(t) = \frac{1}{2\sqrt{2}}e^{i(2t+\frac{\pi}{4})}.$$

In (5), let $x = x_{re} + ix_{im}$ where x_{re} and x_{im} are the real and imaginary parts of x respectively. By equating the real and imaginary parts of (5), the two equations

$$\ddot{x}_{re} - \dot{x}_{re} + 6x_{re} = \cos 2t$$

and

$$\ddot{x}_{im} - \dot{x}_{im} + 6x_{im} = \sin 2t$$

are then obtained. It thus follows that the imaginary part of $\phi_p(t)$ satisfies the original equation (4).

STEP 3: *Choose the imaginary part of*

$$\phi_p(t) = \frac{1}{2\sqrt{2}}\left(\cos(2t + \frac{\pi}{4}) + i\sin(2t + \frac{\pi}{4})\right)$$

to get the particular solution

$$\frac{1}{2\sqrt{2}}\sin(2t + \frac{\pi}{4}).$$

of (4).

The real particular solution just obtained, of the differential equation (4), describes an oscillation with amplitude $A = \dfrac{1}{2\sqrt{2}}$. One of the advantages of using the complex formulation is that we can easily obtain this amplitude as the modulus of the complex solution. Thus, in the above example,

$$\left\{\begin{array}{c}\text{modulus of}\\\text{complex solution}\end{array}\right\} = |ae^{2it}| = |a| = \left|\frac{1+i}{4}\right| = \frac{1}{2\sqrt{2}} = \left\{\begin{array}{c}\text{amplitude of}\\\text{real solution}\end{array}\right\}.$$

The general proof of this result is left to the exercises.

Exercises 15.1

1. Find the real and imaginary parts of the following expressions involving complex numbers.

   (a) $i(3i + 1)$

   (b) $(2i + 1)/(-2i + 1)$

   (c) $(4i + 1)/i$

2. Find the complex solutions of the following quadratic equations.

    (a) $\lambda^2 + 2 = 0$

    (b) $\lambda^2 - 3\lambda + 3 = 0$

    (c) $2\lambda^2 + \lambda + 1 = 0$

3. Differentiate the following functions involving complex numbers.

    (a) $f(t) = ie^{3it}$

    (b) $f(t) = te^{it}$

4. Find the general solutions of each of the following differential equations.

    (a) $\ddot{x} + 5\dot{x} + 6x = 0$

    (b) $2\ddot{x} + 4\dot{x} + 2x = 0$

    (c) $\ddot{x} + \dot{x} + x = 0$

    (d) $2\ddot{x} + \dot{x} + x = 0$

5. Use the complex exponential to find a particular solution of each of the following inhomogeneous equations.

    (a) $\ddot{x} + \dot{x} + x = 2\cos 3t$

    (b) $2\ddot{x} + \dot{x} + x = \frac{1}{2}\sin 2t$

    (c) $\ddot{x} + 4x = \sin \omega t, \qquad \omega \in \mathbb{R}, \quad \omega \neq 0$

    (d) $\ddot{x} + \mu x + 4x = \sin 2t, \qquad \mu \in \mathbb{R}, \quad \mu \neq 0$

6. (a) Find the complex number a so that $z = ae^{3it}$ is a solution of the differential equation

$$\ddot{z} - \dot{z} + 12z = 2e^{3it}.$$

    (b) Hence find a real solution of the differential equation

$$\ddot{x} - \dot{x} + 12x = 2\cos(3t).$$

    (c) Check that the amplitude of the real solution obtained in part (b) is equal to the modulus of the complex solution found in part (a).

7. Let a be a complex number whose polar form is $|a|e^{i\theta}$. Verify that if

$$z = ae^{i\omega t}, \qquad \omega, t \in \mathbb{R},$$

then the real and imaginary parts of $z = x + iy$ are given by

$$x = |a|\cos(\omega t + \theta) \qquad \text{and} \qquad y = |a|\sin(\omega t + \theta).$$

(Thus the real and imaginary parts of z give oscillations whose amplitude equals the modulus of z.)

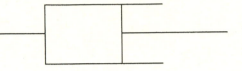

Fig. 15.2.1. Cross-section of a dashpot. This symbol is used to represent ideal damping in a spring.

15.2 Damped oscillations

In Chapter 6, a study was made of the motion executed by a particle attached to a light spring. The magnitude of the force experienced by the particle due to the restoring nature of the spring was taken as being directly proportional to the extension of the spring (Hooke's law). The subsequent motion of the particle was shown to be periodic, with the period of oscillation determined by the spring stiffness and the mass of the particle.

Everyday experience tells us that the amplitude of the oscillations will not remain the same forever. Rather it will decrease with time until the motion eventually stops altogether. Physically, this can be understood as a consequence of the frictional (damping) forces present within both the spring and the attachments to the spring.

A common device which produces significant damping in a spring system is known as a *dashpot*. A dashpot consists of a cylinder filled with a fluid, sealed at one end, with a loose plunger at the other end (see Figure 15.2.1). Examples can be found in car shock absorbers and on some 'self closing' doors. The damping in a dashpot is due to the viscous effect of the fluid. From the discussion in Chapter 14, the magnitude of the damping force F_D will thus be a complicated function of the velocity $\dot{x}$, which increases as $|\dot{x}|$ increases.

The simplest choice, from the viewpoint of the resulting equation of motion, is to have the magnitude of F_D directly proportional to the speed of the mass m. Thus

$$|F_D| = \gamma|\dot{x}| \qquad (1)$$

where $\gamma > 0$ is a proportionality constant called the *damping constant*. This choice will be only approximately correct in general. However (1) allows the equation of motion to be solved exactly, and it is found that the type of motion obtained from the solution of the equation of motion corresponds to that observed in real spring systems. The choice (1) will

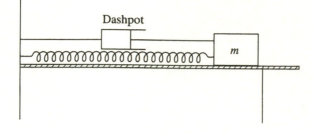

Fig. 15.2.2. A spring and dashpot system.

be referred to as *ideal* damping, and the dashpot will be assumed light so that its mass can be neglected in the equation of motion.

Equation of motion

The equation of motion for a particle on an ideally damped spring can be determined by modifying the procedure of Section 6.2 for the case without damping.

Example 1. *Obtain and classify the equation of motion for the damped spring system of Figure 15.2.2. The table is assumed smooth, the spring light and the damping ideal. The natural length of the spring is ℓ metre and its spring constant is k newton/metre. The particle has mass m kg. Suppose the coordinate x is measured as the extension of the spring beyond its natural length.*

Solution. *In the horizontal direction there are two forces acting on the particle: F_S due to the spring, and F_D due to the dashpot. An application of Hooke's law, as detailed in Section 6.2, gives*

$$F_S = -kx.$$

The formula (1) gives the magnitude of F_D. Its sign is determined from the criteria that the damping force is always opposite in direction to the velocity. Hence

$$F_D = -\gamma \dot{x}.$$

Newton's second law applied to the particle says

$$F_D + F_S = m\ddot{x}$$

and thus

$$-\gamma \dot{x} - kx = m\ddot{x}.$$

Rearrangement gives

$$\ddot{x} + \frac{\gamma}{m}\dot{x} + \frac{k}{m}x = 0, \tag{2}$$

which is a homogeneous linear second-order differential equation with constant coefficients.

Solution of the equation of motion

Let us suppose the spring in Example 1 is initially stretched a distance x_0 metres and released from rest. It has already been remarked that the expected consequence of a damping force in a spring system is the gradual diminishing of the amplitude of the oscillations of the particle about the equilibrium point. Does the solution of the differential equation (2) make this prediction?

The solution of (2) can be obtained using the method of Section 15.1 for homogeneous equations. Recall that the first step is to try a solution of the form $e^{\lambda t}$ and hence obtain the characteristic equation

$$\lambda^2 + \frac{\gamma}{m}\lambda + \frac{k}{m} = 0.$$

It is found that as the parameters k, γ and m vary, *three distinct types of solutions are possible* for the damped spring system, corresponding to the nature of the roots of the characteristic equation. These cases are termed *underdamped*, *overdamped* and *critically damped* harmonic motion. The underdamped motion corresponds to the gradual diminishing of the amplitude mentioned above. Each of the three cases will now be discussed separately. Some of the mathematical details are left for the exercises.

Underdamped motion

It is to be shown in Exercise 15.2.1 that for $\gamma < 2(mk)^{\frac{1}{2}}$ the solution of (2) with the initial conditions $x = x_0$, and $\dot{x} = 0$, at $t = 0$ is

$$x = x_0 e^{-t/\tau} \cos(\omega t) + \frac{x_0}{\omega \tau} e^{-t/\tau} \sin(\omega t) \tag{3}$$

where

$$\tau = 2m/\gamma$$

and

$$\omega = \left(\frac{k}{m}\right)^{\frac{1}{2}} \left(1 - \frac{\gamma^2}{4mk}\right)^{\frac{1}{2}}.$$

The solution (3), with $x_0 = 1$, $\omega = \pi$ and $\tau = 5$, is plotted in Figure 15.2.3. The curve begins at x_0 and oscillates with decreasing amplitudes bounded by the *envelope curves* $x = \pm x_0 e^{-t/\tau}$. The time interval which elapses between any two successive maxima of the displacement is $2\pi/\omega$. The quantity τ is the characteristic time for the decay of the amplitude.

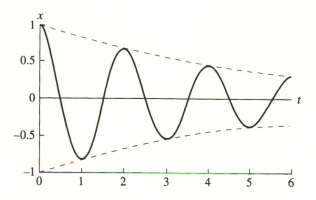

Fig. 15.2.3. Underdamped harmonic motion with $x = 1, \dot{x} = 0$ at $t = 0$ and $\omega = \pi, \tau = 5$.

The motion described by (3) is said to be *underdamped*. A feature of underdamped motion is that the ratio of any two successive maxima is independent of the initial conditions and the position of the maxima. Thus if we denote the first maximum by x_1, the second by x_2, etc, then

$$\frac{x_1}{x_2} = \frac{x_2}{x_3} = \ldots = \frac{x_n}{x_{n+1}}.$$

The *logarithmic decrement*, denoted by Δ, is defined as

$$\Delta = \ln \frac{x_n}{x_{n+1}} = \ln \frac{x_1}{x_2}.$$

From the solution (3), after some algebra, this can be written in terms of γ, k and m as

$$\Delta = 2\pi \left(\frac{4km}{\gamma^2} - 1 \right)^{-1/2}.$$

A plot of Δ as a function of the dimensionless parameter $\gamma(km)^{-1/2}$ is given in Figure 15.2.4.

For fixed values of k and m we see that the ratio of the first maximum to the second maximum increases as the damping constant γ increases, and becomes infinite as the value $\gamma = 2(km)^{\frac{1}{2}}$ is reached. For γ larger than this critical value the motion is no longer underdamped.

Overdamped motion

If the damping force is sufficiently strong, it might be expected that a particle on an extended spring, when released, will return to the

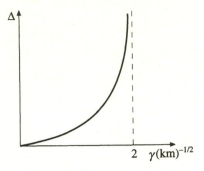

Fig. 15.2.4. The logarithmic decrement Δ as a function of the dimensionless damping parameter $\gamma(km)^{-1/2}$ for underdamped motion.

equilibrium point without any oscillations at all. This can be predicted from the solution of (2) when

$$\gamma > 2(mk)^{\frac{1}{2}}.$$

The characteristic equation then has real, negative and distinct roots which we denote by $\lambda = -\lambda_1$ and $\lambda = -\lambda_2$. The general solution is of the form

$$x = c_1 e^{-\lambda_1 t} + c_2 e^{-\lambda_2 t}$$

where c_1 and c_2 are determined by the initial conditions. If the initial displacement is x_0 and the initial velocity is zero, then

$$x = \frac{x_0 \lambda_2}{\lambda_2 - \lambda_1} e^{-\lambda_1 t} + \frac{x_0 \lambda_1}{\lambda_2 - \lambda_1} e^{-\lambda_2 t}. \tag{4}$$

Although x is never exactly zero for any positive time t, it will approach zero in the limit as t approaches infinity. In practice therefore, the distance of the particle from the origin will become imperceptible after a finite time.

Plots of (4) corresponding to the initial condition $x = 1$ and $\dot{x} = 0$, at $t = 0$, for various values of the dimensionless damping parameter $\gamma(mk)^{-1/2}$, are given in Figure 15.2.5. A feature of these plots is that for fixed values of m and k, the greater the value of the damping constant γ, the slower the return to the equilibrium position $x = 0$.

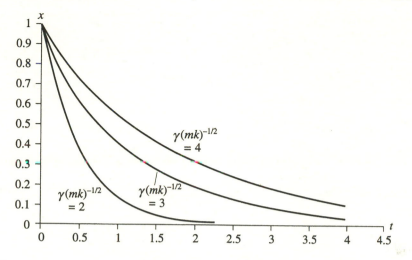

Fig. 15.2.5. Overdamped motion with the initial conditions $x = 1, \dot{x} = 0$ for various values of the dimensionless damping parameter $\gamma(mk)^{-1/2}$ where $k/m = 1$.

Critically damped motion

As suggested by Figure 15.2.5, the value of the damping constant γ which corresponds to the fastest return to the equilibrium point is

$$\gamma = 2(mk)^{1/2}. \tag{5}$$

As γ decreases to values less than $2(mk)^{1/2}$, the motion switches from overdamped to underdamped. For this reason, damped harmonic motion with the condition (5) is said to be critically damped. When (5) holds, the characteristic equation has only a single root $-\lambda_1$ which is negative and The general solution of (2) is

$$x = c_1 e^{-\lambda_1 t} + c_2 t e^{-\lambda_1 t},$$

The constants c_1 and c_2 are determined by the initial conditions, $x = x_0$ and $\dot{x} = 0$ at $t = 0$. This gives

$$x = x_0 e^{-\lambda_1 t} + x_0 t e^{-\lambda_1 t}. \tag{6}$$

Once again, although x is never exactly zero for any $t > 0$, it approaches zero as the time t approaches infinity. Hence the distance of the particle from the origin will be imperceptible after some finite time. In most applications of overdamping, it is desired to return to the equilibrium point as fast as possible. The damping constant is thus chosen to have the critical value (5).

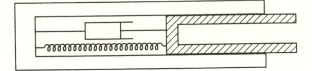

Fig. 15.2.6. Essential features of the artillery gun. The shaded region is the barrel.

Example 2. *The barrel of an artillery gun (see Figure 15.2.6) weighs 500 kg and has a recoil spring of stiffness 40 000 N/m. If the gun is fired horizontally and the barrel recoils 1 m, determine the critical damping coefficient of an attached dashpot with ideal damping.*

Solution. *Critical damping occurs when $\gamma = 2(mk)^{\frac{1}{2}}$. Substituting for m and k gives*

$$\gamma = 4\sqrt{5} \times 10^3 \text{ kg m/s.}$$

as the required damping constant.

Example 3. *For the problem in Example 2 find the time required for the barrel to return to a position 5 cm from its initial position.*

Solution. *To answer this question, it is necessary to set up and solve the appropriate equation of motion. Let ℓ metres denote the natural length of the spring, and let the displacement x metres be the extension of the spring as measured from ℓ. Since the displacement has reached a minimum at $x = -1$ The initial conditions are*

$$x = -1, \qquad \dot{x} = 0, \qquad at \quad t = 0.$$

It is desired to calculate t such that

$$x = -0.05.$$

With this choice of coordinate system, as explained in Example 1, the equation of motion is given by the differential equation (2). Substituting the numerical values of γ, m and k gives

$$\ddot{x} + 8\sqrt{5}\dot{x} + 80 = 0.$$

The corresponding characteristic equation has a double root at $-4\sqrt{5}$. The general solution is thus

$$x = c_1 e^{-4\sqrt{5}t} + c_2 t e^{-4\sqrt{5}t}, \qquad c_1, c_2 \in \mathbb{R}.$$

Substituting both initial conditions gives

$$-1 = c_1$$

and

$$0 = -4\sqrt{5}c_1 + c_2.$$

Hence $c_1 = -1$ and $c_2 = -4\sqrt{5}$ so that the solution of the equation of motion and the initial condition is

$$x = -e^{-4\sqrt{5}t} - 4\sqrt{5}t e^{-4\sqrt{5}t}.$$

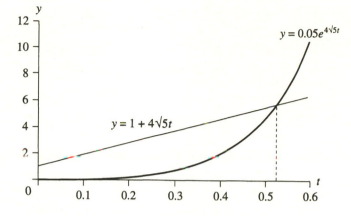

Fig. 15.2.7. Graphical solutions of $0.05e^{4\sqrt{5}t} = 1 + 4\sqrt{5}t$.

From this expression for the displacement, it follows that $x = -0.05$ when

$$0.05e^{4\sqrt{5}t} = 1 + 4\sqrt{5}t.$$

It is not possible to solve this equation using only algebra. An approximate solution can be obtained, however, by plotting both sides of the equation and reading off the point at which the graphs intersect (see Figure 15.2.7). It is thus found that the required time is

$$t \simeq 0.53 \quad seconds.$$

Exercises 15.2

1. The differential equation

$$\ddot{x} + \frac{\gamma}{m}\dot{x} + \frac{k}{m}x = 0$$

describes ideal damped harmonic motion.

    (a) Identify the symbols in this equation, and indicate the choice of coordinate system on a diagram.

    (b) Use the theory of Section 15.1 to show that there are three distinct types of solutions depending on whether

$$\gamma < 2(mk)^{\frac{1}{2}}, \gamma = 2(mk)^{\frac{1}{2}} \text{ or } \gamma > 2(mk)^{\frac{1}{2}},$$

and obtain the general solution in each case. When the roots are complex, write the solution in terms of the exponential and cosine function only.

    (c) In each of these cases, determine the arbitrary constants in the general solution so that the initial conditions $x = x_0$ and $\dot{x} = 0$ when $t = 0$ are satisfied. Compare your answers with (3), (4) and (6) in the text.

(d) Give the name of the type of motion in each case.

2. Consider the spring system of Figure 15.2.2. The table is assumed smooth, the spring light and the damping ideal. Suppose the natural length of the spring is 0.5 m and its stiffness is 10 N/m. Let the mass of the attached particle be 5 kg and the damping constant 10 kg/s.

(a) If the displacement x metres is measured as the extension of the mass from its equilibrium point, write down the equation of motion.

(b) With the initial conditions

$$x = 0.1 \quad \text{and} \quad \dot{x} = 0.2, \quad \text{at} \quad t = 0,$$

obtain the solution of the equation of motion in terms of the exponential and cosine functions only.

(c) Suppose instead that the displacement x metres is measured from the wall. Show that the equation of motion in this case is

$$5\ddot{x} = -10\dot{x} + 10(0.5 - x).$$

Use the workings of (b) to write down the general solution of the homogenized version of this equation. Find a particular solution and thus obtain the solution of the equation of motion when the initial conditions are as in (b).

(d) Compare your answers obtained in parts (b) and (c) and comment. What is the time interval between any two successive maxima?

3. (a) Prove that for ideal underdamped harmonic motion, the ratio of the height of any two successive maxima is a constant. Use this fact to show that the logarithmic decrement Δ can be written as

$$\Delta = \frac{1}{n} \ln \frac{x_1}{x_{n+1}}$$

where x_n is the nth maxima.

(b) Suppose the ratio of any two successive maxima for a car shock absorber is designed to be 0.1. If $m = 1000$ kg and $k = 5000$ N/m, use an appropriate formula in the text to calculate the required value of the damping constant γ.

4. Consider a particle of mass M suspended from a light spring and an ideal dashpot, which hang from a ceiling.

(a) If the spring has natural length ℓ and the displacement x is measured as the distance of the particle below that ceiling, obtain the equation of motion

$$M\ddot{x} = -\gamma\dot{x} + k(\ell - x) + Mg.$$

(b) Explain from the equation of motion why the criteria for the types of motion is the same as in Exercise 1.

(c) Find a particular solution and then use your answer to 1(b) to write down the general solution in each case.

5. For the system of Exercise 4, suppose $\ell = 0.2\,\text{m}$ and $M = 5\,\text{kg}$.

   (a) Calculate γ so that the motion is critical.

   (b) If the spring is compressed so that the particle is $0.1\,\text{m}$ from the ceiling and then released from rest, calculate the subsequent displacement of the particle, assuming γ is as in (a).

   (c) Plot the displacement–time graph in (b).

   (d) Calculate how long it takes the particle to return $1\,\text{mm}$ from its equilibrium point.

6. If a hydrometer is floated in water (recall Figure 14.2.2) and displaced vertically from its equilibrium point, it will undergo damped harmonic motion.

   (a) To understand this effect, suppose the stem of the hydrometer, of cross-sectional area α (metre)^2, is displaced a distance x metres below its floating level. Use Archimedes' principle to show that a net force F newton, given by

   $$F = -1000\,\alpha x,$$

   acts on the body, where the downwards direction has been taken as positive. Show that the same formula applies if the hydrometer is displaced a distance x metres above its floating point.

   (b) Assuming a resistive force proportional to the velocity of the hydrometer, obtain the equation of motion. For what values of the damping constant will the motion be underdamped?

15.3 Forced harmonic motion

A feature of the three types of solutions for damped harmonic motion, from the previous section, is that the amplitude progressively decreases. For this reason, the motion is referred to as *transient* (temporary). To maintain the amplitude of oscillation in a damped system, it is necessary to have an external force acting on the system. This is known to us all from experience on a swing. Motion of the legs and body at regular intervals provides a force capable of maintaining a steady amplitude of oscillation, even though significant friction is present.

   The key requirement for the external force $F_f(t)$ (the subscript f indicates that this term forces the oscillations) is that it must be periodic in time. Perhaps the simplest choices with this feature are

$$F_f(t) = F_0 \sin \omega_f t \qquad \text{and} \qquad F_f(t) = F_0 \cos \omega_f t \qquad (1)$$

where F_0 and ω_f are constants.

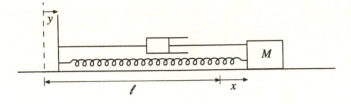

Fig. 15.3.1. The spring system and coordinates for Example 1.

Equation of motion

An equation of motion with a forcing term of the form (1) can be obtained in a spring system by periodically varying the position of the attachment of the spring to the wall.

Example 1. *Suppose a light spring and ideal dashpot are attached at one end to a particle of mass M which moves on a smooth table. The other end of the spring and dashpot are forced to oscillate with displacement given by*

$$y = a \sin \omega_f t$$

as shown in Figure 15.3.1.

If x is the displacement of the particle as measured from the point with coordinate ℓ (the natural length of the spring), and $z = x - y$, show that

$$\ddot{z} + \frac{\gamma}{M}\dot{z} + \omega^2 z = F_0 \sin \omega_f t \tag{2}$$

where

$$F_0 = a(\omega_f)^2, \qquad \omega^2 = \frac{k}{M}$$

and k is the stiffness of the spring and γ is the damping constant.

Solution. *Let the direction to the right of the origin in Figure 15.3.1 be positive. There are two forces acting on the particle: F_S due to the spring, and F_D due to the dashpot. To calculate F_S, note from Figure 15.3.1 that the extension of the spring beyond its natural length is $x - y$, assuming $x > y$. Hence, from Hooke's law, since the positive direction is to the right,*

$$F_S = -k(x - y).$$

This formula holds if $y > x$ also, since then the spring is compressed so the force is positive.

The force due to an ideal dashpot is directly proportional to the difference between the velocity of the two ends of the dashpot. Since the rightmost end is moving at velocity $\dot{x}$, and the leftmost end at velocity $\dot{y}$, this difference is $\dot{x} - \dot{y}$. If $\dot{x} > \dot{y}$, the force is in the opposite direction to the particle motion, so

$$F_D = -\gamma(\dot{x} - \dot{y}).$$

This formula also applies if $\dot{x} < \dot{y}$.

Since the origin for the x-coordinate is fixed, the acceleration of the particle is $\ddot{x}$. *Hence, by Newton's second law,*

$$M\ddot{x} = F_S + F_D.$$

Changing from x to the relative coordinate $z = x - y$ *gives*

$$F_S = -kz, \qquad F_D = -\gamma\dot{z} \qquad and \qquad \ddot{x} = \ddot{z} + \ddot{y}.$$

Hence the equation of motion $M(\ddot{z} + \ddot{y}) = F_S + F_D$ *becomes*

$$M\ddot{z} = -kz - \gamma\dot{z} - M\ddot{y}$$

and so

$$M\ddot{z} = -kz - \gamma\dot{z} + Ma(\omega_f)^2 \sin\omega_f t,$$

since $y = a\sin(\omega_f t)$. *Hence*

$$\ddot{z} + \frac{\gamma}{M}\dot{z} + \frac{k}{M}z = F_0 \sin(\omega_f t) \tag{3}$$

as required.

Note that if, in Example 1 of Section 15.2, an additional force $F_0 \sin(\omega_f t)$ were applied directly to the particle of mass M, then the equation of motion would become

$$\ddot{x} + \frac{\gamma}{M}\dot{x} + \frac{k}{M}x = F_0 \sin(\omega_f t).$$

This equation is exactly the same as (3), but with x in place of z.

The non-transient solution

The solution of the equation of motion (2) can be obtained by using the method of Section 15.1. The general solution is the sum of the homogenized solution, which for $\gamma \neq 0$ is a transient function, and a particular solution. The particular solution has the same period as the forcing term, whereas the amplitude depends on all the parameters as illustrated in the following example.

Example 2. *Consider the equation of motion*

$$\ddot{z} + 0.1\dot{z} + 9z = \sin(\omega_f t) \tag{4}$$

which from (2) describes forced harmonic motion with $\gamma/M = 0.1$, $\omega = 3$ *and* $F_0 = 1$. *Find the amplitude of the particular solution and plot its dependence on* ω_f *for* $0 \le \omega_f \le 4$.

Solution. *To find a particular solution of (4) we use the method of Section 15.1. We consider the complex equation*

$$\ddot{z} + 0.1\dot{z} + 9z = e^{i\omega_f t}, \tag{5}$$

noting that $\sin(\omega_f t)$ *is the imaginary part of* $e^{i\omega_f t} = \cos(\omega_f t) + i\sin(\omega_f t)$. *We try a particular solution of the form*

$$z_p = a e^{i\omega_f t}$$

where a *is a complex constant, to be determined.*

Substituting z_p *into the complex differential equation (5) gives*

$$\left[a(i\omega_f)^2 + 0.1ai\omega_f + 9a \right] e^{i\omega_f t} = e^{i\omega_f t}$$

from which it follows that

$$a = \frac{1}{9 - (\omega_f)^2 + 0.1i\omega_f}.$$

Hence

$$z_p = \frac{1}{9 - (\omega_f)^2 + 0.1i\omega_f} e^{i\omega_f t} \tag{6}$$

is the complex solution of (5). To get the real solution of (5) we need to take the imaginary part of (5). (Note that if the right-hand side of (4) had been $\cos(\omega_f t)$ *then we would then need to take the real part of the particular solution (6).)*

However we are only interested here in the amplitude A of the real solution to (4). This can be obtained (see the discussion at the end of Section 15.1) directly from the complex solution (6) as $A = |a|$. *Hence the required amplitude A is*

$$A = \left| \frac{1}{9 - (\omega_f)^2 + 0.1i\omega_f} \right| = \frac{1}{|9 - (\omega_f)^2 + 0.1i\omega_f|}.$$

Hence

$$A = \left((9 - \omega_f^2)^2 + 0.01\omega_f^2 \right)^{-1/2}.$$

A plot of the amplitude A against the forcing frequency ω_f *is given in Figure 15.3.2.*

Figure 15.3.2 shows that the amplitude of the oscillations of the particular solution peaks very close to $\omega_f = 3$ (i.e. the frequency $\omega_f/2\pi = 3/2\pi$). The value $3/2\pi$ is the frequency of the system with damping and forcing removed — that is, the *natural frequency* of the system. The frequency $\omega_f/2\pi$ at which this peak occurs is called the *resonant frequency* of the system. In general, for small damping, when the frequency of the forcing term approximately equals the natural frequency of the spring system, the amplitude of the non-transient solution peaks. The phenomenon is known as *resonance*.

Resonance is responsible for some surprising happenings. A well-known example is a singer holding a note at a certain frequency and shattering a glass. This occurs when the motion of the natural frequency

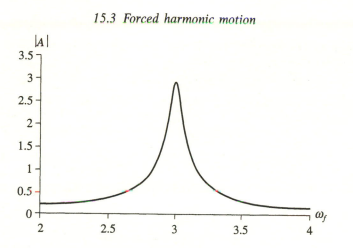

Fig. 15.3.2. Displacement amplitude against frequency near resonance: $\omega_f = 2.9$.

of vibration of the molecules in the glass coincides with the frequency of the note. Another example is the massive tides in the Bay of Fundy on the Canadian east coast. In the oceans, the tidal rise is approximately 0.3 m, but in the bay it is 11m. This occurs because the characteristic period of water as it moves back and forth across the bay is about 13 hours, which is close to the period of 12.4 hours between successive high tides. Since the ocean tide can be regarded as a driving force, a large resonant amplitude in the bay results.

Resonance, induced by periodic fluctuations in the wind, has also been used as an explanation of the collapse of a bridge in the State of Washington, USA. An entertaining discussion of this is given by Braun (1983).

Exercises 15.3

1. For the equation of motion

$$\ddot{z} + 9z = \cos(\omega_f t),$$

which describes forced harmonic motion without damping, find the amplitude of the particular solution and sketch its dependence on ω_f for $0 \le \omega_f \le 6$, $\omega_f \ne 3$.

2. (a) Recall that the natural frequency of a system is the frequency of the oscillations when the damping force and the external force are removed. What is the natural frequency of the system in Exercise 1?

  (b) What is the resonant frequency for the system in Exercise 1?

3. Repeat Exercise 1 for the equation

$$\ddot{z} + 0.05\dot{z} + 4z = \cos \omega_f t,$$

here sketching the amplitude in the range $0 \le \omega_f \le 4$.

4. Suppose a particle rests on a light spring. If the weight of the particle causes the spring to be compressed a distance of 5 cm, use Newton's and Hooke's laws to deduce the ratio of the mass of the particle to the spring constant, and thus determine the resonant frequency, assuming small damping.

5. Consider a particle resting on a smooth table and attached to a spring of stiffness 10 N/m and an ideal dashpot. The spring and dashpot are attached to a wall.

   (a) Suppose the spring is extended and released. You are given that the time interval between successive maxima is 1.5 s and that the ratio of the amplitude of successive maxima is 2/7. Use appropriate formulae in the text to calculate the mass of the particle and the damping constant.

   (b) Determine the amplitude and phase of the non-transient part of the motion if an external force $F_f(t) = 2 \sin(4t)$ acts on the system.

6. The diagram below shows the essential elements of a vibration-measuring device (seismometer or accelerometer). If in the y-direction there is a displacement of the form

$$y = Y \sin(\omega_f t)$$

(due to an earthquake for example), the aim is to deduce Y from the recorded amplitude of the relative displacement

$$z = x - y + \frac{mg}{k}$$

of the particle.

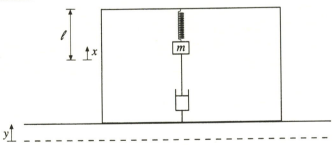

   (a) By following the working of Example 1, show that the equation of motion for the particle can be written

$$m\ddot{z} + \gamma\dot{z} + kz = mY\omega_f^2 \sin(\omega_f t) - mg.$$

   (b) Show that the amplitude A of the non-transient part of the solution to this equation is

$$A = \frac{Y(\omega_f/\omega)^2}{\sqrt{[1 - (\omega_f/\omega)^2]^2 + (2\zeta\omega_f/\omega)^2}}$$

where $\omega = (k/m)^{\frac{1}{2}}$ and $2\zeta = \gamma/(mk)^{\frac{1}{2}}$.

(c) For a seismometer, ω is very small in comparison to ω_f, so that ω_f/ω is large. Show from your answer to (b) that in this circumstance

$$Y \simeq A.$$

What is the motion of the particle?

(d) For the accelerometer, the ratio ω_f/ω is small. Show that here

$$Y \simeq (\omega/\omega_f)^2 A.$$

16

Motion in a plane

This chapter introduces some problems from particle mechanics in which the action takes place in a plane rather than along a line. Typical problems are

- motion down an inclined plane,
- motion of a projectile,

where in each case the particle is assumed to move in a vertical plane. The first problem is suggested by Galileo's famous experiment, but it also has relevance to the motion of a skier down a slope. The second problem has relevance to ball games as well as to warfare.

Our overall approach to these problems is to define the acceleration of the particle as a vector $\ddot{X}$. After writing the net force acting on the particle as a vector F, we then apply Newton's second law in its vector form

$$F = m\ddot{X},$$

m being the mass of the particle. This leads to differential equations, from which the motion of the particle can be predicted.

The basic ideas of vector algebra will be assumed, but ideas about vector-valued functions and their derivatives will be explained fully.

16.1 Kinematics in a plane

In this section the ideas of velocity and acceleration will be formulated in sufficient generality to be applicable to particles moving in a plane. Our definitions of these quantities will be closely related to the method by which they are calculated. An important geometrical interpretation of velocity in terms of tangent vectors to curves will be explained later.

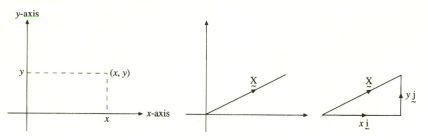

Fig. 16.1.1. Cartesian coordinates and position vector of a particle.

Position and displacement

A familar way of specifying the position of a particle moving in a plane is to give its cartesian coordinates x and y relative to a pair of axes, as as Figure 16.1.1. The axes, which should be fixed relative to an inertial frame, are chosen in two mutually perpendicular directions which fit in with the geometry of the mechanical problem. These directions will often (but not always) be horizontal and vertical respectively.

An alternative way to specify the position of the point is to give the *position vector* $\underset{\sim}{X}$ of the particle relative to the origin O. This vector is just the directed line segment from O to the point in question. As can be seen from the triangle on the right in Figure 16.1.1, the position vector is given in terms of the cartesian coordinates of the point by the formula

$$\underset{\sim}{X} = x\underline{i} + y\underline{j} \qquad (1)$$

where $\underline{i}$ and $\underline{j}$ are the vectors of unit length pointing in the directions of the x- and y-axes respectively. Note that the cartesian coordinates of a point determine its position vector uniquely, and conversely.

Curves

As the time varies, the particle traces out a curve in the (x, y)-plane, as illustrated in Figure 16.1.2.

The coordinates of the particle are thus functions of time given by, say,

$$x = \phi(t) \quad \text{and} \quad y = \psi(t) \qquad (2)$$

where ϕ and ψ are functions mapping times into displacements along the axes. The position vector of the particle is also a function of the time

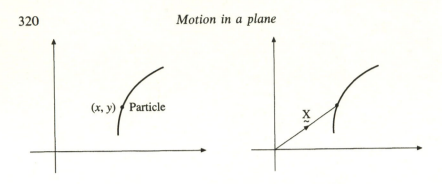

Fig. 16.1.2. Curve traced out in the (x, y)-plane.

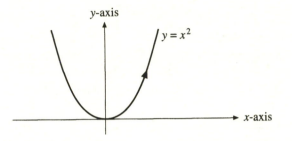

Fig. 16.1.3. Particle moving on a parabola in the (x, y)-plane.

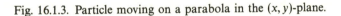

which, from (1) and (2), is given by

$$\underset{\sim}{X} = \phi(t)\underset{\sim}{i} + \psi(t)\underset{\sim}{j}. \tag{3}$$

The following examples show how the curve on which the particle is moving may be obtained once these functions are known.

Example 1. *The cartesian coordinates of a particle are given as functions of the time by*

$$x = t \quad and \quad y = t^2.$$

Find the curve in the (x, y)-plane on which the particle moves.

Solution. *Eliminating t from these two equations gives the single equation relating x and y,*

$$y = x^2.$$

This is the equation of a parabola in the (x, y)-plane, as shown in Figure 16.1.3. Since x increases as t increases, the particle moves along the parabola in the direction shown by the arrow.

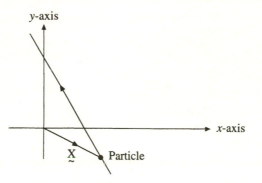

Fig. 16.1.4. Particle moving along a straight line in the (x, y)-plane.

In the above example, the equation $y = x^2$ is called the *equation* of the curve along which the particle moves. The pair of equations $x = t$ and $y = t^2$, however, are called the *parametric equations* of this curve because they involve the parameter t.

While the equation of the curve gives a purely 'static' picture, the parametric equations tell which point on the curve the particle occupies at any given time. Thus, in eliminating t, we lose some information about the motion of the particle.

The following example shows how to get the equation of the curve when the position vector, rather than each coordinate, is given as a function of time.

Example 2. *Describe the curve along which the particle moves in the (x, y)-plane when its position vector is given as a function of time by*

$$\underline{X} = -t\underline{i} + (2t + 1)\underline{j}.$$

Solution. *The cartesian coordinates x and y of the particle are just the respective coefficients of $\underline{i}$ and $\underline{j}$ on the RHS of the formula for $\underline{X}$. Hence*

$$x = -t \text{ and } y = 2t + 1.$$

Elimination of t gives

$$y = -2x + 1$$

and so the curve along which the particle moves is a straight line, as shown in Figure 16.1.4. The particle moves along the line in the direction indicated by the arrow because the parametric equations show that x decreases as t increases.

Velocity and acceleration

Our approach to these ideas is motivated by Galileo's work on projectiles. He imagined the motion in the vertical plane to consist of two independent components: one vertical and the other horizontal. Vertically the motion was as for free fall, while horizontally the motion was with constant velocity. By arguing in this way, Galileo was able to deduce the correct result that the path traced out by the projectile was a parabola.

We now apply this idea quite generally to a particle moving in a plane in such a way that its cartesian coordinates x and y are twice differentiable functions of time. The velocity consists essentially of two components: $\dot{x}$ along the direction of the x-axis and $\dot{y}$ along the direction of the y-axis. Just as we combined the two coordinates in equation (1) to give a single position vector

$$\underline{X} = x\underline{i} + y\underline{j},$$

we now combine the two velocity components to get a single *velocity vector*

$$\dot{\underline{X}} = \dot{x}\underline{i} + \dot{y}\underline{j}. \tag{4}$$

Likewise we combine the two components of acceleration $\ddot{x}$ and $\ddot{y}$ to get a single *acceleration vector*

$$\ddot{\underline{X}} = \ddot{x}\underline{i} + \ddot{y}\underline{j}. \tag{5}$$

The formulae (4) and (5) may be regarded as the definitions of velocity and acceleration for a particle moving in a plane, with position vector $\underline{X}$, given by (1), at time t. The *speed* of the particle is defined as the magnitude of the velocity vector, is given by

$$\|\dot{\underline{X}}\| = \sqrt{\dot{x}^2 + \dot{y}^2}.$$

The advantage of combining the components to produce a single vector will be apparent later when we give the geometrical interpretation of the velocity vector. Meanwhile, the following example shows how to calculate the velocity and acceleration vectors using familiar rules of differentiation.

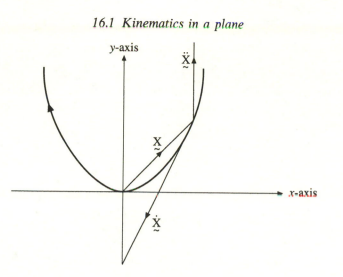

Fig. 16.1.5. Position, velocity and acceleration vectors.

Example 3. *A particle moving in the (x, y)-plane has position vector at time t given by*

$$\underset{\sim}{X} = -t\underset{\sim}{i} + t^2\underset{\sim}{j}.$$

Find the velocity and acceleration vectors $\dot{\underset{\sim}{X}}$ and $\ddot{\underset{\sim}{X}}$ at time t. When $t = -1$, show $\underset{\sim}{X}$, $\dot{\underset{\sim}{X}}$ and $\ddot{\underset{\sim}{X}}$ on the curve traced out by the particle. Find also the speed of the particle at this time, if distance is in metres and time in seconds.

Solution. *By our definitions, the velocity and acceleration are given as functions of t by*

$$\dot{\underset{\sim}{X}} = -\underset{\sim}{i} + 2t\underset{\sim}{j},$$
$$\ddot{\underset{\sim}{X}} = \qquad 2\underset{\sim}{j}.$$

If $t = -1$ then $\underset{\sim}{X} = \underset{\sim}{i} + \underset{\sim}{j}$, $\dot{\underset{\sim}{X}} = -\underset{\sim}{i} - 2j$ and $\ddot{\underset{\sim}{X}} = 2\underset{\sim}{j}$.

These vectors are shown in Figure 16.1.5 together with the curve along which the particle moves, which has the equation $y = x^2$. Thus

$$\|\dot{\underset{\sim}{X}}\| = \sqrt{(-1)^2 + (-2)^2} = \sqrt{5} \quad \text{m/s}$$

is the speed of the particle when $t = -1$.

Velocity and tangency

Note that in Figure 16.1.5 the velocity vector $\dot{\underset{\sim}{X}}$ seems to be tangent to the curve along which the particle is moving. The following argument

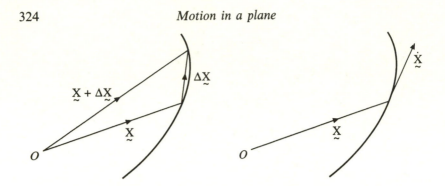

Fig. 16.1.6. Tangency of the velocity vector.

makes it seem plausible that this will always be the case: by definition

$$\dot{X} = \dot{x}\underset{\sim}{i} + \dot{y}\underset{\sim}{j}$$

$$= \left(\lim_{\delta t \to 0} \frac{\delta x}{\delta t}\right)\underset{\sim}{i} + \left(\lim_{\delta t \to 0} \frac{\delta y}{\delta t}\right)\underset{\sim}{j}$$

where δx and δy are the changes in the coordinates produced by a change δt in the time. Hence, when δt is sufficiently small, $\dot{X}$ will be close to the vector

$$\frac{\delta x}{\delta t}\underset{\sim}{i} + \frac{\delta y}{\delta t}\underset{\sim}{j} = \frac{\delta x\underset{\sim}{i} + \delta y\underset{\sim}{j}}{\delta t} = \frac{\delta X}{\delta t}$$

where δX is the change in the position vector produced by changes δx and δy in the coordinates. Now the vector

$$\frac{\delta X}{\delta t}$$

has the same direction as the vector δX. Hence Figure 16.1.6 makes it plausible that

$$\lim_{\delta t \to 0} \frac{\delta X}{\delta t} = \dot{X}$$

has the same direction as the tangent to the curve along which the particles moves, at the point with position vector X.

The notations

$$\frac{dX}{dt} \quad \text{and} \quad \frac{d^2X}{dt^2}$$

are often used for the velocity and the acceleration.

Exercises 16.1

1. In each case state what is wrong with the equation involving vectors:

(a) $a = \underset{\sim}{b} + \underset{\sim}{c}$ (b) $\underset{\sim}{a} = \underset{\sim}{b} + c$

(c) $\underset{\sim}{a} = \underset{\sim}{b} \cdot \underset{\sim}{c}$ (d) $\underset{\sim}{X} = t + 2t^3\underset{\sim}{j}$.

2. Plot each of the following points in the (x, y)–plane and then express the position vector of the point as a linear combination of the vectors $\underset{\sim}{i}$ and $\underset{\sim}{j}$.

(a) $(0, 1)$ (b) $(1, 0)$

(c) $(1, 1)$ (d) $(\cos(\pi/6), \sin(\pi/6))$.

3. In each case sketch the curve traced out in the (x, y)-plane by a particle when its coordinates are given as functions of time by the stated formulae. Show the direction of motion along the curve.

(a) $x = t^2$ and $y = t$ (b) $x = t^2$ and $y = t^2$

(c) $x = 1$ and $y = t$ (d) $x = e^t$ and $y = e^{-t}$.

4. In each case sketch the curve traced out in the (x, y)-plane by a particle when its position vector is given as a function of time by the stated formula. Calculate the velocity and acceleration vectors and the speed of the particle.

(a) $\underset{\sim}{X} = 2t\underset{\sim}{i} + 2t\underset{\sim}{j}$ (b) $\underset{\sim}{X} = -t\underset{\sim}{i} - t^2\underset{\sim}{j}$

(c) $\underset{\sim}{X} = t\underset{\sim}{i} + e^t\underset{\sim}{j}$ (d) $\underset{\sim}{X} = e^t\underset{\sim}{i} + e^{2t}\underset{\sim}{j}$.

5. The position vector $\underset{\sim}{X}$ of a particle at time t is given by the formula

$$\underset{\sim}{X} = t(\cos(\pi/3)\underset{\sim}{i} + \sin(\pi/3)\underset{\sim}{j}).$$

(a) Sketch the curve which the particle traces out and indicate the direction of motion.

(b) Guess the velocity and acceleration vectors and the speed.

(c) Calculate the velocity and acceleration vectors by differentiation and hence check your answer to (b).

6. Repeat Exercise 5, but now assume

$$\underset{\sim}{\dot{X}} = t(\cos(\alpha)\underset{\sim}{i} + \sin(\alpha)\underset{\sim}{j}) \qquad \text{where} \quad 0 \leq \alpha < \pi/2.$$

7. A particle moves around a circle. Its position and velocity vectors at time t are $\underset{\sim}{X}$ and $\underset{\sim}{\dot{X}}$. What is the angle between $\underset{\sim}{X}$ and $\underset{\sim}{\dot{X}}$? (A careful sketch should reveal the answer!) What is the value of the dot product $\underset{\sim}{X} \cdot \underset{\sim}{\dot{X}}$?

8. Suppose that a particle moves in such a way that its position vector at time t is given by

$$\underset{\sim}{X} = e^t\underset{\sim}{i} + e^{-t}\underset{\sim}{j}.$$

Calculate the velocity $\underset{\sim}{\dot{X}}$ and verify that it is perpendicular to $\underset{\sim}{X}$ when $t = 0$.

16.2 Motion down an inclined plane

Galileo tested his hypothesis that bodies fall towards the Earth with uniform acceleration by rolling a brass ball down a plane inclined to the horizontal at various angles. The inclined plane 'diluted' the effect of gravity and, by slowing the motion, made it possible to measure accurately the times at which the balls passed through various points along its path. Galileo found that, for a given inclination, the distance which a ball rolled down the plane when released from rest was proportional to the square of the time taken – just as would be expected if the acceleration were uniform.

The aim here is to study motion down an inclined plane using Newton's laws. To simplify our model we shall replace the brass ball by a particle and shall ignore the effect of friction. The problems which arise in trying to make a more realistic mathematical model of Galileo's experiment will be discussed later.

To find the equations of motion for the particle, we shall perform the following steps:

STEP 1: *Choose a coordinate system for the vertical plane containing the line along which the particle moves.*

STEP 2: *Express the forces acting on the particle as a linear combination of the vectors $\underline{i}$ and $\underline{j}$.*

STEP 3: *Apply Newton's second law in its vector form to get the equations of motion of the particle.*

Similar steps apply to other problems involving motion of a particle in a plane.

Example 1.   *A particle is released from rest on a smooth plane making an angle α with the horizontal, where $0 < \alpha < \frac{1}{2}\pi$. Show that the particle moves down the plane with constant acceleration. Find also the magnitude of the normal reaction of the plane on the particle.*

Solution.

STEP 1: Choose coordinates. *Let the x-axis run straight down the plane and let the y-axis point up perpendicularly to it with the origin at the point where the particle is initially released, as in Figure 16.2.1. Thus the particle moves along the line in the (x, y)-plane with $y = 0$.*

The unit vectors $\underline{i}$ and $\underline{j}$ point along the directions of the x- and y-axes, as usual. We let

$$\underline{X} = x\underline{i} + y\underline{j}$$

be the position vector of the particle at time t, hence

$$\underline{\ddot{X}} = \ddot{x}\underline{i} + \ddot{y}\underline{j}.$$

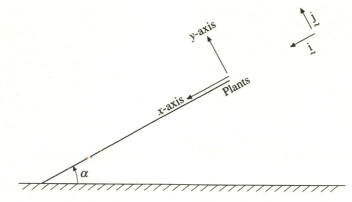

Fig. 16.2.1. Choosing the *x*- and *y*-axes.

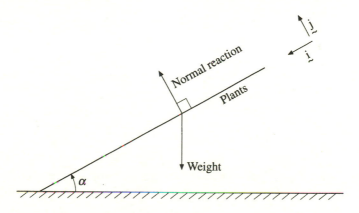

Fig. 16.2.2. Forces on the particle.

Since we assume the particle stays on the plane, it follows that y = 0 throughout the motion. Hence ÿ = 0.

STEP 2: Write the force as a vector. *Suppose the particle has mass m. There are two forces which act on it: its* weight, *of magnitude mg, and the* normal reaction *of the plane, of magnitude N, say. These forces are shown in Figure 16.2.2.*

These forces are to be expressed in terms of the vectors $\underset{\sim}{i}$ *and* $\underset{\sim}{j}$. *Since* $\underset{\sim}{j}$ *is the unit vector with the same direction as the normal reaction,*

$$\{\text{normal reaction}\} = N\underset{\sim}{j}.$$

To express the weight as a vector, we first find a unit vector pointing vertically downwards. So consider the right angle triangle in Figure 16.2.3 with one side parallel to the inclined plane and with hypotenuse of length 1.

The angle at the bottom of the triangle is $\pi/2 - \alpha$; *hence the angle at the top is* $\pi/2 - (\pi/2 - \alpha) = \alpha$. *Thus the remaining sides of the triangle have lengths* $\sin(\alpha)$

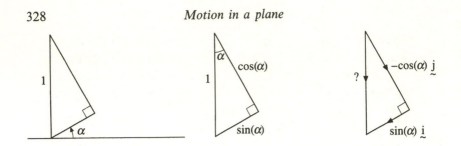

Fig. 16.2.3. Unit vector vertically downwards.

and $\cos(\alpha)$ *as shown. Hence the unit vector in the vertically downwards direction is*

$$\sin(\alpha)\underset{\sim}{i} - \cos(\alpha)\underset{\sim}{j}.$$

Hence, in vector form,

$$\{\text{weight}\} = mg(\sin(\alpha)\underset{\sim}{i} - \cos(\alpha)\underset{\sim}{j}).$$

The net force $\underset{\sim}{F}$ *acting on the particle is now given by*

$$\underset{\sim}{F} = \{\text{normal reaction}\} + \{\text{weight}\}$$
$$= N\underset{\sim}{j} + mg(\sin(\alpha)\underset{\sim}{i} - \cos(\alpha)\underset{\sim}{j})$$
$$= mg\sin(\alpha)\underset{\sim}{i} + (N - mg\cos(\alpha))\underset{\sim}{j}$$

STEP 3: Apply Newton's second law. *The vector form of this law,* $\underset{\sim}{F} = m\underset{\sim}{\ddot{X}}$, *now shows that*

$$mg\sin(\alpha)\underset{\sim}{i} + (N - mg\cos(\alpha))\underset{\sim}{j} = m\ddot{x}\underset{\sim}{i} + m\ddot{y}\underset{\sim}{j}.$$

Equating coefficients of $\underset{\sim}{i}$ *and* $\underset{\sim}{j}$ *on each side and using* $\ddot{y} = 0$ *gives*

$$\ddot{x} = g\sin(\alpha), \tag{1}$$
$$0 = N - mg\cos(\alpha). \tag{2}$$

Equation (1) shows that the acceleration down the plane is the constant $g\sin(\alpha)$. *Equation (2) shows that the normal reaction has magnitude* $mg\cos(\alpha)$.

Note that (1) is a differential equation for x as a function of t. Since the RHS is a constant, the differential equation can be solved by antidifferentiation, as explained in Section 2.4. In this way a complete description of the motion of the particle can be obtained.

Effect of friction

It is relatively easy to include in the above model the effect of friction between the plane and the particle. (See Chapter 4 for the basic ideas of friction.) Exercises to work out the details are set later.

Even when friction is included, however, the particle model is still inadequate as a description of Galileo's experiment with rolling balls. For example, a ball will start rolling down the plane even for very small inclinations α to the horizontal, but for a particle with friction there is a certain threshold inclination below which the particle will remain stationary.

The main obstacle to obtaining a simple model for the rolling balls is that the laws of friction merely give an inequality for the friction acting on a rolling body, as explained in Section 4.3. (This difficulty can be overcome by using the principle of conservation of energy for a rigid body, but this topic lies beyond the scope of this book.) It turns out that the actual acceleration of the balls is considerably less than that predicted by the particle on the smooth plane, at least for small inclinations of the plane.

Exercises 16.2

1. Recall from the text that, for a particle moving down a smooth plane inclined at an angle α to the horizontal, the acceleration is given by the formula

$$\ddot{x} = g \sin(\alpha) \qquad 0 < \alpha < \pi/2.$$

(a) Find the limit of the RHS as $\alpha \to 0$.

(b) Find the limit of the RHS as $\alpha \to \pi/2$.

(c) What are the physical interpretations of your answers to (a) and (b)?

(d) Suggest a further check which can be applied to the answer in the text.

2. Recall from the text that, for a particle moving down a smooth plane inclined at an angle α to the horizontal, the normal reaction on the particle has magnitude given by the formula

$$N = mg \cos(\alpha).$$

(a) Find the limit of the RHS as $\alpha \to 0$.

(b) Find the limit of the RHS as $\alpha \to \pi/2$.

(c) What are the physical interpretations of your answers to (a) and (b)?

3. Solve the differential equation

$$\ddot{x} = g \sin(\alpha),$$

which arises in Example 1 in the text, after imposing the relevant initial conditions. Does your answer show that the displacement of the particle from the origin is proportional to the square of the time taken?

4. Consider a vertical circle of radius $r > 0$. Suppose that a smooth plank is placed so that it forms a chord of the circle with lower end at the point where the circle touches the ground, as shown below. Let the plank have inclination α radians with respect to the ground.

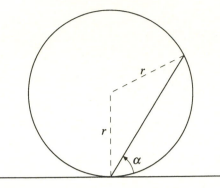

(a) Find the length of the plank in terms of r and α.

(b) Show that the time taken by a particle to roll from rest down the length of the plank is the same no matter what its inclination α happens to be. [This result is Galileo's famous *circle theorem*.]

5. Consider a particle lying at rest on a rough plane with coefficient of static friction μ_s between the plane and the particle. Suppose the inclination α of the plane to the horizontal is gradually increased until the particle is on the point of sliding down the plane. At this stage:

(a) show all the forces acting on the particle,

(b) use the fact that the acceleration is zero to show that

$$\mu_s = \tan(\alpha),$$

(c) and hence devise a practical method of estimating the coefficient of static friction.

6. Suppose a particle is sliding down a rough inclined plane making an angle α with the horizontal and let the coefficient of kinetic friction between the plane and the particle be μ_k.

(a) Show that if the particle moves a distance x down the plane in time t then

$$\ddot{x} = g(\sin(\alpha) - \mu_k \cos(\alpha)).$$

(b) Deduce there is an angle α_0 such that the particle

(i) accelerates *up* the plane if $\alpha < \alpha_0$,

(ii) maintains constant velocity if $\alpha = \alpha$,

(iii) accelerates *down* the plane if $\alpha > \alpha_0$.

7. Solve the differential equation in Exercise 6(a), assuming that the particle was initially at rest.

8. Investigate how the results stated in Exercise 6 are affected if the particle has been initially projected so as to slide *up* the plane.

9. Find the time taken by a skier to ski 1 km down a slope making an angle of 30° to the horizontal

(a) in the absence of friction,

(b) if the coefficient of kinetic friction between the skis and the snow is 0.04.

16.3 Projectiles

In the study of projectiles, the aim is to set up a mathematical model for the motion of a body projected near the Earth's surface, like a cricket ball or an artillery shell. In the simple model investigated in this section, the Earth is regarded as an inertial frame and hence the projectile is assumed to move in the verticle plane containing its initial direction of projection. While this simplification is appropriate for a cricket ball, it is less so for a shell. The only force taken into account, moreover, is gravity (assumed constant in magnitude and direction).

This model predicts that the path traced out by the particle (its *trajectory*) is a parabola — as was already known to Galileo, who realized that the horizontal and vertical components of the motion could be analysed separately. We achieve the same end by the use of vectors, obtaining a pair of uncoupled simultaneous differential equations to describe the motion.

If the model is modified to include the effect of air-resistance, the use of vectors makes it easy to obtain the modified equations of motion, as will be seen from a later exercise. The modified equations are coupled and, while it is possible to uncouple them by a suitable change of coordinates, the easiest way to get useful information about the behaviour of the solutions is by approximate numerical techniques.

The steps to follow in deriving the equations of motion are much the same as in the previous section.

Example 1. *Find the equations of motion in suitable coordinates, of a particle moving in a vertical plane subject only to the force of gravity, assumed constant.*

Solution.

STEP 1: Choose a coordinate system. *Since the only force on the particle acts vertically, we choose the y-axis vertically upwards and hence the x-axis horizontal. The origin is chosen at the point of projection, assumed to be at ground level.*

STEP 2: Write the force as a vector. *Let the mass of the particle be m. The only force acting on the particle is its weight. This has magnitude mg and direction that of the unit vector $-\underset{\sim}{j}$, as shown in Figure 16.3.1. Thus*

$$\{\text{weight}\} = -mg\underset{\sim}{j}.$$

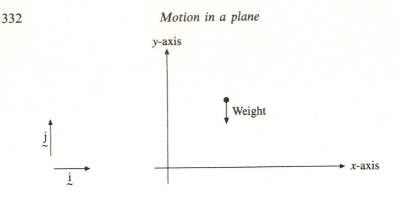

Fig. 16.3.1. Weight as a vector.

STEP 3: Apply Newton's second law. *The vector form of this law shows that*

$$-mg\underset{\sim}{j} = m\ddot{\underset{\sim}{X}} = m\ddot{x}\underset{\sim}{i} + m\ddot{y}\underset{\sim}{j}.$$

Equating the coefficients of $\underset{\sim}{i}$ *and of* $\underset{\sim}{j}$ *on each side gives, after cancellation of* m,

$$\ddot{x} = 0, \qquad\qquad\qquad\qquad\qquad (1)$$

$$\ddot{y} = -g. \qquad\qquad\qquad\qquad\qquad (2)$$

The simultaneous differential equations (1) and (2) are the desired equations of motion.

The subsequent motion of the particle can be found from these differential equations once the point from which the particle is projected and the velocity of projection are specified. The initial velocity can be obtained from the initial speed and angle of projection by elementary trigonometry, as in the following example.

Example 2. *Let the x-axis and the y-axis be horizontal and vertically upwards respectively, as in the previous example. Suppose that a particle is projected up from the origin at ground level with an initial speed of* 10 m/s *at an angle of* $\pi/4$ *radians to the x-axis. Find the coordinates x and y of the particle as functions of the time t after projection, and describe the trajectory of the particle in the (x, y)-plane. (Assume x and y are in metres, t in seconds.)*

Solution. *To write the initial vector in terms of* $\underset{\sim}{i}$ *and* $\underset{\sim}{j}$*, we first get a vector of unit length in the direction of projection.*

By the construction given in Figure 16.3.2, this unit vector is

$$\cos\left(\frac{\pi}{4}\right)\underset{\sim}{i} + \sin\left(\frac{\pi}{4}\right)\underset{\sim}{j} = \frac{1}{\sqrt{2}}\underset{\sim}{i} + \frac{1}{\sqrt{2}}\underset{\sim}{j}.$$

To get the correct magnitude for the initial velocity vector, we now multiply this unit vector by 10 giving

$$\dot{\underset{\sim}{X}} = \dot{x}\underset{\sim}{i} + \dot{y}\underset{\sim}{j} = \frac{10}{\sqrt{2}}\underset{\sim}{i} + \frac{10}{\sqrt{2}}\underset{\sim}{j} \qquad \text{for } t = 0.$$

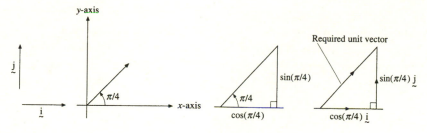

Fig. 16.3.2. Unit vector in direction of projection.

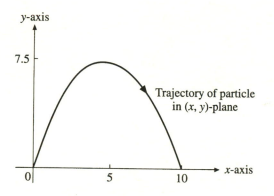

Fig. 16.3.3. The trajectory of the projectile.

Equating coefficients of $\underset{\sim}{i}$ and $\underset{\sim}{j}$ in this vector equation and using the fact that the particle starts at the origin gives

$$\left.\begin{array}{c} x=0 \\ y=0 \end{array}\right\} \ and \ \left.\begin{array}{c} \dot{x}=5\sqrt{2} \\ \dot{y}=5\sqrt{2} \end{array}\right\} \qquad when \ t=0. \tag{3}$$

Solving the differential equations (1) and (2) subject to the initial conditions (3) gives the solutions

$$x = 5\sqrt{2}t,$$
$$y = -\frac{1}{2}gt^2 + 5\sqrt{2}t, \tag{4}$$

which are valid while $y \geq 0$.

Eliminating t between these equations gives as the equation of the trajectory

$$y = -\left(\frac{1}{100}g\right)x^2 + x.$$

Thus the trajectory is part of a parabola, as shown in Figure 16.3.3.

One of the main applications of the mathematics of projectile motion is to military ballistics (see Hart and Croft, 1988; and Farrar and Leeming, 1983). Some interesting stroboscopic photographs of a moving projectile are given in Cohen (1985). To model accurately the motion of shells our model needs to be extended to take into account rotation of the Earth, air-resistance, and the spinning of the shell.

Applications of the mathematics of projectiles to sport are also interesting; a very accessible account is given in Chapter 8 of De Mestre (1990). This book also discusses the effect of air-resistance on projectile motion.

Exercises 16.3

1. Verify the claims made in the solution of Example 2 in the text concerning the solutions of the simultaneous differential equations

$$\ddot{x} = 0$$
$$\ddot{y} = -g$$

with the relevant initial conditions.

2. Use the solutions obtained in the text to Example 2 to determine:

    (a) how long the particle takes to return to ground level,

    (b) how far it had then moved horizontally,

    (c) the maximum height reached by the particle.

3. Let the x- and y-axes be horizontal and vertically upwards respectively, with the origin at ground level. Suppose a particle is projected from the origin in the (x, y)-plane with the speed $u > 0$ at an angle α to the x-axis $(0 < \alpha < \pi/2)$.

    (a) By following the method used in Example 2, show that the equation of the trajectory of the particle is

$$y = -\frac{g}{2\cos^2(\alpha)u^2}x^2 + \tan(\alpha)x \qquad (y \geq 0).$$

    (b) Verify that this equation is dimensionally correct.

4. In each of the following cases find the limiting form of the equation given in Exercise 3 and describe the limiting trajectory:

    (a) when u approaches ∞

    (b) when α approaches 0.

    (c) Do your results seem plausible physically? Give reasons.

5. In Exercise 3, show that the particle returns to ground level at a point whose distance from the origin is

$$\sin(2\alpha)u^2/g.$$

Apply some checks to this answer. For which angle of projection is this distance a maximum?

6. Show that, if $0 < \beta < \alpha$, the time t which elapses before the particle in Exercise 3 again crosses through the line through 0 making an angle β with the horizontal is given by

$$t = \frac{2u}{g} \cos(\alpha)(\tan(\alpha) - \tan(\beta)).$$

Apply some checks to this answer.

7. A projectile of mass m moves in a vertical plane subject to the forces of gravity and air-resistance. The magnitude of the air-resistance is assumed to be proportional to the square of the speed of the particle. The direction in which it acts is opposite to the velocity vector. Suppose that the cartesian coordinates of the particle at time t are x and y where the x-axis is horizontal and the y-axis is vertically upwards.

(a) In terms of $\dot{x}$ and $\dot{y}$ find

(i) the magnitude of the velocity vector $\dot{\underline{X}} = \dot{x}\underline{i} + \dot{y}\underline{j}$,

(ii) a vector of unit length with the direction of $\dot{\underline{X}}$,

(iii) the speed of the projectile,

(iv) the force due to air-resistance.

(b) Deduce that the equations of motion of the projectile are

$$m\ddot{x} = -k\dot{x}\sqrt{\dot{x}^2 + \dot{y}^2},$$
$$m\ddot{y} = -g - k\dot{y}\sqrt{\dot{x}^2 + \dot{y}^2},$$

for some constant $k > 0$.
[These differential equations can be uncoupled, and hence solved in closed form (see Synge and Griffiths, 1959); however, the resulting formulae are complicated. Alternatively, numerical approximations to the solutions can be obtained, on a computer, and hence graphs can be sketched.]

17

Motion on a circle

This chapter explains the mathematical ideas needed to study the motion of a particle moving on a circle. These ideas enable us to formulate mathematical models for a number of interesting problems. For example, as an introduction to the mechanics of the solar system you are shown how to derive one of Kepler's laws for the special case of a planet which moves in a circular orbit. In another application you are shown how to set up the equation of motion for a pendulum, from which useful information can be obtained by linearizing the differential equation.

17.1 Kinematics on a circle

The key step in solving many problems in mechanics is to introduce coordinates which fit in with the geometry of the problems. Up till now, the only coordinates which we have used in mechanics have been cartesian coordinates, which measure distances along axes. The coordinate which seems most suited to problems involving circles, however, measures an angle rather than a distance.

The angular coordinate

Consider a particle P moving around a circle of centre O in the (x, y)-plane with radius $a > 0$, as in Figure 17.1.1. The *angular coordinate* of the particle is defined as the angle θ radians ($-\pi < \theta \leq \pi$) which the position vector $\overrightarrow{OP}$ makes with the x-axis.

The values of the angular coordinate at various points on the circle are shown in Figure 17.1.2. If, for example, the particle moves around the circle anti-clockwise then θ assumes the value 0 when the particle passes through the point $(a, 0)$ and then increases to the value π when it

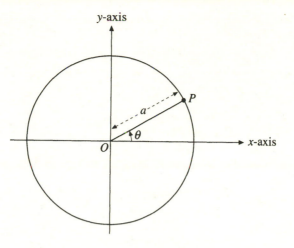

Fig. 17.1.1. The angular coordinate θ.

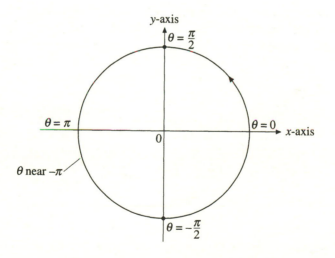

Fig. 17.1.2. Values of θ at various points.

reaches $(-a, 0)$. After this, θ drops to values near $-\pi$ and then increases up to 0 as the particle passes through $(a, 0)$ once more. Note the sudden jump, or discontinuity, in the values of θ as the particle passes through $(-a, 0)$.

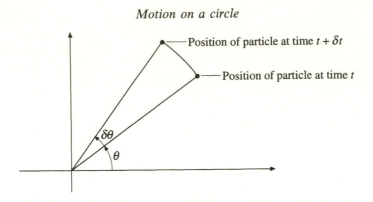

Fig. 17.1.3. Change in angular coordinate over a small time interval δt.

Angular velocity and acceleration

Imagine now a particle moving around the circle, so that its angular coordinate θ is a function of the time, say $\theta = f(t)$. We assume that the function f is twice differentiable. As the time changes from t to $t + \delta t$, let the angular coordinate change from θ to $\theta + \delta\theta$, as in Figure 17.1.3.

Thus $\delta\theta$ is the angle swept out by the ray joining the origin to the particle during the time interval from t to $t + \delta t$, where $\delta t \neq 0$. We assume, for the present, that $\theta \neq \pi$ so as to avoid the discontinuity in the value of θ.

The ratio

$$\frac{\delta\theta}{\delta t} = \frac{f(t + \delta t) - f(t)}{\delta t}$$

is called the *average angular velocity* of the particle during this interval. The *angular velocity* at time t is defined to be

$$\lim_{\delta t \to 0} \frac{\delta\theta}{\delta t} = f'(t)$$

and we denote it by $\dot{\theta}$. Thus the angular velocity is the rate of increase of the angular coordinate with respect to time. It measures the rate at which the particle is spinning around the origin. If the angular velocity is positive, the particle is moving anti-clockwise around the circle; if it is negative, however, the particle is moving clockwise. If $\dot{\theta}$ is constant, the motion of the particle around the circle is said to be *uniform*.

In a similar way, the ratio

$$\frac{\delta\dot{\theta}}{\delta t} = \frac{f(t + \delta t) - f(t)}{\delta t}$$

is called the *average angular acceleration* of the particle during the time

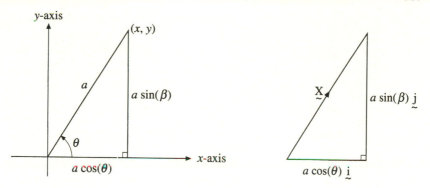

Fig. 17.1.4. Cartesian coordinates in terms of the angular coordinate θ.

interval from t to $t + \delta t$. The *angular acceleration* at time t is defined to be

$$\lim_{\delta t \to 0} \frac{\delta \dot{\theta}}{\delta t} = f(t)$$

and we denote it by $\ddot{\theta}$. The angular acceleration is the rate of increase of the angular velocity with respect to the time.

In spite of the jump in the value of the angular coordinate whenever the particle crosses the negative x-axis it turns out that, in the problems of interest to us, there is no such jump in the angular velocity or acceleration.

We now relate the angular quantities to their cartesian counterparts.

Cartesian coordinates

As can be seen from Figure 17.1.4, the cartesian coordinates of the particle are given in terms of its angular coordinate by the formulae

$$x = a \cos \theta, \qquad y = a \sin \theta.$$

Hence the position vector of the particle is given in terms of its angular coordinate by the formula

$$\underset{\sim}{X} = a \cos(\theta)\underset{\sim}{i} + a \sin(\theta)\underset{\sim}{j}. \tag{1}$$

In accordance with the definitions in Section 16.1, the velocity $\dot{\underset{\sim}{X}}$ of the particle is obtained from (1) by differentiating each component with respect to the time t. Now, by the rule for differentiating composite functions,

$$\frac{d}{dt}a\cos(\theta) = \frac{d}{d\theta}a\cos(\theta)\frac{d\theta}{dt} = -a\sin(\theta)\dot{\theta},$$

$$\frac{d}{dt}a\sin(\theta) = \frac{d}{d\theta}a\sin(\theta)\frac{d\theta}{dt} = a\cos(\theta)\dot{\theta}$$

and hence from (1)

$$\dot{\underset{\sim}{X}} = -a\sin(\theta)\dot{\theta}\underset{\sim}{i} + a\cos(\theta)\dot{\theta}\underset{\sim}{j}. \tag{2}$$

In a similar way the acceleration vector $\ddot{\underset{\sim}{X}}$ is obtained from (2) by differentiating each component with respect to the time t. This involves using both the product and the composite rules for differentiation. The result is

$$\ddot{\underset{\sim}{X}} = \left(-a\cos(\theta)\dot{\theta}^2 - a\sin(\theta)\ddot{\theta}\right)\underset{\sim}{i} + \left(-a\sin(\theta)\dot{\theta}^2 + a\cos(\theta)\ddot{\theta}\right)\underset{\sim}{j}. \tag{3}$$

Tangent and normal vectors

The equations for the velocity and acceleration will now be rewritten so as to show their geometrical significance. In equation (2) take out the common factor $a\dot{\theta}$ to get

$$\dot{\underset{\sim}{X}} = a\dot{\theta}\left(-\sin(\theta)\underset{\sim}{i} + \cos(\theta)\underset{\sim}{j}\right) \tag{4}$$

while in equation (3) expand and then interchange the second and third terms to get

$$\ddot{\underset{\sim}{X}} = -a\cos(\theta)\dot{\theta}^2\underset{\sim}{i} + -a\sin(\theta)\dot{\theta}^2\underset{\sim}{j} - a\sin(\theta)\ddot{\theta}\underset{\sim}{i} + a\cos(\theta)\ddot{\theta}\underset{\sim}{j}$$

and hence

$$\ddot{\underset{\sim}{X}} = -a\dot{\theta}^2\left(\cos(\theta)\underset{\sim}{i} + \sin(\theta)\underset{\sim}{j}\right) + a\ddot{\theta}\left(-\sin(\theta)\underset{\sim}{i} + \cos(\theta)\underset{\sim}{j}\right). \tag{5}$$

Notice that equation (5) expresses $\ddot{\underset{\sim}{X}}$ as a linear combination of the two vectors

$$\cos(\theta)\underset{\sim}{i} + \sin(\theta)\underset{\sim}{j} \qquad \text{and} \qquad -\sin(\theta)\underset{\sim}{i} + \cos(\theta)\underset{\sim}{j},$$

which we shall denote by $\underset{\sim}{\nu}$ ('nu') and $\underset{\sim}{\tau}$ ('tau') respectively.

We imagine the vectors $\underset{\sim}{\nu}$ and $\underset{\sim}{\tau}$ as sitting with their tails at the point of the circle occupied by the particle. The geometrical significance of these vectors is described in the following proposition, which is illustrated in Figure 17.1.5(b).

Proposition. The vectors

$$\underset{\sim}{\nu} = \cos(\theta)\underset{\sim}{i} + \sin(\theta)\underset{\sim}{j} \qquad \text{and} \qquad \underset{\sim}{\tau} = -\sin(\theta)\underset{\sim}{i} + \cos(\theta)\underset{\sim}{j} \tag{6}$$

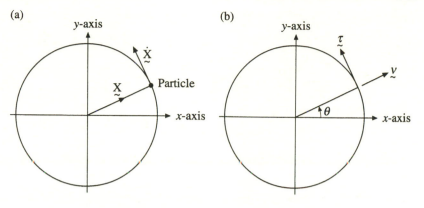

Fig. 17.1.5. Unit tangent and normal vectors.

have length 1 and, at the point on the circle with angular coordinate θ,

$\underset{\sim}{v}$ is normal to the circle, pointing outwards from the circle, while

$\underset{\sim}{\tau}$ is tangent to the circle and points anti-clockwise.

In terms of these vectors, the velocity $\dot{\underset{\sim}{X}}$ and the acceleration $\ddot{\underset{\sim}{X}}$ may be written very simply as

$$\dot{\underset{\sim}{X}} = a\dot{\theta}\underset{\sim}{\tau} \tag{7}$$

and

$$\ddot{\underset{\sim}{X}} = -a\dot{\theta}^2\underset{\sim}{v} + a\ddot{\theta}\underset{\sim}{\tau}. \tag{8}$$

Proof The vectors $\underset{\sim}{v}$ and $\underset{\sim}{\tau}$ have length 1; this follows directly from (6). The direction of $\underset{\sim}{v}$ is the same as that of $\underset{\sim}{X}$, as comparison of (1) and (6) shows. Thus $\underset{\sim}{v}$ points out radially from the circle.

On the other hand it follows directly from (6) that the dot product of $\underset{\sim}{v}$ and $\underset{\sim}{\tau}$ is zero. Hence $\underset{\sim}{\tau}$ is perpendicular to $\underset{\sim}{v}$ and so must be tangent to the circle.

Finally, the formulae (7) and (8) follow immediately from (4) and (5) on use of (6). □

The geometric significance of the formulae (7) and (8) for the velocity and acceleration is now clear. Since $\underset{\sim}{v}$ and $\underset{\sim}{\tau}$ are the unit vectors normal and tangent to the circle, these formulae express the velocity and acceleration in terms of normal and tangential components, as illustrated in Figure 17.1.6.

Thus (7) shows that the velocity is in the direction of the tangent to

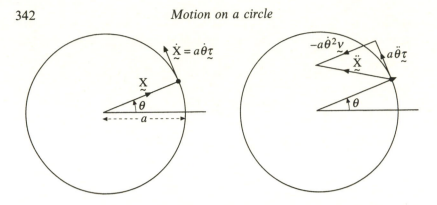

Fig. 17.1.6. Velocity and acceleration in terms of tangent and normal vectors.

the circle and has magnitude $|a\dot{\theta}| = a|\dot{\theta}|$ which is also called the *speed* of the particle.

On the other hand, (8) shows that the acceleration has both a tangential component of magnitude $|a\ddot{\theta}| = a|\ddot{\theta}|$ and a normal component, of magnitude $a\dot{\theta}^2$, directed towards the centre of the circle.

The formulae (7) and (8) are the key to solving problems involving motion in a circle and they should be remembered.

Exercises 17.1

1. Each of the following points lies on the circle of radius 2 and centre $(0,0)$ in the (x, y)-plane:

$$(0, -2), \quad (2, 0), \quad (\sqrt{2}, \sqrt{2}), \quad (0, 2), \quad (-1, \sqrt{3}), \quad (-2, 0).$$

Plot each of these points on the circle and state the value of its angular coordinate.

2. A particle moves around the circle in Exercise 1 in such a way that its position vector at time t is given by the formula

$$\underset{\sim}{X} = 2\cos(t)\underset{\sim}{i} + 2\sin(t)\underset{\sim}{j}.$$

(a) Express the angular coordinate θ as a function of t when $-\pi < t \leq \pi$. Hence find the values of the angular velocity $\dot{\theta}$ and angular acceleration $\ddot{\theta}$ at time t. Find also the speed of the particle.

(b) Now use appropriate formulae in the text to express the velocity $\dot{\underset{\sim}{X}}$ and the acceleration vector $\ddot{\underset{\sim}{X}}$ as functions of t for $-\pi < t \leq \pi$.

(c) State briefly why the formulae obtained in (b) are still valid when $\pi < t \leq 3\pi$.

3. By using the formulae given in the text for the unit tangent and unit normal vectors, verify that they each have length 1.

4. By using the formulae given in the text for the unit tangent and unit normal vectors $\underset{\sim}{\tau}$ and $\underset{\sim}{v}$ at a point on the circle, verify that their dot product $\underset{\sim}{\tau} \cdot \underset{\sim}{v}$ is zero. To which geometrical fact does this correspond?

5. The bob P of a simple pendulum oscillates in a vertical plane along the arc of a circle between two points A and B (at the same height). Let θ be the angle which the pendulum makes with the vertical at time t, as shown below.

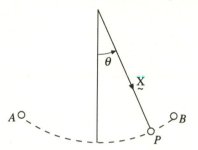

(a) Which point does the particle P occupy when its angular coordinate θ assumes (i) its maximum value? (ii) its minimum value?

(b) What is the value of the angular velocity $\dot{\theta}$ at these points?

(c) Show the direction of the acceleration vector $\underset{\sim}{\ddot{X}}$ at each of these points.

6. In Exercise 5, which point does the particle P occupy when its angular acceleration is zero? Show the direction of its acceleration vector $\underset{\sim}{\ddot{X}}$ at this point.

7. Complete the details given in the text of the derivation of formula (3) for $\underset{\sim}{\ddot{X}}$ from formula (2) for $\underset{\sim}{\dot{X}}$.

17.2 Uniform circular motion

A number of interesting problems in physics can be modelled by a particle moving around a circle with constant angular velocity. In this special case, the formula for the acceleration of the particle simplifies. As in the previous section, we assume the particle is moving on a circle of radius a and centre at the origin and we let θ and $\underset{\sim}{X}$ denote the angular coordinate and the position vector of the particle at time t.

The assumption of constant angular velocity means that

$$\dot{\theta} = \omega \tag{1}$$

for some constant $\omega \neq 0$ and hence that $\ddot{\theta} = 0$. Hence in the formula (8) of Section 17.1 the tangential component drops out and the formula simplifies to

$$\begin{aligned} \underset{\sim}{\ddot{X}} &= -a\dot{\theta}^2 \underset{\sim}{v}, \\ &= -a\omega^2 \underset{\sim}{v}. \end{aligned} \tag{2}$$

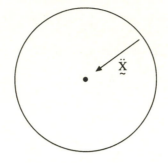

Fig. 17.2.1. Acceleration vector points to centre of circle.

This shows that, when circular motion is uniform, the acceleration always points in the direction from the particle to the origin, as shown in Figure 17.2.1.

We assume that the particle starts on the x-axis so that $\theta = 0$ when $t = 0$. The differential equation (1) together with this initial condition has the solution

$$\theta = \omega t \tag{3}$$

provided t is in the interval for which $-\pi < \theta \leq \pi$. If t lies outside this interval, then a multiple of 2π must be added to the RHS of (3). Hence the formula

$$\underset{\sim}{X} = a\cos(\theta)\underset{\sim}{i} + a\sin(\theta)\underset{\sim}{j}$$

gives in all cases

$$\underset{\sim}{X} = a\cos(\omega t)\underset{\sim}{i} + a\sin(\omega t)\underset{\sim}{j}. \tag{4}$$

This formula shows that $\underset{\sim}{X}$ is a periodic function of t with period $2\pi/|\omega|$. Thus the particle makes one complete revolution around the circle in time $2\pi/|\omega|$. The distance which the particle moves during this time is the length $2\pi a$ of the circumference of the circle. Hence the speed u of the particle is given by

$$u = \frac{\{\text{distance moved}\}}{\{\text{time taken}\}} = \frac{2\pi a}{2\pi/|\omega|} = a|\omega|. \tag{5}$$

This result can also be obtained from the more generally applicable formula (4) of Section 17.1 (in which the angular velocity need not be constant). The formula (5) enables us to deduce the speed of the particle from its angular velocity, and vice versa.

The following example gives a simple application of the above kinematical formulae to astronomy. In this example the Earth's acceleration is calculated from other data about its orbit.

Example 1. *The Earth moves around the sun in an orbit which is nearly circular of radius* 1.49×10^8 *km. Find*

(a) *the angular velocity of the Earth around the sun,*
(b) *the speed of the Earth,*
(c) *the magnitude of its acceleration towards the sun.*

Solution. *We choose the centre of the sun as the origin and let*

$$\underset{\sim}{X} = \left\{ \begin{array}{c} \text{position vector} \\ \text{of the Earth} \\ \text{at time } t \end{array} \right\}$$

$$\underset{\sim}{v} = \left\{ \begin{array}{c} \text{outward unit normal} \\ \text{to the Earth's orbit} \\ \text{at this point} \end{array} \right\}$$

$$\omega = \left\{ \begin{array}{c} \text{angular velocity} \\ \text{of the Earth} \\ \text{around the sun} \end{array} \right\}$$

$$u = \left\{ \begin{array}{c} \text{speed} \\ \text{of the} \\ \text{Earth} \end{array} \right\}.$$

(a) *The position vector of the Earth sweeps out an angle of 2π radians in 1 siderial year, of length $365\frac{1}{4}$ days. Thus the angular velocity ω is given by*

$$\omega = \frac{\{\text{angle swept out}\}}{\{\text{time taken}\}}$$

$$= \frac{2\pi}{365.25 \times 24 \times 60 \times 60}$$

$$= 1.99 \times 10^{-7} \text{ rad/s}.$$

(b) *The speed of the Earth is given by (5) as*

$$u = a\omega$$

$$= 1.49 \times 10^8 \times 1.99 \times 10^{-7}$$

$$= 29.7 \text{ km/s}.$$

(c) *From (2), the acceleration of the Earth is directed towards the sun and has magnitude*

$$a\omega^2 = 1.49 \times 10^8 \times (1.99 \times 10^{-7})^2$$

$$= 5.90 \times 10^{-6} \text{ km/s}^2.$$

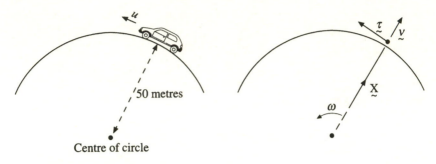

Centre of circle

Fig. 17.2.2. A car crossing a bridge. See Example 2.

Similar calculations to those in the above example can be made for other planets and their satellites by using data about their orbits obtained, for example, from the *CRC Handbook of Chemistry and Physics*.

While the above example involves only kinematics, the following example requires the use of the Newton's second law of motion in its vector form, $\underset{\sim}{F} = m\underset{\sim}{\ddot{X}}$.

Example 2. *A car is crossing a bridge whose vertical cross-section has the form of an arc of a circle of radius 50 metres. Show that, if the car is to maintain contact with the bridge at the highest point, then its speed u must satisfy the inequality*

$$u \leq 22 \quad \text{m/s}.$$

Solution. *We regard the car as a particle. The solution follows the usual steps for a dynamical problem:* introduce notation; draw a force diagram; apply Newton's second law.

We choose the centre of the circle as origin and let

$$\underset{\sim}{X} = \{\text{position vector of car at time } t\},$$
$$\omega = \{\text{angular velocity of car around origin}\},$$
$$\underset{\sim}{v} = \{\text{outward unit normal at top of circle}\},$$
$$\underset{\sim}{\tau} = \{\text{unit tangent at top of circle}\},$$
$$N = \{\text{magnitude of normal reaction on car}\},$$
$$f = \{\text{magnitude of friction force on car}\},$$
$$m = \{\text{mass of car}\}.$$

We need only consider the forces acting on the car when it is at the highest point of the road. The directions of the individual forces are then either horizontal or vertical, as shown in Figure 17.2.3. Thus the net force acting on the car, at the top of the road, is given by

$$\underset{\sim}{F} = (N - mg)\underset{\sim}{v} - f\underset{\sim}{\tau}.$$

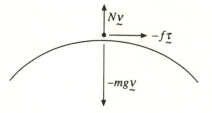

Fig. 17.2.3. Forces on the car at highest point of road.

Now, while the car remains in contact with the road, it is moving around a circle of radius a = 50 metres. Hence the formula (2) gives

$$\ddot{X} = -a\omega^2 \underset{\sim}{v}$$

where a = 50 metres. Application of Newton's second law $\underset{\sim}{F} = m\ddot{X}$ now gives

$$(N - mg)\underset{\sim}{v} - f\underset{\sim}{\tau} = -ma\omega^2 \underset{\sim}{v}.$$

Equating coefficients on each side of this equation gives

$$N - mg = -ma\omega^2$$

and hence

$$a\omega^2 = g - N/m.$$

But, since $N \geq 0$, this implies that

$$a\omega^2 \leq g.$$

Hence by (5)

$$u^2 \leq ag,$$

and so $u \leq \sqrt{50 \times 9.8} = 22$ m/s.

It can be shown that, for the car to maintain contact with the road at points near the top of the bridge, an even stricter speed limit must be applied than that obtained in the above example.

Exercises 17.2

1. The moon orbits the Earth once every 27.3 days in a nearly circular orbit of radius 3.83×10^8 metres. Find the acceleration of the moon.

2. A car of mass 1000 is travelling at a constant speed of 20 m/s as it moves down one hill and up the next as in the diagram below. Near the lowest point, the vertical cross-section of the road may be regarded as part of the arc of a circle of radius 200 metres.

Find (i) the normal reaction of the road on the car and (ii) the net frictional force acting on the tyres as the car passes over the lowest point on the road.

3. The breaking strength of a 50 cm long string of a simple pendulum is 30 newton. What is the maximum mass that may be used for the bob of the pendulum if the speed of the bob at the lowest point of the string is 1 m/s.

4. An unbanked curve on a highway has the shape of a circular arc. What is the minimum safe radius for the arc if the coefficient of static friction between the tyres and the road is 0.6 and the speed limit is 100 km/h?

INTRODUCTION TO ASTRONOMY: Exercise 5 below shows how to derive one of Kepler's three famous laws of planetary motion:

$$T^2 \quad \text{is proportional to} \quad r^3$$

where T is the period of revolution of a planet around the sun and r is the radius of its orbit. To simplify the mathematics, it will be assumed that the orbits are circular, although it would be more realistic to take them as ellipses. You will need to assume Newton's Law of Universal Gravitation, which states:

- *each pair of bodies in the universe attract each other with*
- *a force which has magnitude GmM/r²*

where m and M are the masses of the two bodies, r is the distance between them, and G is the universal constant of gravitation.

---

5. Assume that a planet moves around the sun of mass M in a circular orbit of radius r, with angular velocity ω.

   (a) Introduce notation for the position vector of the planet at time t and the unit normal to the circle at the point with this position vector.

   (b) From Newton's law of universal gravitation, write down the force acting on the planet, at time t.

   (c) Apply Newton's second law of motion to derive the formula

$$\omega^2 = GM/r^3.$$

Hence obtain Kepler's third law.

17.3 The pendulum and linearization

When Galileo performed his famous experiment of rolling balls down an inclined plane, the device he used to measure equal intervals of time was similar to an hour-glass, but involving the flow of water rather than sand. The accuracy obtainable with such a device was very limited and

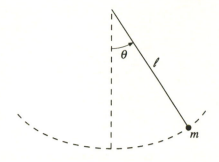

Fig. 17.3.1. Angular coordinate for the pendulum.

developments in physics and astronomy during the seventeenth century called for a more accurate way of measuring time.

This need was met by the invention of the pendulum clock in 1657. It was invented by the Dutch physicist and mathematician Christiaan Huygens, to whom the mathematical theory of the pendulum is due. Although the equation of motion for the pendulum is non-linear, Huygens was able to obtain useful information about its solutions, by using the technique of linearization.

The model

Our model for the pendulum consists of a light rod of length ℓ pivoted smoothly at one end and with a particle of mass m attached to the other end. The particle is called the *bob* of the pendulum. Motion is possible in a vertical plane. We choose θ to be the angle which the pendulum makes with the vertical at time t, as shown in Figure 17.3.1.

Thus the bob of the pendulum is free to move in a vertical circle under gravity. The bob is held on the circle by the force due to the rod, which is assumed to act along the direction of the rod.

The equation of motion

The derivation of the equation of motion of the bob of the pendulum uses methods from earlier in this chapter. Working through the details provides useful revision of these methods and is set below as Exercise 1. The equation of motion so obtained is

$$\ddot{\theta} = -\frac{g}{\ell}\sin(\theta). \qquad (1)$$

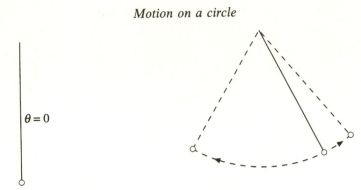

$\theta = 0$

Fig. 17.3.2. The bob resting and oscillating.

This differential equation is second order, but non-linear. Hence the methods of solution explained earlier in this book do not apply. Since, however, the RHS is a smooth function of θ, the existence–uniqueness theorem of Section 5.2 is still applicable and tells us that there are solutions of (1) (even though we cannot express them in terms of functions which we already know).

Some types of solutions

Because we cannot solve the equation of motion (1) in terms of familiar functions, we reverse the usual procedure: instead of using the solutions of (1) to tell us about the physics, we use our physical intuition to tell us about the behaviour of the solutions. In particular, physical intuition leads us to expect the following types of motion to be possible.

A stable equilibrium point. The lowest point on the circle is where $\theta = 0$, as shown on the left of Figure 17.3.2. If the bob starts at this point with zero velocity it will stay there for ever. This means that the equation of motion (1) should have the constant solution $\theta = 0$. This point is called an *equilibrium point* for the bob. The equilibrium point is said to be *stable* because, if you give the bob a small push, it stays close to the equilibrium point.

Small oscillations. If, before being released, the bob is displaced away from the equilibrium point, then it oscillates. Information about these oscillations is not so easy to obtain directly from the equation of motion (1).

In this special case where the oscillations have small amplitude, however, Huygens was able to obtain useful information about them by

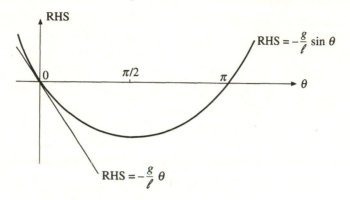

Fig. 17.3.3. Linear approximation based at $\theta = 0$.

approximating the differential equation (1) by a linear differential equation — which he could solve easily. The approximation process which he used is now called *linearization*. As it is very important in the study of differential equations, it will now be explained in some detail. Those who have already read Section 9.4 will notice that the ideas are much the same as in linearising difference equations.

Linearization near $\theta = 0$

Huygens knew that during small oscillations the values of θ stay close to 0. So instead of using the actual RHS of (1), given by

$$\text{RHS} = -\frac{g}{\ell} \sin \theta,$$

he looked for a simpler RHS which approximates this closely when θ is small. As we see from the graph in Figure 17.3.3, the tangent at the point $(0,0)$ follows the graph closely when θ is small. So following the tangent rather than the original graph gives a good approximation and it has the advantage of being linear in θ.

Now the tangent to the graph is the straight line through the origin with the same slope as the graph at $(0,0)$. We use differentiation to get the slope. Thus the new RHS is given by

$$\text{RHS} = -\frac{g}{\ell} \sin'(0)\theta,$$
$$= -\frac{g}{\ell} \cos(0)\theta,$$
$$= -\frac{g}{\ell}\theta.$$

In the differential equation (1) we now substitute this new RHS in place of the old one. This gives the new, and simpler, differential equation

$$\ddot{\theta} = -\frac{g}{\ell}\theta. \tag{2}$$

The differential equation (2) is called the *linearization near $\theta = 0$ of the differential equation* (1).

The linearized differential equation (2) is like (1) in that it is a second-order differential equation, but is unlike (1) in that it is linear and it can be solved easily. It is, in fact, just the SHM equation from Chapter 5, and so we know its solutions are given by

$$\theta = A \sin\left(\sqrt{\frac{g}{\ell}}t + \varepsilon\right) \tag{3}$$

where A and ε are arbitrary constants. These solutions are oscillatory with period given by

$$T = 2\pi\sqrt{\frac{\ell}{g}}, \tag{4}$$

which is independent of the amplitude A.

Exercises 17.3

1. Show that $\theta = 0$ is an equilibrium point of the differential equation
$$\ddot{\theta} = \theta - \sin(2\theta)$$
and then linearise the differential equation near this equilibrium point.

Part five
Coupled Models

18

Models with linear interactions

In this chapter we look at simple models in which two quantities interact with each other. The models lead to pairs of simultaneous differential equations for the two quantities and, because of the interaction, the equations are coupled.

The models considered will be simple enough to produce differential equations which are first-order linear, with constant coefficients. A systematic method for uncoupling — and hence solving — such equations will be explained. It involves eliminating one of the two quantities to give a second-order differential equation of the type studied in Chapter 15.

The ideas will be illustrated by a mixing model, similar to the one in Chapter 13, but involving a pair of vats. Two models from physiology are then presented: the first models the glucose–insulin homeostasis in the bloodstream, while the second models the mother–fetus exchange of nutrients via the placenta.

18.1 Two-compartment mixing

From now on we shall be considering models which lead to a *pair* of simultaneous differential equations, rather than a single differential equation. The equations will involve a *pair* of quantities, rather than a single quantity, which are to be expressed as functions of the time, say. In this section we show how an extension of the mixing problem discussed in Chapter 13 leads to such a pair of equations.

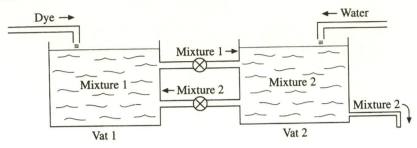

Fig. 18.1.1. Two interconnected vats.

Mixing with two vats

Consider two interconnected vats, each containing a mixture of dye and water, as in Figure 18.1.1. Dye runs into the first vat. Mixture from the first vat is pumped into the second vat and vice versa. As part of the model we assume the mixture in each vat is well stirred and hence is homogeneous (of uniform concentration throughout).

Given the various rates of flow of the liquids into and out of the two vats, we shall show how to set up simultaneous differential equations for the concentrations of the two mixtures.

Formulating the equations: an example

The following problem illustrates the procedure for setting up the differential equations.

> *Assume that initially the two vats each contain 100 litres of pure water. Pure dye is then pumped into the first vat at a fixed rate of 1 litre/minute, while pure water is pumped into the second vat at the same rate. Pumps exchange the mixtures between the two vats — at a rate of 4 litres/minute from vat 1 into vat 2, and 3 litres/minute from vat 2 into vat 1. The diluted mixture is drawn off from vat 2 at a rate of 2 litres/minute. Derive a pair of differential equations for the concentrations of dye in each vat.*

The steps involved in the solution follow closely those used in the example in Section 13.1. First we draw a diagram and show the rates of flow of the liquids into the vats, as in Figure 18.1.2.

We then note that, for vat 1, liquid (pure dye and mixture 2) flows in at the rate of 4 litres/minute. This is exactly balanced by the outflow

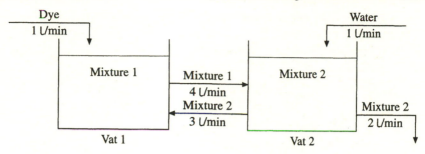

Fig. 18.1.2. Rates of flow of the liquids.

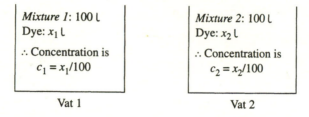

Fig. 18.1.3. Volumes and concentrations at time $t \geq 0$.

of mixture 1 at 4 litres/minute. Hence the liquid in vat 1 maintains its initial volume of 100 litres. A similar argument shows that the volume of the mixture in vat 2 also stays fixed.

The next step is to introduce notation for the quantities involved and to show them on the diagram.

Let x_1 and x_2 litres be the volumes of dye in each tank at time $t \geq 0$. Hence the concentrations are

$$c_1 = \frac{\left\{\begin{array}{c}\text{volume of dye}\\\text{in vat 1}\end{array}\right\}}{\left\{\begin{array}{c}\text{volume of mixture}\\\text{in vat 1}\end{array}\right\}} = \frac{x_1}{100},$$

$$c_2 = \frac{\left\{\begin{array}{c}\text{volume of dye}\\\text{in vat 2}\end{array}\right\}}{\left\{\begin{array}{c}\text{volume of mixture}\\\text{in vat 2}\end{array}\right\}} = \frac{x_2}{100}.$$

(1)

These volumes and concentrations are shown in Figure 18.1.3.

We now consider what happens to the volumes of the dye in the vats during a small time interval from t to $t + \delta t$. The volumes of dye in the respective vats at time t are denoted by x_1 and x_2; at time $t + \delta t$ they become $x_1 + \delta x_1$ and $x_2 + \delta x_2$.

Thus δx_1 is the change in volume of the dye in vat 1 during this time interval and hence

$$\delta x_1 = \left\{ \begin{array}{c} \text{volume of dye} \\ \text{flowing} \\ \text{into vat 1} \end{array} \right\} - \left\{ \begin{array}{c} \text{volume of dye} \\ \text{flowing} \\ \text{out of vat 1} \end{array} \right\}.$$

From Figure 18.1.2 it now follows that

$$\delta x_1 = \left\{ \begin{array}{c} \text{volume of} \\ \text{pure dye} \\ \text{flowing} \\ \text{into vat 1} \end{array} \right\} + \left\{ \begin{array}{c} \text{volume of dye} \\ \text{in mixture 2} \\ \text{flowing} \\ \text{into vat 1} \end{array} \right\} - \left\{ \begin{array}{c} \text{volume of dye} \\ \text{in mixture 1} \\ \text{flowing out} \\ \text{of vat 1} \end{array} \right\}$$

$$\simeq (1\,\delta t) \times 1 + (3\,\delta t) \times \frac{x_2}{100} - (4\,\delta t) \times \frac{x_1}{100}. \tag{2}$$

Here the first factor in each term is the volume of the appropriate mixture flowing into or out of vat 1 during the time δt. These factors are each multiplied by the *fraction of dye* (concentration) in that mixture to give the *volume of the dye*. The final result is only an approximation because the concentrations change slightly during the time interval.

Similarly, δx_2 is the change in volume of the dye in vat 2 during this time interval and hence

$$\delta x_2 = \left\{ \begin{array}{c} \text{volume of dye} \\ \text{flowing} \\ \text{into vat 2} \end{array} \right\} - \left\{ \begin{array}{c} \text{volume of dye} \\ \text{flowing} \\ \text{out of vat 2} \end{array} \right\}$$

$$\simeq (4\,\delta t) \times \frac{x_1}{100} - (3\,\delta t) \times \frac{x_2}{100} - (2\,\delta t) \times \frac{x_2}{100}. \tag{3}$$

As in Section 13.1, we can show that the errors involved in (2) and (3) are small compared with δt. Hence dividing each of (2) and (3) by δt, and then letting δt approach 0 gives

$$\begin{aligned} \frac{dx_1}{dt} &= 1 - \frac{4}{100}x_1 + \frac{3}{100}x_2 \\ \frac{dx_2}{dt} &= \frac{4}{100}x_1 - \frac{5}{100}x_2. \end{aligned} \tag{4}$$

Thus we have derived a pair of simultaneous differential equations for x_1 and x_2 as functions of t. The initial conditions

$$x_1 = 0 \quad \text{and} \quad x_2 = 0 \quad \text{when} \quad t = 0 \tag{5}$$

must also be satisfied, since there is no dye in either vat initially.

We could solve the initial-value problem (4)–(5) for x_1 and x_2 as functions of t, and then obtain the concentrations c_1 and c_2 from (1).

Alternatively, we can use (1) to get

$$x_1 = 100c_1 \quad \text{and} \quad x_2 = 100c_2.$$

Substitution in (4) then gives the following pair of simultaneous differential equations for c_1 and c_2 as functions of time,

$$
\begin{aligned}
\frac{dc_1}{dt} &= \frac{1}{100} - \frac{4}{100}c_1 + \frac{3}{100}c_2 \\
\frac{dc_2}{dt} &= \frac{4}{100}c_1 - \frac{5}{100}c_2,
\end{aligned}
\tag{6}
$$

while the initial condition (5) now becomes

$$c_1 = 0 \quad \text{and} \quad c_2 = 0 \quad \text{when} \quad t = 0. \tag{7}$$

Thus we have obtained the desired pair of simultaneous differential equations for the concentrations, together with the relevant initial conditions.

Steady-state solutions

How to solve pairs of simultaneous differential equations like (4) and (6) will be explained in the next section. One thing we can do straight away, however, is to look for the *constant solutions* of the differential equations (which correspond to *steady-states* or *equilibrium states* of the mixing problem). These solutions can be found by setting

$$\frac{dc_1}{dt} = 0 \quad \text{and} \quad \frac{dc_2}{dt} = 0$$

in (6) to get the simultaneous pair of linear algebraic equations

$$
\begin{aligned}
-4c_1 + 3c_2 &= -1, \\
4c_1 - 5c_2 &= 0.
\end{aligned}
\tag{8}
$$

These equations have the solution

$$c_1 = \frac{5}{8} \quad \text{and} \quad c_2 = \frac{1}{2},$$

which are therefore the steady-state concentrations for the mixing problem.

The next question to ask is whether the actual concentrations approach these steady-state concentrations as time goes on. We will be able to answer this in the next section, by taking limits as $t \to \infty$, after we have found the time-dependent solutions of the intial-value problem (6)–(7).

Exercises 18.1

1.   Initially two vats each contain 50 litres of pure water. Pure dye is then pumped into the first vat at a fixed rate of 2 litres/minute, while pure water is pumped into the second vat at a rate of 2 litres/minute. Pumps exchange the mixtures between the two vats — at a rate of 6 litres/minute from vat 1 into vat 2, and 4 litres/minute from vat 2 into vat 1. The diluted mixture is drawn off from vat 2 at a rate of 4 litres/minute.

Show that the concentrations of dye in the vats at time t minutes after the start satisfy the following simultaneous differential equations:

$$\frac{dc_1}{dt} = \frac{1}{25} - \frac{3}{25}c_1 + \frac{2}{25}c_2,$$
$$\frac{dc_2}{dt} = \frac{3}{25}c_1 - \frac{4}{25}c_2.$$

What are the initial conditions for these differential equations?

2.   Find the steady-state concentrations for the mixing problem in Exercise 1.

3.   As in Exercise 1 except that vat 1 contains 50 litres of pure water initially and vat 2 contains 100 litres of a 25% mixture of dye initially.

4.   As in Exercise 1 except that dye is now pumped into the first vat at the rate of 3 litres/minute. Show now that the concentrations satisfy the following differential equations until such time as the first vat overflows:

$$\dot{c}_1 = (3 - 7c_1 + 4c_2)/(50 + t)$$
$$\dot{c}_2 = (6c_1 - 8c_2)/50.$$

18.2 Solving constant-coefficient equations

In this section we introduce a procedure which is capable of solving the simultaneous pair of differential equations from the previous section. Let A, B, C, D, P and Q be given functions of t. A pair of simultaneous differential equations of the form

$$\frac{dx}{dt} = Ax + By + P$$
$$\frac{dy}{dt} = Cx + Dy + Q$$
(1)

is said to be *first order* because it involves only the first derivatives of x and y, and *linear* because of the linear way in which x and y appear on the RHS. If $P = Q = 0$ then it is said to be *homogeneous*.

When A, B, C, D are constants, the pair of equations is called *constant coefficient*. This is the case to which the solution procedure is applicable, and it includes the pair of differential equations of the previous section, for the concentrations of the dye.

Example 1. *Determine in each case if the pair of differential equations is linear, constant coefficient, or homogeneous.*

(a)　$\dfrac{dx}{dt} = 3x - 2xy$　　　　　(b)　$\dfrac{dx_1}{dt} = 2x_1 + 2tx_2$

　　　$\dfrac{dy}{dt} = 2x - y$　　　　　　　　$\dfrac{dx_2}{dt} = x_1 + x_2$

(c)　$\dfrac{dx}{dt} = 7x - 2y + t^2$　　　(d)　$\dfrac{dx}{dt} = 7x - 2y$

　　　$\dfrac{dy}{dt} = 3x - y + 1$　　　　　　$\dfrac{dy}{dt} = 3x - y$

Solution. *The pair (a) is not linear because of the 2xy term in the first equation. All other pairs are linear. The pair (b) is not constant coefficient, because of the coefficient 2t in the first equation, but (c) and (d) are constant coefficient. The pairs (b) and (d) are homogeneous.*

A *solution* of the pair of differential equations (1) is a pair of functions ϕ and ψ such that when we substitute

$$x = \phi(t) \quad \text{and} \quad y = \psi(t)$$

into (1) we get LHS = RHS for each of the pair of differential equations. The domain of each function is assumed to be an interval, chosen as large as possible.

The solution procedure

We now give a step-by-step procedure for solving a pair of simultaneous differential equations of the form

$$\dot{x} = ax + by \tag{2a}$$
$$\dot{y} = cx + dy \tag{2b}$$

where a, b, c, d are constants with $b \neq 0$. To simplify the algebra, we will explain the solution procedure only for a pair of homogeneous equations. The extension to inhomogeneous ones is left as Exercise 3. (You should try to understand why the procedure works. Do not try to memorize the detailed results of each step.)

STEP 1: *Find a second-order differential equation for x alone by differentiating (2a) to get*

$$\ddot{x} = a\dot{x} + b\dot{y}$$
$$= a\dot{x} + b(cx + dy) \qquad \text{(removing } \dot{y} \text{ with (2a))}$$
$$= a\dot{x} + b\left(cx + d\frac{\dot{x} - ax}{b}\right) \qquad \text{(removing } y \text{ with (2b)).}$$

Thus

$$\ddot{x} = (a + d)\dot{x} + (bc - ad)x. \tag{3}$$

This is the desired second-order differential equation involving x and its derivative (but not y). It is linear with constant coefficients; hence it can be solved as in Chapter 15.

STEP 2: *Solve (3) for x as a function of t (involving two arbitrary constants a_1 and a_2).*

STEP 3: *Use (2a) to get y as a function of t (involving a_1 and a_2).* By (1)

$$y = \frac{1}{b}(\dot{x} - ax).$$

We now use the solution obtained for x as a function of t in Step 2 to replace the RHS of this formula by a function of t.

Remarks about the procedure

- If at Step 3 we had used (2) instead of (1), we would have obtained a *differential* equation for y rather than an *algebraic* equation. This would have made a lot of extra work for us.
- The solution obtained at Steps 2 and 3 for x and y involves two arbitrary constants A and B. Specific values of these constants can be found so as to satisfy initial conditions of the form

$$x = x_0 \quad \text{and} \quad y = y_0 \quad \text{when} \quad t = 0.$$

- The assumption $b \neq 0$ is needed for the validity of the above procedure. (In Exercise 5 you are asked to determine how to proceed if $b = 0$.)
- The above procedure is not the only one which can be used to solve simultaneous linear differential equations with constant coefficients. A particularly efficient method (using more matrix theory and linear algebra than this book assumes) may be seen in Braun (1975), for example.

Application to the mixing problem

By using the above procedure, we can now solve the differential equations for the mixing problem modelled in the previous section.

Example 2. *Find the solution of the differential equations*

$$\dot{c}_1 = \frac{1}{100} - \frac{4}{100}c_1 + \frac{3}{100}c_2 \qquad (4)$$

$$\dot{c}_2 = \frac{4}{100}c_1 - \frac{5}{100}c_2 \qquad (5)$$

which satisfies the initial conditions $c_1 = c_2 = 0$ when $t = 0$.

Solution. *We follow the steps in the above procedure.*

STEP 1: *We find a second-order equation for c_1, by differentiating (4) to get*

$$\ddot{c}_1 = -\frac{4}{100}\dot{c}_1 + \frac{3}{100}\dot{c}_2$$

$$= -\frac{4}{100}\dot{c}_1 + \frac{3}{100}\left(\frac{4}{100}c_1 - \frac{5}{100}c_2\right) \qquad (by\ (5))$$

$$= -\frac{4}{100}\dot{c}_1 + \frac{3}{100}\left(\frac{4}{100}c_1 - \frac{5}{100}\frac{100}{3}\left(\dot{c}_1 - \frac{1}{100} + \frac{4}{100}c_1\right)\right) \qquad (by\ (4)).$$

Thus

$$\ddot{c}_1 = -\frac{9}{100}\dot{c}_1 - \frac{8}{100^2}c_1 + \frac{5}{100^2}$$

or

$$\ddot{c}_1 + \frac{9}{100}\dot{c}_1 + \frac{8}{100^2}c_1 = \frac{5}{100^2} \qquad (6)$$

which is a second-order linear constant-coefficient differential equation.

STEP 2: *We solve (6) for c_1 as a function of t, by using the method from Chapter 15. Substituting*

$$c_1 = e^{\lambda t}$$

into the homogenized version of (6) gives the characteristic equation

$$\lambda^2 + \frac{9}{100}\lambda + \frac{8}{100^2} = 0$$

which has the roots $\lambda_1 = -\frac{8}{100}$ and $\lambda_2 = -\frac{1}{100}$. Looking for a constant solution of (6) itself gives $c_1 = \frac{5}{8}$. Hence, by the method given in Chapter 15, we obtain

$$c_1 = a_1 e^{-\frac{8}{100}t} + a_2 e^{-\frac{1}{100}t} + \frac{5}{8} \qquad (7)$$

where a_1 and a_2 are constants whose values are to be found from the initial conditions.

STEP 3: *We find c_2 as a function of t, by substituting (7) back into (4). This gives*

$$c_2 = \frac{100}{3}\left(\dot{c}_1 + \frac{4}{100}c_1 - \frac{1}{100}\right)$$

$$= \frac{100}{3}\left(-\frac{8}{100}a_1 e^{-\frac{8}{100}t} - \frac{1}{100}a_2 e^{-\frac{1}{100}t} + \frac{4}{100}\right.$$

$$\left. + \left(a_1 e^{-\frac{8}{100}t} + a_2 e^{-\frac{1}{100}t} + \frac{5}{8}\right) - \frac{1}{100}\right).$$

Thus

$$c_2 = -\frac{4}{3}a_1e^{-\frac{8}{100}t} + a_2e^{-\frac{1}{100}t} + \frac{1}{2}. \tag{8}$$

STEP 4: *Finally the initial conditions, $c_1 = c_2 = 0$ when $t = 0$, imply by (7) and (8) that*

$$a_1 + a_2 = -\frac{5}{8},$$

$$\frac{4}{3}a_1 - a_2 = \frac{1}{2}.$$

These equations have the solutions $a_1 = -\frac{3}{56}$ and $a_2 = -\frac{4}{7}$. Hence

$$c_1 = -\frac{3}{56}e^{-\frac{8}{100}t} - \frac{4}{7}e^{-\frac{1}{100}t} + \frac{5}{8},$$

$$c_2 = \frac{1}{14}e^{-\frac{8}{100}t} - \frac{4}{7}e^{-\frac{1}{100}t} + \frac{1}{2}. \tag{9}$$

It is easy to verify that c_1 and c_2, given as functions of t by (9), satisfy both the differential equations (4) and (5) and the initial conditions.

Discussion of the solutions

From (9) it follows that $c_1 \to \frac{5}{8}$ and $c_2 \to \frac{1}{2}$ as $t \to \infty$. This means that the concentrations approach the steady-state solutions in the previous section as the time becomes arbitrarily large. The graphs of c_1 and c_2 against time are shown in Figure 18.2.1. The graphs show that it takes nearly 4 hours for the concentration c_2 of the mixture being drawn off to reach 90% of its steady-state value.

Exercises 18.2

1. In each of the following cases, state if the pair of simultaneous differential equations is linear, constant-coefficient, or homogeneous.

(a) $\dfrac{dx}{dt} = 3x - 7y$

$\dfrac{dy}{dt} = x - ty^2$

(b) $\dfrac{dx}{dt} = 7x - 2y$

$\dfrac{dy}{dt} = x - ty^2$

(c) $\dfrac{dx_1}{dt} = 2x_1 - 2x_2$

$\dfrac{dx_2}{dt} = 2x_1 - 2x_2$

(d) $\dfrac{du}{dt} = 2u + v + \cos(t)$

$\dfrac{dv}{dt} = v + 2u + \sin(t)$

2. Use the procedure described in the text to find the solution of the pair of simultaneous linear constant-coefficient differential equations

$$\dot{x} = x + 2y$$
$$\dot{y} = 2x + y$$

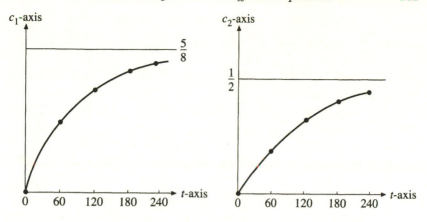

Fig. 18.2.1. Concentrations approach the steady-state.

which satisfy the initial conditions $x = y = 1$ when $t = 0$.

3. Show that the procedure given in the text may be used to solve pairs of simultaneous differential equations of the form

$$\dot{x} = ax + by + p$$
$$\dot{y} = cx + dy + q$$

where a, b, c, d are constants and p, q are given functions of t (thereby extending the procedure to linear constant-coefficient equations which may not be homogeneous).

4. In Exercise 1 of Section 18.1 the following system of differential equations was obtained:

$$\frac{dc_1}{dt} = \frac{1}{25} - \frac{3}{25}c_1 + \frac{2}{25}c_2,$$
$$\frac{dc_2}{dt} = \frac{3}{25}c_1 - \frac{4}{25}c_2.$$

Using appropriate initial conditions, solve this system to find c_1 as a function of t.

5. The procedure in the text shows how to solve the pair of linear constant coefficient differential equations

$$\dot{x} = ax + by$$
$$\dot{y} = cx + dy$$

when $b \neq 0$. Explain how you would solve these equations in each of the remaining cases (i) $b = c = 0$ and (ii) $b = 0$ but $c \neq 0$.

18.3 A model for detecting diabetes

Glucose, an end product of cabohydrate digestion, is converted into energy in the cells of the body. A hormone *insulin*, secreted by the pancreas, facilitates the absorption of glucose by cells other than those of the brain and nervous system.

A delicate balance is normally maintained between the amounts of glucose and insulin in the bloodstream. If the insulin concentration is too low, then too little glucose is absorbed from the bloodstream; the unabsorbed glucose is then lost in the urine along with other nutrients. If, on the other hand, the insulin concentration is too high, then too much glucose is absorbed by cells other than those of the brain and nervous system; lack of glucose available to the cells of the brain then impairs its function. The end result in either case, whether too little or too much insulin, can be coma and even death.

In the medical disorder *Diabetes Mellitus*, not enough insulin is secreted by the pancreas. People suffering from this require supplements of insulin in the form of regular injections, together with a modification of their diet to regulate glucose input. In this section a *simple* model of the interaction between glucose and insulin in the body is presented; we then use this model to discuss a clinical test for the detection of mild forms of diabetes.

The model

The main features that a model of the glucose–insulin regulation system must take into account are as follows.

(a) *A rise in the concentration of glucose in the bloodstream results in the liver absorbing more of the glucose, which it converts and stores as glycogen; a drop in the concentration of glucose reverses the process.*

(b) *A rise in the concentration of insulin in the bloodstream enables the glucose to pass more readily through the membranes of the cells in skeletal muscle, resulting in greater absorption of glucose from the bloodstream.*

(c) *A rise in the concentration of glucose in the bloodstream stimulates the pancreas to produce insulin at a faster rate; a drop in the glucose concentration lowers the rate of insulin production.*

(d) *Insulin, produced by the pancreas, is constantly being degraded by the liver.*

The model omits details of the biochemistry involved and ignores the effects of other hormones. It treats the bloodstream, moreover, as if it were contained in a single compartment throughout which concentrations of glucose and insulin are uniform at each instant. In spite of these simplifications, the model is nonetheless suitable as a basis for understanding what is, in reality, a complicated situation.

Provided there has been no recent digestion, glucose and insulin concentrations will be in equilibrium. We are interested in how the system responds to a change in that equilibrium. Thus we put

$$g = \{\text{excess glucose concentration}\},$$
$$h = \{\text{excess insulin concentration}\},$$

at time t. We use 'h' because insulin is a hormone. Equilibrium occurs for $g = h = 0$. Positive values of g or h corresponds to concentrations greater than the equilibrium values and negative values to concentrations less than the equilibrium values.

If either of g or h is given a non-zero value, then the body tries to restore the equilibrium. We assume that the rates of change of these quantities depend only on the values of g and h so that

$$\frac{dg}{dt} = F_1(g, h),$$
$$\frac{dh}{dt} = F_2(g, h),$$

for some functions F_1 and F_2.

The simplest way to construct a model is to assume that these differential equations are linear with constant coefficients. Since $g = h = 0$ are equilibrium solutions, it now follows that the linear differential equations must be homogeneous. Hence we assume them to be of the form

$$\frac{dg}{dt} = -ag - bh, \tag{1}$$

$$\frac{dh}{dt} = cg - dh, \tag{2}$$

where a, b, c and d are constants.

Signs of the coefficients

To determine how the solutions of the differential equations behave, it is necessary to have some more information about the coefficients which occur in them. A particularly useful fact is that each of a, b, c and d is positive.

To illustrate why this is so we show that $d > 0$. We do this by looking at what must happen if initially $g = 0$ and $h > 0$. From (2)

$$\frac{dh}{dt} = -dh$$

at the initial instant. But the liver will immediately start to degrade the insulin, as noted in (d) above, since the concentration of the insulin has exceeded its equilibrium value. Thus its concentration starts to drop so that initially

$$\frac{dh}{dt} < 0$$

Hence the previous equation shows that d must be positive.

In Exercise 1 it is left for you to derive in a similar way that a, b and c must be positive too. The signs found for the coefficients will be used later to predict some features of the behaviour of the model.

Testing for diabetes

In a *glucose tolerance test* a patient is asked to fast overnight and the following morning is given an injection of glucose. Blood samples are then taken at subsequent times and the concentration of glucose measured, to test the response of the glucose–insulin regulatory system. We might expect the glucose concentration to return after a time to the equilibrium level and to take longer in diabetic patients than in normal ones.

In modelling this test we suppose that during the short time interval while the glucose is being injected, the insulin concentration stays zero. Thus if g_0 is the total amount of glucose injected, then

$$g = g_0, \qquad h = 0, \qquad \text{at} \qquad t = 0. \tag{3}$$

Our model for the glucose and insulin concentrations is thus the solution of the pair of differential equations (1), (2) with the initial conditions (3).

To solve the differential equations (1), (2) we apply the procedure from the previous section and so obtain the second-order equation

$$\ddot{g} + (a + d)\dot{g} + (ad + bc)g = 0 \tag{4}$$

The solution for g as a function of t can then be substituted into

$$h = -\frac{1}{b}(\dot{g} + ag). \tag{5}$$

Table 18.3.1. *Different types of solutions for the glucose–insulin model.*

Solutions of characteristic equation	Solutions of differential equation
Two real solutions λ_1 and λ_2:	$g = g_0 \dfrac{\left((a + \lambda_2)e^{\lambda_1 t} - (a + \lambda_1)\,e^{\lambda_2 t}\right)}{(\lambda_2 - \lambda_1)}$
One real solution λ:	$g = g_0\left(1 - (a + \lambda)t\right)e^{\lambda t}$
To unreal solutions $\lambda = -\alpha \pm i\omega$:	$g = g_0\left(\cos(\omega t) - \dfrac{a}{\omega}\sin(\omega t)\right)e^{-\alpha t}$

The initial conditions for (4), which can be found from (3) and (5), are

$$g = g_0, \quad \dot{g} = -ag_0 \quad \text{at} \quad t = 0. \tag{6}$$

To solve the second-order equation (4) we use the procedure of Chapter 15. The type of solution obtained will depend on the number of real solutions of the characteristic equation

$$\lambda^2 + (a + d)\lambda + (ad + bc) = 0. \tag{7}$$

This in turn will depend on the values of a, b, c and d. The solutions for g in the various cases are listed in Table 18.3.1

From the fact that a, b, c and d are all positive it can be shown that the solutions of (7) for λ are negative or have negative real part. Hence the factors $e^{\lambda_1 t}, e^{\lambda_2 t}, e^{\lambda t}$ and $e^{\alpha t}$ which occur in the above table must all decay exponentially with time. Thus our model predicts that *the glucose concentration will approach its original undisturbed value with sufficient lapse of time.* This is just as we would expect.

In these solutions g is a linear function of g_0. This is a consequence of our assumption that the differential equations are linear. It can be shown that these solutions are good approximations to those of any smooth non-linear model of the problem, if g_0 is sufficiently small.

Experimental results

The model can be used to make numerical predictions about the concentration of glucose at various times once the constants occurring in the differential equtions (1) (2) are known. In order to determine these constants, Bolie (1961) used three different methods, based on data from previous experiments with dogs, which he extrapolated to humans. Measuring glucose in grams, insulin in 'units' and time in hours, he obtained the following averaged values for normal individuals:

$$a = 2.92, \quad b = 4.34, \quad c = 0.208, \quad d = 0.780,$$

measured in units corresponding to grams for mass and hours for time. Substituting these values into the characteristic equation (7) and then solving for λ gives

$$\lambda_1 = -1.36, \qquad \lambda_2 = -2.34.$$

Thus the characteristic equation has two real roots. Hence we can substitute the numerical values into the first row of Table 18.3.1 to get g as a function of t and then use (5) to get h. This gives

$$
\begin{aligned}
g &= g_0(-0.56e^{\lambda_1 t} + 1.56e^{\lambda_2 t}) \\
h &= 0.202g_0(e^{\lambda_1 t} - e^{\lambda_2 t}).
\end{aligned}
\tag{8}
$$

By putting $g = 0$ we can find the time at which the glucose concentration returns to normal (and then slightly undershoots before coming back up again to approach the equilibrium value exponentially). The value that we obtain is about 1 hour, for a normal individual. The graph of g against t has the general shape shown in Figure 18.3.1.

Possible refinements of the model are discussed in Bolie (1961) and Edelstein-Keshet (1988).

Orally administered glucose

In an alternative version of the glucose-tolerance test, the glucose is administered orally, rather than by injection. The differential equations which model this test are no longer homogeneous. In particular, the differential equation (4) is replaced by

$$\ddot{g} + (a + d)\dot{g} + (ad + bc)g = S(t)$$

where $S(t)$ is a 'forcing' which takes account of the glucose coming in through the digestive system.

In the approach given by Ackerman, Rosevar and McGuckin (1964) no attempt is made to determine the parameters a, b, c and d. Instead they

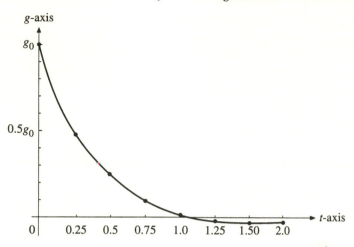

Fig. 18.3.1. Glucose concentration returning to normal after glucose tolerance test.

work directly with the above second-order differential equation, writing it in the form

$$\ddot{g} + 2\alpha\dot{g} + \omega_0{}^2 g = S(t) \qquad (9)$$

where $\alpha = a + d$ and $\omega_0{}^2 = ad + bc$ and where the forcing term $S(t)$ is chosen on the basis of their modelling assumptions. They then put $\omega^2 = \omega_0{}^2 - \alpha^2$ and assume $\omega^2 > 0$ (so that the characteristic equation has two unreal solutions) and they deduce that the solutions of (9) are

$$g = A\sin(\omega t)e^{-\alpha t} \qquad (10)$$

where A is an arbitrary constant.

During the test, the concentration of glucose (c_g) in the individual's bloodstream is measured at regular intervals. Ackerman *et al.* then chose the parameters in (10) so as to get a curve which fits the measured data. Typical examples of data and the fitted curves are shown in Figure 18.3.2(a–c). As a result of data obtained in this way from hundreds of individuals, Ackerman *et al.* concluded that the value of the parameter ω_0 is a reliable guide as to whether or not an individual is diabetic. They called $T_0 = 2\pi/\omega_0$ the 'resonant period' and claimed that *normal individuals have resonant periods of less than 4 hours, whereas diabetics have periods greater than 4 hours.*

An account of this work is included in Middleman (1972). Note that

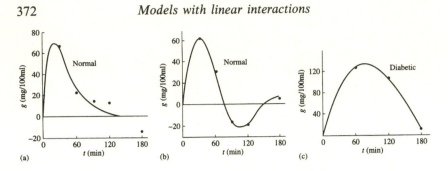

Fig. 18.3.2. Responses of individuals to an oral glucose-tolerance test. Case (c) suggests diabetes because the period of oscillation is greater than 4 hours, indicating that the response to the excess glucose is too slow. Graphs are from Ackerman *et al.* (1964).

Burghes and Borrie (1981) switch the rôles of ω and ω_0 and then state a conclusion which is different from that given in Ackerman *et al.* (1964).

Exercises 18.3

1. The differential equations for the glucose and insulin concentrations are given in the text as

$$\frac{dg}{dt} = -ag - bh$$

$$\frac{dh}{dt} = cg - dh$$

where a, b, c and d are constants. We showed in the text that $d > 0$. Show in a similar way, by referring to the relevant principles from physiology, that each of a, b and c is positive.

2. Derive, from the equations in Exercise 1, the following equations.

$$\ddot{g} + (a + d)\dot{g} + (ad + bc)g = 0, \qquad h = -\frac{1}{b}(\dot{g} + ag).$$

(Apply the solution procedure for simultaneous linear differential equations with constant coefficients from Section 18.2.)

Show, furthermore, that the initial conditions

$$g = g_0, \quad h = 0 \quad \text{at} \quad t = 0$$

are equivalent to

$$g = g_0, \quad \dot{g} = -ag_0 \quad \text{at} \quad t = 0.$$

3. Given that a, b, c and d are all positive, show that the solutions of the characteristic equation

$$\lambda^2 + (a + d)\lambda + (ad + bc) = 0$$

are either negative or have negative real part.

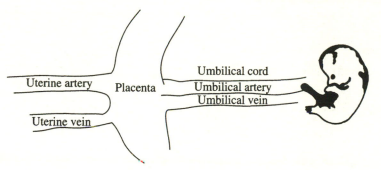

Fig. 18.4.1. Placenta provides interface between the bloodstreams of mother and fetus.

4. On the basis of Bolie's model for the glucose-tolerance test, we obtained the following formula for the insulin concentration t hours after the glucose injection:

$$h = 0.202g_0(e^{\lambda_1 t} - e^{\lambda_2 t})$$

where $\lambda_1 = -1.36$ and $\lambda_2 = -2.34$. Choose any positive value for g_0 and then sketch the graph of h against t and find the time at which h reaches its maximum value. (For any other choice of g_0, the graph can be obtained simply by scaling in the vertical direction.)

5. In an insulin-tolerance test, an injection of insulin is given to an individual after fasting and the level of glucose in the blood is measured at subsequent times. Assume that that the concentrations of glucose and insulin in the bloodstream satisfy the differential equations of Exercise 1, together with the initial conditions

$$g = 0, \quad h = h_0 \qquad at \quad t = 0.$$

Show that, on the basis of Bolie's estimates for the coefficients in the differential equations,

$$g = 4.35h_0(-e^{\lambda_1 t} + e^{\lambda_2 t}),$$
$$h = h_0(1.57e^{\lambda_1 t} - .57e^{\lambda_2 t}).$$

Sketch the graph of h as a function of t.

18.4 Nutrient exchange in the placenta

In the placenta nutrients pass from mother to fetus, while waste products from the fetus go the other way. During this exchange the blood of the mother and the fetus do not mix but are separated by a membrane across which nutrients and waste must pass. The nutrients flow from a high concentration in the maternal blood to a lower concentration in the fetal blood. For both mother and fetus, blood flows to the placenta along an artery and returns via a vein, as shown in Figure 18.4.1.

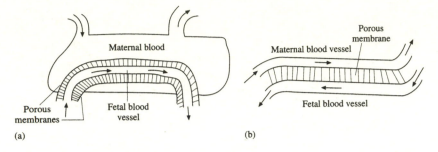

Fig. 18.4.2. Two different types of blood flow in a placenta: (a) maternal pool, (b) countercurrent flow.

There is some variation among species in the arrangement of blood vessels within the placenta. In humans, fetal blood vessels are bathed in maternal blood, as in Figure 18.4.2a. In rabbits and sheep, on the other hand, there is a system of maternal blood vessels adjacent to the fetal ones. The blood which they contain is believed to flow in opposite directions as shown in Figure 18.4.2b.

Simple mathematical models can be used to compare the advantages and disadvantages of the different types of arrangements of blood vessels within the placenta. The model to be described here is for the type of placenta shown in Figure 18.4.2b, appropriate to sheep and rabbits.

The nutrient concentrations in both maternal and fetal blood vessels can be expected to vary with the distance along the blood vessels since, as nutrients are transferred, the concentrations change. It will be assumed that the concentrations have reached a steady-state, so that they depend only on the distance along the blood vessels.

The model

The model of the placenta we will describe is illustrated in Figure 18.4.3, our notation being as follows. We use Q_1 and Q_2 to denote the rates of flow of the maternal and fetal blood respectively. We suppose that the blood vessels of the mother and fetus stay in contact with the membrane along a total length L.

We choose as coordinate the distance x of a typical point along the blood vessels from the point where they first make contact with the membrane.The concentration of nutrient in each blood vessel is then a

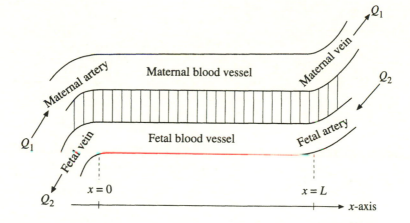

Fig. 18.4.3. Schematic view of countercurrent bloodflow in neighbouring blood vessels in a rabbit or sheep placenta.

function of x and we put

$$c_1 = \left\{ \begin{array}{c} \text{nutrient concentration} \\ \text{in the maternal blood} \\ \text{at a distance } x \end{array} \right\} = \phi_1(x),$$

$$c_2 = \left\{ \begin{array}{c} \text{nutrient concentration} \\ \text{in the fetal blood} \\ \text{at a distance } x \end{array} \right\} = \phi_2(x).$$

(1)

Thus c_1 and c_2 are functions of x while L, Q_1 and Q_2 are constants, independent of x.

We now consider the amount of nutrient contained within two planes, perpendicular to the blood vessels, which pass through the points with coordinates x and $x + \delta x$ respectively. These planes are shown in Figure 18.4.4a.

In the maternal bloodstream, nutrient enters at the first plane and leaves at the second. It also leaves, to enter the fetal bloodstream, via the membrane. Thus, as the mass of nutrient is conserved,

$$\left\{ \begin{array}{c} \text{mass entering} \\ \text{through plane} \\ \text{at } x \end{array} \right\} = \left\{ \begin{array}{c} \text{mass leaving} \\ \text{through plane} \\ \text{at } x + \delta x \end{array} \right\} + \left\{ \begin{array}{c} \text{mass leaving} \\ \text{across} \\ \text{membrane} \end{array} \right\}.$$

(2)

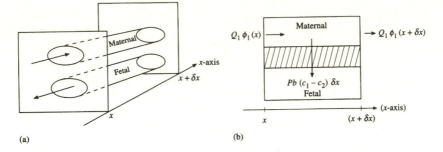

(a)   (b)

Fig. 18.4.4. (a) Planes through the blood vessels. (b) Nutrient flow in maternal blood vessels.

Now, in any time interval of length δt,

$$\left\{\begin{array}{c} \text{mass entering} \\ \text{through plane} \\ \text{at } x \end{array}\right\} = \left\{\begin{array}{c} \text{rate of} \\ \text{flow} \\ \text{of blood} \end{array}\right\} \times \left\{\begin{array}{c} \text{concentration} \\ \text{of} \\ \text{nutrient} \end{array}\right\} \times \{\text{time}\} \tag{3}$$

$$= Q_1\phi_1(x)\,\delta t$$

and similarly

$$\left\{\begin{array}{c} \text{mass leaving} \\ \text{through plane} \\ \text{at } x+\delta x \end{array}\right\} = Q_1\phi_1(x+\delta x)\,\delta t. \tag{4}$$

The principle which enables us to estimate the amount of nutrient transported across the membrane, known as *Fick's law*, states that if the concentrations on either side of the membrane were homogeneous then

$$\left\{\begin{array}{c} \text{rate of} \\ \text{transport through} \\ \text{the membrane} \end{array}\right\} = P \times \left\{\begin{array}{c} \text{area of} \\ \text{membrane} \end{array}\right\} \times \left\{\begin{array}{c} \text{difference between} \\ \text{concentrations} \\ \text{on either side} \end{array}\right\}$$

where P is a constant called the *permeability* of the membrane.

In our problem, we wish to apply Fick's law across the portion of membrane cut off by the two planes, which has area $b\,\delta x$ where b is the width of the membrane, assumed constant. The difference in concentration across this portion of membrane is approximately $\phi_1(x) - \phi_2(x) = c_1 - c_2$, with an error which approaches 0 as δx approaches 0. Thus Fick's law gives

$$\left\{\begin{array}{c} \text{rate of} \\ \text{flow through} \\ \text{membrane} \end{array}\right\} \simeq P \times b\,\delta x \times (c_1 - c_2)$$

and hence

$$\left\{ \begin{array}{c} \text{mass of} \\ \text{nutrient leaving} \\ \text{across membrane} \end{array} \right\} \simeq P \times b\,\delta x \times (c_1 - c_2) \times \delta t. \tag{5}$$

Now substitute (3), (4) and (5) into (2) and then divide by δt and rearrange to get

$$Q_1(\phi_1(x + \delta x) - \phi_1(x)) + Pb(c_1 - c_2)\,\delta x \simeq 0,$$

the error involved in the approximation being small compared with δx. Hence, dividing by δx and then letting δx approach 0 gives

$$\frac{dc_1}{dx} = -\alpha_1(c_1 - c_2) \tag{6}$$

where $\alpha_1 = Pb/Q_1$.

A similar derivation for the fetal bloodstream gives

$$\frac{dc_2}{dx} = -\alpha_2(c_1 - c_2) \tag{7}$$

where $\alpha_2 = Pb/Q_2$.

Solving the differential equations

The pair of differential equations (6), (7) are linear with constant coefficients and hence may be solved by the procedure given in Section 18.2. This leads to the second-order differential equation

$$\frac{d^2 c_1}{dx^2} = (-\alpha_1 + \alpha_2)\frac{dc_1}{dx}. \tag{8}$$

From this we can deduce that, if $\alpha_1 \neq \alpha_2$, then the differential equations (6), (7) have the solutions

$$\begin{aligned} c_1 = \phi_1(x) = \phi_1(0) - \frac{\alpha_1}{\alpha_2 - \alpha_1}\,(\phi_1(0) - \phi_2(0))\left(e^{(-\alpha_1 + \alpha_2)x} - 1\right), \\ c_2 = \phi_2(x) = \phi_2(0) - \frac{\alpha_2}{\alpha_2 - \alpha_1}\,(\phi_1(0) - \phi_2(0))\left(e^{(-\alpha_1 + \alpha_2)x} - 1\right). \end{aligned} \tag{9}$$

Comparisons

The placenta modelled above is called a countercurrent type of placenta because the two bloodstreams flow in opposite directions. Middleman (1972) gives further details, and he models other types of placenta, obtaining solutions analogous to (9) for the concentrations of the nutrients. On the basis of such solutions, he is able to make some comparisons between the efficiency of the various types of placenta in exchanging nutrients.

Models analogous to that of the countercurrent nutrient exchange system also occur in other applications. These include simple models of an artificial kidney machine, which is described in Burghes and Borrie (1981), and oxygen exchange in the swim bladders of deep sea fish, described in Rodin and Jacques (1989).

Exercises 18.4

1. In the text, the differential equation for the concentration c_1 of nutrient in the maternal bloodstream at a distance x along the placental membrane was shown to be

$$\frac{dc_1}{dx} = -\alpha_1(c_1 - c_2)$$

for a suitable constant α_1.

By arguing in a similar way, derive the corresponding differential equation

$$\frac{dc_2}{dx} = -\alpha_2(c_1 - c_2)$$

for the concentration of nutrient in the fetal bloodstream.

Under which physical condition does $\alpha_1 = \alpha_2$?

2. Why can the procedure given in Section 18.2 be used to solve the differential equation in Exercise 1 ? Use this procedure to solve these equations in the case $\alpha_1 \neq \alpha_2$. Check your answers against those given in the text.

3. Repeat Exercise 2 but with $\alpha_1 = \alpha_2$. Interpret your answers.

4. Use the solutions (9) in the text to establish each of the following statements.

   (a) If the concentration of nutrient entering via the maternal artery is equal to that leaving via the fetal vein, then the concentrations must be constant along the entire length of the placental membrane.

   (b) If the concentration of nutrient entering via the maternal artery exceeds that leaving via the fetal vein, then at each point along the membrane the concentration in the maternal bloodstream exceeds that in the fetal bloodstream.

5. Modify the model given in the text if, instead of flowing in opposite directions, the maternal and fetal bloodstreams flow in the same direction. (This is called concurrent exchange.)

19

Non-linear coupled models

Further models of two interacting quantities are studied in this chapter. These models are distinguished from those of the previous chapter because they lead to coupled non-linear, rather than linear, differential equations. Such equations usually cannot be solved explicitly. Instead, a method of eliminating the independent variable, called the phase-plane technique, is developed for studying properties of the solution.

Non-linear equations often exhibit unusual or non-intuitive behaviour. A classic example considered in this chapter is the Lotka–Volterra equations which describe the interactions between predators and their prey. Applied to the study of fish and shark populations they predict that an increase in fishing will actually lead to an increase in the fish population.

In addition to the predator–prey system, models of combat and epidemics are presented.

19.1 Predator–prey interactions

Our concern in this chapter is with models which lead to coupled *non-linear* differential equations. In this section it will be shown how a model describing the interaction of a predator (shark) and its prey (fish) leads to coupled non-linear differential equations for the shark and fish populations as functions of time.

Some unusual data

The Italian mathematician Vito Volterra (1860–1940) developed a model of predator–prey interaction in response to some unusual data. Extensive records had been kept of the yearly catches of fish and sharks at an Italian sea port (Fiume, 1914–1923). Table 19.1.1 shows the percentage ratio

Table 19.1.1. *Percentage catch of sharks at Fiume, Italy, 1914–1923.*

Year	Sharks (% total catch)
1914	11.9
1915	21.4
1916	22.1
1917	21.2
1918	36.4
1919	27.3
1920	16.0
1921	15.9
1922	14.8
1923	10.7

of sharks to total catch. It reveals that this ratio, and thus the relative number of sharks, increased substantially during a time of reduced fishing (1915 to 1919 corresponding to the First World War).

How can these data be understood? Volterra set up the following model to try to explain why the decrease in fishing increased the percentage catch of sharks.

The model

Following Volterra, our objective is to formulate some differential equations describing the shark and fish populations, which will be done using the small time-interval method introduced in Chapter 13. The following influencing factors will be taken into account.

(i) Natural births and deaths of the sharks and fish in isolation from each other.
(ii) Decline of the fish population due to the fish being the prey of the sharks.
(iii) Increase in the shark population due to the presence of more fish.
(iv) Fishing of both sharks and fish.

For the populations, we introduce the notation

$$x = \left\{ \begin{array}{c} \text{number of} \\ \text{fish at time } t \end{array} \right\}$$

$$y = \left\{ \begin{array}{c} \text{number of} \\ \text{sharks at time } t \end{array} \right\}$$

and denote by δx and δy the change in the corresponding populations during a small time interval δt. From the points (i)–(iv) it follows that

$$\delta x = \left\{\begin{array}{c} \text{fish} \\ \text{born in} \\ \text{isolation} \end{array}\right\} - \left\{\begin{array}{c} \text{fish} \\ \text{deaths in} \\ \text{isolation} \end{array}\right\} - \left\{\begin{array}{c} \text{fish} \\ \text{eaten by} \\ \text{sharks} \end{array}\right\} - \left\{\begin{array}{c} \text{fish} \\ \text{caught by} \\ \text{fishermen} \end{array}\right\} \quad (1)$$

and

$$\delta y = \left\{\begin{array}{c} \text{sharks} \\ \text{born in} \\ \text{isolation} \end{array}\right\} - \left\{\begin{array}{c} \text{shark} \\ \text{deaths in} \\ \text{isolation} \end{array}\right\} + \left\{\begin{array}{c} \text{extra sharks} \\ \text{surviving} \\ \text{due to} \\ \text{fish} \end{array}\right\} - \left\{\begin{array}{c} \text{sharks} \\ \text{caught by} \\ \text{fishermen} \end{array}\right\} \quad (2)$$

where each of the quantities on the right-hand side refers to the number of sharks and fish which are born or die during the time interval δt. We need to relate the various quantities in (1) and (2) to the shark and fish populations. To do this we will make the following modelling assumptions.

(a) *The change in the shark and fish populations, in isolation, is proportional to the present population of sharks and fish, respectively.* (This is the same assumption as for the linear model of population growth made in Section 9.1.) The proportionality constant for the shark population is negative, indicating that the shark population would decrease if isolated from the fish population. This is a consequence of the fish being the food supply of the sharks.

(b) *The number of sharks and fish caught by fishermen is directly proportional to the present population of the shark and fish populations respectively.* The proportionality constant is the same in both cases, which means that the fishing methods do not discriminate between sharks or fish.

(c) *The number of fish eaten by sharks is directly proportional to the product of the number of fish present and the number of sharks present.* This is equivalent to assuming that *each* shark eats a constant fraction of the fish population.

(d) *The additional number of sharks surviving is directly proportional to the number of fish eaten.*

In the exercises you are asked to show that, when written as mathematical equations and substituted into equations (1) and (2), the assumptions (a)–(d) imply the differential equations

$$\frac{dx}{dt} = (r - f)x - \alpha xy,$$

$$\frac{dy}{dt} = -(s + f)y + \beta xy,$$

$$(3)$$

where r, s, f, α and β are all positive constants. The constants r and $-s$ are the growth rates for the fish and shark populations if they existed in isolation, the constant α is the rate at which fish are eaten by a single shark and β/α gives the fraction of a shark surviving by eating one fish. These equations are called the Lotka–Volterra equations after Volterra, and Lotka — another mathematician who formulated them independently.

The Lotka–Volterra equations are coupled since the unknowns x and y appear in both of the equations. Furthermore they are non-linear due to the presence of the terms αxy and βxy. Because of the non-linear terms, it is not possible to solve these equations explicitly for general initial conditions. We therefore seek ways of obtaining useful information about the solutions. The steady-state solutions provide an informative beginning.

Steady-state solutions

The steady-state solutions (i.e. time-independent constant solutions) are the easiest particular solutions to find for an autonomous differential equation. To find these solutions set

$$x = X \quad \text{and} \quad y = Y,$$

where X and Y are constants, and substitute into the Lotka–Volterra equations (3) to give two non-linear simultaneous equations

$$(r - f)X - \alpha XY = 0,$$

$$-(s + f)Y + \beta XY = 0.$$

In the exercises, you are asked to show that the constant solutions of the equations are

$$(X, Y) = (0, 0)$$

and

$$(X, Y) = \left(\frac{s + f}{\beta}, \frac{r - f}{\alpha}\right).$$

The first of these solutions corresponds to extinction of both species, which we know doesn't happen in practice. The second solution contains the more interesting information. One interesting thing about this solution of the Lotka–Volterra equations is that the steady-state fish population depends on the shark growth rate parameter s but not on the fish growth rate parameter r. Similarly, the steady-state shark population depends on the fish growth rate parameter r but not on the shark growth rate parameter s.

It will be shown in Section 19.2 via a phase-plane analysis that the second steady-state solution gives the value about which the shark and fish populations oscillate. This value may thus be regarded as the average population of the shark and fish populations. The effect of fishing can therefore be deduced by simply considering this steady-state solution.

Effect of fishing

Consider the circumstance that there is less fishing, as in the First World War. In the Lotka–Volterra model this means that f is decreased but all other parameters remain the same. The second constant solution above then gives that the fish population actually decreases while the shark population increases. Hence the ratio of sharks to fish increases, which is consistent with the data of Table 19.1.1. The mechanism for this result will be discussed in the next section.

Exercises 19.1

1. Consider the statements (a)–(d) of the text which relate the quantities influencing the shark and fish populations to the populations themselves. Suppose that the proportionality constants are r and $-s$ in (a) for the fish and shark populations respectively, f in (b), α in (c) and β in (d). Write the statements as mathematical equations, keeping in mind that each quantity on the right-hand side of equations (1) and (2) is also directly proportional to the time interval δt. Substitute these equations in (1) and (2) and thus derive the Lotka–Volterra equations.

2. Factorize and thus solve the two non-linear equations given in the text which specify the steady-state solutions of the Lotka–Volterra equations.

3. (a) Suppose there are two different species of prey x and z, which are the food of a single predator y. Derive differential equations for the changes in the populations of the species by using modelling assumptions analogous to those used to derive the Lotka–Volterra equations.

   (b) Find all the steady-state solutions of the differential equations found in part (a), and thus show that at least one species of prey becomes extinct in this instance.

4. An orchard is infested by a population of aphids. The aphids are preyed upon by a type of beetle. The owner of the orchard decides to use a pesticide which kills a fixed fraction of both aphid and beetle. Use the steady-state solution of the Lotka–Volterra equation to decide whether or not this is a wise move.

5.   Modify the Lotka–Volterra equations to account for the fish population growing logistically in the absence of sharks (recall Section 11.3). Do the original conclusions regarding the effect of fishing alter?

19.2 Phase-plane analysis

The Lotka–Volterra equations are examples of first-order, autonomous, coupled differential equations of the form

$$\dot{x} = f(x, y) \quad \text{and} \quad \dot{y} = g(x, y), \tag{1}$$

where f and g are non-linear function of x and y. Although in general it is not possible to solve these equations exactly, as seen in the previous section, useful information can be obtained from the steady-state solution. This information can be further supplemented by use of a *phase-plane analysis*.

    Instead of solving (1) for x and y as functions of t, we obtain a single differential equation for y as a function of x, by use of the chain rule. The resulting curves in the (x, y)-plane are called *phase-plane trajectories*.

The method

There are three main steps in applying a phase-plane analysis, which we will illustrate by the following example.

Example 1. *Find, and sketch, the phase-plane trajectories for the coupled system*

$$\dot{x} = -y \quad \text{and} \quad \dot{y} = x. \tag{2}$$

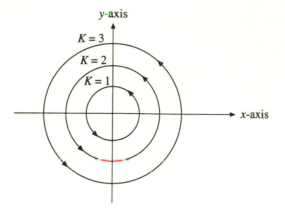

Fig. 19.2.1. Phase-plane trajectories for the coupled system (2) in Example 1.

Solution.

STEP 1: Write down the chain rule and substitute the values of $\dot{x}$ and $\dot{y}$ from the given equations. *Hence obtain a single first-order differential equation in y and x by using the chain rule and then substituting for the derivatives from the original coupled differential equations.*

The chain rule says that

$$\frac{dy}{dx}\frac{dx}{dt} = \frac{dy}{dt},$$

or equivalently,

$$\frac{dy}{dx}\dot{x} = \dot{y}.$$

Thus, since $\dot{x} = -y$ and $\dot{y} = x$, the differential equation

$$\frac{dy}{dx} = -\frac{x}{y}$$

is obtained. This equation is of the first-order separable type studied in Chapter 11.

STEP 2: Solve the differential equation for x and y and sketch the solution in the (x, y)-plane (the phase-plane) for various initial values.

Solving the differential equation (3) (using the methods explained in Section 11.1) we obtain

$$x^2 + y^2 = x_0^2 + y_0^2$$

where x_0 and y_0 are the initial values of x and y. Since x_0 and y_0 are constants, we can write

$$x_0^2 + y_0^2 = K^2$$

where K is a constant. The relationship between x and y defines a circle of radius K, and varying K gives a family of circles as sketched in Figure 19.2.1.

STEP 3: Determine the directions of the trajectories by finding if x and y increase or decrease with t at a few selected regions in the phase-plane.

*From the original differential equations, for the region $x > 0$, $y > 0$, we see that
$\dot{x} < 0$, which means x decreases with t;
$\dot{y} > 0$, which means y increases with t.
Thus the trajectories are anti-clockwise around the circles, as indicated in Figure 19.2.1.*

Features of the phase-plane trajectories

Some features exhibited by the phase-plane trajectories of Figure 19.2.1 are common to all phase-plane trajectories. These features are as follows.

- Through any point (x, y) in the phase-plane there is at most one trajectory. This follows from the uniqueness theorem for the solution of coupled first-order equations.
- Each constant solution of the equations is given by a single point in the phase-plane.
- Closed trajectories in the phase-plane correspond to oscillatory solutions of the coupled equations.

The last result is illustrated by the exact solution of the coupled system (1), which is

$$x = A\cos(t + \delta) \quad \text{and} \quad y = A\sin(t + \delta).$$

These solutions oscillate with period 2π.

The Lotka–Volterra equations

In the exercises you are asked to apply the phase-plane technique of Example 1 to the Lotka–Volterra equations,

$$(r - f)x - \alpha xy = 0$$
$$-(s + f)y + \beta xy = 0.$$

This gives, for the phase-plane trajectories, the equation

$$\bar{r}\ln y - \alpha y + \bar{s}\ln x - \beta x = K \tag{3}$$

where

$$K = \bar{r}\ln y_0 - \alpha y_0 + \bar{s}\ln x_0 - \beta x_0$$

and K is a constant, which depends on the initial populations x_0 and y_0. The notation $\bar{r} = r - f$ and $\bar{s} = s + f$ has been used, and x_0 and y_0 are the initial values of x_0 and y_0.

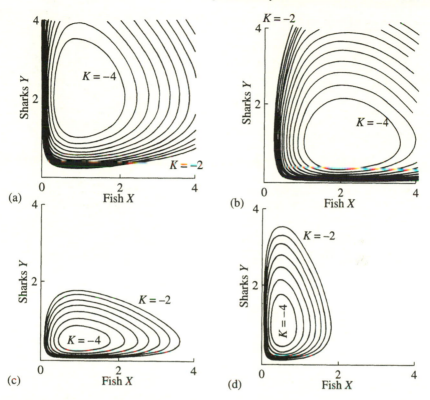

Fig. 19.2.2. Computer-generated phase-plane trajectories for Lotka–Volterra model. (a) $\bar{r} = 2$, $\bar{s} = 1$, $\alpha = 1$, $\beta = 1$; (b) $\bar{r} = 1$, $\bar{s} = 2$, $\alpha = 1$, $\beta = 1$; (c) $\bar{r} = 1$, $\bar{s} = 1$, $\alpha = 2$, $\beta = 1$; (d) $\bar{r} = 1$, $\bar{s} = 1$, $\alpha = 1$, $\beta = 2$. Values of K for some trajectories are marked.

Unfortunately, it is not possible to solve (3) for y in terms of x. However, phase-plane trajectories can be obtained using a contouring software package. This is a computer program which evaluates the left-hand side of (3) for a large number of values of x and y and then joins up those points for which the left-hand side equals K. The procedure is the same as that which produces isobars on a weather map. Some typical plots are shown in Figure 19.2.2. In each plot the phase-plane trajectories are closed curves. In fact, it turns out that the trajectories for $x > 0$ and $y > 0$ are always closed curves, for all positive values of the parameters r, s, α, and β. A detailed proof of this result is given in Braun (1975), Section 4.9.

The direction of the plots can be deduced by writing the Lotka–

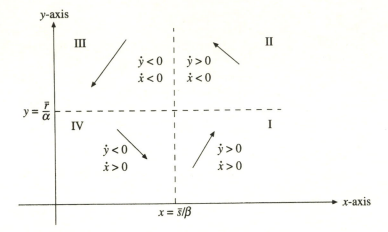

Fig. 19.2.3. Direction of the trajectories in each of the four quadrants.

Volterra equations in the form

$$\dot{x} = x(\bar{r} - \alpha y)$$
$$\dot{y} = -y(\bar{s} - \beta x).$$

As indicated in Figure 19.2.3, the phase-plane can be divided into four quadrants about the fixed point (X, Y) according to the sign of $\dot{x}$ and $\dot{y}$. If $x > \bar{s}/\beta$ and $y > \bar{r}/\alpha$ (in region I), then $\dot{x} < 0$ and $\dot{y} > 0$. This means that, as we move along a trajectory, x is increasing and y is decreasing. This implies that the trajectories are traversed in the anti-clockwise direction.

Interpretation

Consideration of Figure 19.2.3 allows aspects of the predator–prey system of sharks and fish to be better understood. Suppose initially both the shark and fish populations are above their steady-state values and thus in region II of Figure 19.2.3. In this circumstance the shark population increases at the expense of the fish population until the fish population drops below its steady-state value into region III. Now there is not enough food for the sharks so the shark population also decreases until it drops below its steady-state value into region IV. The fish population can now recover as there are fewer sharks. It eventually increases to above its steady-state value into region I. Now there is sufficient food for the shark population to recover, so it begins to increase until region II is again entered and the cycle repeated.

Exercises 19.2

1. Elimination of the independent variable from the Lotka-Volterra equations leads to the differential equation

$$\left(\frac{\bar{r}}{y} - \alpha\right) \frac{dy}{dx} = \left(\beta - \frac{\bar{s}}{x}\right)$$

where $\bar{r} = r - f$ and $\bar{s} = s + f$. Given $x = x_0$ and $y = y_0$ solve this equation to show that

$$\bar{r} \ln y - \alpha y + \bar{s} \ln x - \beta x = K$$

where K is a constant depending on the initial conditions.

2. Consider the system of equations

$$\dot{x} = 2y, \quad \dot{y} = -x.$$

(a) Use the chain rule to show that

$$2y \frac{dy}{dx} = -x.$$

(b) Hence obtain the equations for the family of phase-plane trajectories.

(c) Sketch the family of phase-plane trajectories in the (x, y)-plane. Remember to indicate by arrows the direction in which the solution moves along these curves.

3. Repeat Exercise 2 for the system

$$\dot{x} = x^2, \quad \dot{y} = y, \quad x > 0, \quad y > 0.$$

4. Suppose that the fish population grows according to the logistic equation, in the absence of sharks (see Exercise 5 of Section 19.1). What would you *expect* the phase-plane trajectories to look like? (If you have access to a differential equation solving program you can use a computer to check your answer.)

19.3 Models of combat

The combat situation of two armies fighting a battle can be modelled, subject to some simplifying assumptions, as a pair of coupled differential equations. Mathematical models of combat can be used to understand what factors can influence the outcome of a battle: some questions which might be asked include which side is the victor, how many survivors remain, how long does the battle take?

In this section we look at one particular combat situation where one army is exposed to fire and the other is hidden. This situation may be used to model guerilla warfare.

Guerilla combat model

The exposed army will be termed the 'home' army and the hidden army the 'enemy' army. We will use the notation

$$x = \left\{ \begin{array}{c} \text{number of} \\ \text{home soldiers} \\ \text{at time } t \end{array} \right\} \qquad y = \left\{ \begin{array}{c} \text{number of} \\ \text{enemy soldiers} \\ \text{at time } t \end{array} \right\}$$

and assume that the number of soldiers can be approximated as continuous variables.

In an isolated battle the major factor reducing the size of each army is the number of soldiers put out of action (either killed, wounded or taken prisoner) by the opposing army. We will assume that

- neither army takes prisoners
- each army is using gunfire against the other.

Thus, with δx and δy denoting the changes in the numbers of the respective armies during a time interval δt,

$$\delta x = - \left\{ \begin{array}{c} \text{number of} \\ \text{home soldiers} \\ \text{hit by gunfire} \\ \text{of enemy army} \end{array} \right\} \quad \text{and} \quad \delta y = - \left\{ \begin{array}{c} \text{number of} \\ \text{enemy soldiers} \\ \text{hit by gunfire of} \\ \text{home army} \end{array} \right\}$$

The number of soldiers hit in a small time interval δt is equal to the product of

(i) the rate at which each soldier shoots (R_x shots per unit time for the home army and R_y for the enemy),

(ii) the probability that a single shot hits its target (P_x for the home army and P_y for the enemy),

(iii) the number of soldiers firing the shots.

Hence

$$\delta x = -R_y P_y y \, \delta t \qquad \text{and} \qquad \delta y = R_x P_x x \, \delta t. \tag{1}$$

The firing rates R_x and R_y are assumed to be constant, while the probabilities P_x and P_y are determined according to whether the target is exposed or hidden.

If the target is exposed (the home army soldiers), it is reasonable to assume that each single shot has a *constant* probability of hitting its target, independent of the number of home soldiers. So P_y is a constant with respect to x and y.

If the target is hidden, however, then the probability of hitting a soldier by a shot fired at random into a given area will depend on the concentration of hidden soldiers in the area. If there are y enemy soldiers, and if each enemy soldier has on average an area α exposed, then

$$P_x = \frac{\alpha}{A}y$$

where A is the total area occupied by the enemy soldiers. Note that αy gives the total area of soldiers available to be hit by random fire.

Substituting this formula into (1) and dividing by δt gives the coupled non-linear equations

$$\frac{dx}{dt} = -ay \quad \text{and} \quad \frac{dy}{dt} = -bxy \tag{2}$$

where $a = R_y P_y$ and $b = R_x \alpha / A$ are positive constants.

Steady-state solutions

Setting $x = X$ and $y = Y$, where X and Y are constants, gives

$$0 = -aY \quad \text{and} \quad 0 = -bXY.$$

Hence the steady-state solution is $Y = 0$, X unspecified, which corresponds to the enemy (guerilla) army being defeated. However, as will be shown below via a phase-plane analysis, certain initial conditions lead to $X = 0$ and thus victory to the enemy army.

Phase-plane analysis

The phase-plane technique of Section 19.2 allows us to eliminate t in the equations (2) to obtain the differential equation

$$\frac{dy}{dx} = \frac{b}{a}x.$$

This equation is first-order separable so the methods of Section 11.1 can be used to show that it has solution

$$y = \frac{b}{2a}x^2 + K, \tag{3}$$

where

$$K = y_0 - \frac{b}{2a}(x_0)^2 \tag{4}$$

is a constant depending on the initial conditions.

The parabolas defined by (3) for varying K are sketched in Figure 19.3.1. Only the region of the phase-plane $x, y \geq 0$ is sketched since

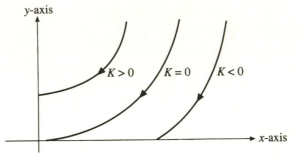

Fig. 19.3.1. Phase-plane trajectories for the guerilla combat model.

the number of soldiers cannot be negative. Note also from (2) that, if $x, y > 0$, then $\dot{x}, \dot{y} < 0$ so all the trajectories point towards the axes.

Suppose that a victory to the home army occurs when $x > 0$ and $y = 0$ (although in practice the defeated army would probably surrender before this stage). From (3) and (4) this occurs when $K < 0$ and thus when

$$y_0 < \frac{b}{2a}(x_0)^2.$$

On the other hand, a victory to the hidden army occurs when $K > 0$ so that

$$y_0 > \frac{b}{2a}(x_0)^2.$$

As an application of these results, let us suppose it appeared that the two armies were heading for mutual annihilation. From (3) and (4) mutual annihilation will occur if

$$K = 0$$

and thus

$$\frac{y_0}{(x_0)^2} = \frac{b}{2a}. \tag{5}$$

However if the enemy army were to double the number of soldiers it had initially we see that the home army would only have to increase the size of its army by a factor of $\sqrt{2}$ to match the enemy army.

Generally, however, the hidden army often has an advantage over the exposed army. If the hidden army is spread over a very large area A then the parameter $b = R_x \alpha / A$ is a small number compared with $a = R_y P_y$. Thus (5) states that the two armies are evenly matched when x_0 is large and y_0 is relatively small.

Other combat models

Models of combat similar to the above were first devised by F.W. Lanchester. Details may be found in a reprinted article in Newman (1956), pp. 2138–2157. Additional discussions may be found in Giordano and Weir (1985) and in Braun (1975).

Other types of combat models include

- hidden armies verses hidden armies,
- exposed armies verses exposed armies.

It is also possible to take account of reinforcements and illness amongst troops. Some of these ideas are considered in the exercises. One particularly interesting situation arises when considering an exposed verses exposed battle, which is discussed in Exercise 5. Here it is shown how an army which would lose a battle by committing all of its soldiers to the battle can gain victory by committing half of the army to a first battle with the whole of the enemy army and then sending the second half to fight the surviving enemy.

These combat models have been applied to some real battles. Braun (1975) discusses an exposed verses exposed model incorporating reinforcements which gives excellent agreement with the battle of Iwo Jima which occurred during the Second World War.

Exercises 19.3

1. In this exercise a model of combat in which both armies are exposed to gun fire is to be developed.

(a) Using equation (1) of the text, obtain the coupled differential equations

$$\frac{dx}{dt} = -py \quad \text{and} \quad \frac{dy}{dt} = -qx$$

and specify the constants p and q in terms of the average firing rates and probability of hitting the target for each army.

(b) Use a phase-plane analysis to determine an inequality relating p and q such that the home army (x) wins. Do the same for a win to the enemy army.

2. (Continuation of Exercise 1) If one side was to double their initial number, by how much should the other side increase their firing rate so that the former action has no net effect on the outcome of the battle?

3. (Continuation of Exercise 1) Suppose a soldier from the home army fires at a rate of two shots per minute with a probability of success of 80% and suppose that the enemy army fires at a rate of three shots per minute with a probability of success of 70%. If initially there are 1000 home soldiers and 900 enemy, who wins the battle? How many survivors are there?

4. The equations obtained in Exercise 1 are linear and can thus be solved using the method of Section 18.2.

    (a) Hence find x and y as functions of t, given that initially $x = x_0$ and $y = y_0$.

    (b) By eliminating t from your solutions in (a) find a relation between y and x. You should thus reproduce the result derived in Exercise 1(b).

5. Suppose that initially there are 80 000 soldiers of the 'home' army and 100 000 soldiers of the enemy, and assume that both armies are exposed as in Exercise 1 above, with soldiers on each side being equally effective. (This means that $p = q$ in the differential equations.)

    (a) If all soldiers take part in the battle, show that the enemy army wins the battle.

    (b) Show that the home army can win by first engaging half of its army against the enemy army and then fighting the other half against the enemy survivors of the first battle.

6.   (a) Argue that a battle with long range artillery will satisfy the differential equations

$$\dot{x} = -\alpha xy \quad \text{and} \quad \dot{y} = -\beta xy$$

    where α and β are positive constants. Express these constants in terms of other meaningful constants such as a firing rate, number of soldiers per missile launcher, and firing rates.

    (b) Sketch the phase-plane trajectories, and give a condition for one of the sides to win.

19.4 Epidemics

A mathematical model of a measles epidemic was presented in Chapter 9. The model was presented as a pair or coupled non-linear *difference* equations. The reason for the applicability of difference equations was the significant latent period between catching the disease and becoming contagious. If this latent period is small (ideally zero) a model of an epidemic involving coupled *differential* equations can be formulated. This will be done in this section. A phase-plane analysis of the model is left to the exercises.

The model

For the purposes of formulating our model for the spread of a disease, the population will be divided into three groups: *susceptibles* (who are not immune to the disease), *infectives* (who are capable of infecting susceptibles) and *removed* (who have previously had the disease and may not be reinfected because they are immune, have been quarantined or have died from the disease). The symbols S, I and R will be used to denote the number of susceptibles, infectives and removed, respectively, in the population at time t.

The following modelling assumptions will be made.

(a) *The disease is transmitted by close proximity or contact between an infective and susceptible.*

(b) *A susceptible becomes an infective immediately after transmission.*

(c) *Infectives eventually become removed.*

(d) *The population of susceptibles is not altered by emigration, immigration, births or deaths.*

(e) *Each infective infects a constant fraction β of the susceptible population per unit time.*

(f) *The number of infectives removed is proportional to the number of infectives present.*

As mentioned earlier, it is assumption (b) which makes a formulation involving differential equations rather than difference equations relevant. Diseases for which this assumption is applicable include diphtheria, scarlet fever and herpes. Assumption (e) is the same as that used in Section 9.4. It is valid provided that the number of infectives is small in comparison to the number of susceptibles.

To set up the differential equations using a δt argument, let $\delta S, \delta I$ and δR denote the changes in the population of susceptibles, infectives and removed during a small time interval δt. By assumptions (a), (c) and (d)

$$\delta S = - \left\{ \begin{array}{c} \text{number of susceptibles} \\ \text{infected in time } \delta t \end{array} \right\}.$$

By assumptions (a), (b) and (c)

$$\delta I = \left\{ \begin{array}{c} \text{number of susceptibles} \\ \text{infected in time } \delta t \end{array} \right\} - \left\{ \begin{array}{c} \text{number of} \\ \text{infectives removed} \end{array} \right\}.$$

By assumptions (a), (c) and (d)

$$\delta R = \left\{ \begin{array}{c} \text{number of infectives} \\ \text{removed in time } \delta t \end{array} \right\}.$$

But from assumptions (e) and (f)

$$\left\{ \begin{array}{c} \text{number of susceptibles} \\ \text{infected in time } \delta t \end{array} \right\} = \beta SI \, \delta t,$$

$$\left\{ \begin{array}{c} \text{number of infectives} \\ \text{removed in time } \delta t \end{array} \right\} = \gamma I \, \delta t.$$

Substituting the last two equations into the previous three, dividing by δt and letting $\delta t \to 0$ gives the coupled differential equations

$$\frac{dS}{dt} = -\beta SI, \tag{1a}$$

$$\frac{dI}{dt} = \beta SI - \gamma I, \tag{1b}$$

$$\frac{dR}{dt} = -\gamma I, \tag{1c}$$

where β and γ are positive constants of proportionality. Here β is known as the *infection rate* which governs how fast the disease is spread from one infective to one susceptible and γ is the removal rate which governs how fast infectives are removed (by dying, becoming immune or by being quarantined).

Equations (1a–c) give three differential equations in three unknowns S, I and R. Notice, however, that the variable R does not occur in (1a) or (1b) hence equation (1c) is not coupled to the system (1a–b).

Note also, that adding the three equations (1a), (1b) and (1c) gives

$$\frac{d}{dt}(S + I + R) = 0$$

which implies that for all times

$$S + I + R = N_0 \tag{2}$$

where N_0 denotes the initial population. Equation (2) is evident from the formulae for δS, δI and δR which in turn follow from assumptions (b), (c) and (d) used in the formulation of the model.

The equations (1a) and (1b) can be analysed using the phase-plane technique, which is to be done in the exercises. An important consequence of the analysis is the following result:

If the initial number of susceptibles is greater than γ/β then the number of infectives will increase and we say that an epidemic has occurred. If it is smaller than γ/β then the number of infectives decreases and thus no epidemic occurs.

Thus knowing β and γ can help decide vaccination strategies: one would try to decrease the number of susceptibles through vaccination to below the value γ/β (called the threshold value).

Alternative versions of the model can be formulated by altering some of the assumptions (a)–(f). These include allowance for births, diseases where immunity is not conferred on sufferers with the disease, and diseases with carriers. Some such models are considered in the exercises. These models are prototypes for much more complicated *stochastic models* which allow for random variation in infectivity and are not restricted to large populations. Braun (1975) and Edelstein-Keshet (1988) pp. 242–253 give fairly comprehensive discussions of epidemics. A more advanced treatment, including stochastic models, is given by Bailey (1975).

Exercises 19.4

1. Consider the equations (1*a*) and (1*b*) in the text, which describe the model for an epidemic with removal.

(a) Eliminate t and solve the resulting differential equation to obtain the formula

$$I = K - S + \frac{\gamma}{\beta} \ln S$$

where K is a constant depending on the initial values of I and K.

(b) From your answer to (a), find the maximum of I regarded as a function of S and sketch some typical trajectories. Thus determine a condition on S_0 for which the number of infectives will keep decreasing with time. Also give the condition on S_0 for which the number of infectives will initially increase with time. In this instance an epidemic is said to occur.

(c) Find a method to determine the number of susceptibles remaining when the disease has run its course and the number of infectives is zero.

2. The differential equations

$$\dot{S} = -\beta SI + \lambda S, \qquad \dot{I} = \beta SI - \gamma I$$

model a disease spread by contact. The same assumptions apply as for the model in the text except that now, in place of assumption (c), we allow birth of susceptibles.

(a) Identify which terms in the RHS of each differential equation arises from birth of susceptibles.

(b) Make a correspondence between these differential equations and the Lotka–Volterra system. Use this correspondence to sketch some phase-plane trajectories. What are the average values of the number of infectives and the number of susceptibles?

3. Generalize the model of the text to include the possibility of those recovering from the disease becoming reinfected.

4. With $\gamma = 0$ and thus zero rate of removal, equations (1a) and (1b) of the text read

$$\dot{S} = -\beta SI \quad \text{and} \quad \dot{I} = \beta SI.$$

(a) Determine, and sketch, the phase-plane trajectories.

(b) Show that

$$\dot{I} = \beta I(N_0 - I)$$

where $N_0 = S_0 + I_0$.

(c) Solve this differential equation, and check that the solution is consistent with the phase-plane trajectories.

(d) Give some examples of diseases which could be modelled by these equations.

5. The following set of coupled non-linear differential equations has been used to model tranmission of AIDS (May, Anderson and McLean, 1988). Here X represents a susceptible population and Y represents those infected with the HIV virus, and $N = X + Y$.

$$\frac{dX}{dt} = \nu N - (\lambda + \mu)X,$$

$$\frac{dY}{dt} = \lambda X - (\nu + \mu)Y,$$

where λ is the probability of acquiring infection from any one partner.

(a) Discuss why it is plausible to write $\lambda = \beta c Y / N$, where β is the probability of acquiring infection from one infected partner and c is the rate at which new partners are acquired.

(b) Which terms in the equation represent

(i) input of new susceptibles,

(ii) HIV carriers who get AIDS,

(iii) those who become infected with HIV?

(c) Show that the total population N satisfies the differential equation

$$\frac{dN}{dt} = \nu N - \mu N - \nu Y.$$

References

Chapter 1

Andrade, E.N. da C. (1979) *Sir Isaac Newton*. Greenwood Press, Westport Connecticut.
Cohen, I.B. (1987) *The Birth of a New Physics*. Penguin Books, Middlesex.
Koestler, A. (1958) *The Sleepwalkers*. Penguin Books, Middlesex.
Westfall, R.S. (1971) *Force in Newtonian Physics*. Elsevier, New York.

Chapter 4

Halliday, D. and Resnick, R. (1974) *Physics Parts I and II*. Wiley, New York.

Chapter 7

Gardner, M. (1981) *Mathematical Circus*. Penguin, Middlesex.

Chapter 8

Archibald, G.C. and Lipsey, R.G. (1973) *An Introduction to a Mathematical Treatment of Economics*. 3rd edition. Weidenfeldt and Nicolson, London.
Ayres, F. (1963) *Theory and Problems of Mathematics of Finance*. Schaum, New York.
Gandolfo, G.C. (1971) *Mathematical Methods and Models in Economic Dynamics*. North Holland, Amsterdam.
Goldberg, S. (1958) *An Introduction to Difference Equations: with illustrated examples from Economics, Psychology and Sociology*. Wiley, New York.
Kenkal, J.L. (1974) *Dynamic Linear Economic Models*. Gordon and Breach, London.
Lipsey, R.G., Langley, P.C. and Mahoney, D.M. (1981) *Positive Economics for Australian Students*. Weidenfeldt and Nicolson, London.
Pfouts, R.W. (1972) *Elementary Economics: a Mathematical Approach*. Wiley, New York.

Chapter 9

Anderson, R. and May, R. (1982) 'The logic of vaccination.' *New Scientist*, November.
Devaney, R.L. (1986) *An Introduction to Chaotic Dynamical Systems*. Benjamin-Cummings, Menlo-Park, California.

Edelstein-Keshet, L. (1988) *Mathematical Models in Biology*. Random House, New York.

Gleick, J. (1987) *Chaos: Making a New Science*. Cardinal, London.

Greenwell, R.N. and Ng, H.K. (1984) 'The Ricker Salmon Model' (Unit 653) UMAP Modules, Comap.

May, R.M. (1975) 'Biological Populations obeying Difference Equations: stable points, stable cycles, and chaos.' *Journal of Theoretical Biology*. **51**, 511–524.

May, R.M. (ed.) (1976) *Theoretical Ecology: Principles and Applications*. Blackwell, Oxford.

May, R.M. (1978) 'Simple mathematical models with very complicated dynamics.' *Nature*, **261**, 459–567.

Maynard-Smith, J. (1968) *Mathematical Ideas in Biology*. Cambridge University Press.

Stewart, I. (1990) *Does God Play Dice? The Mathematics of Chaos*. Penguin, Middelsex.

Tuck, E. and de Mestre N. (1991) *Computer Ecology and Chaos: an Introduction to Mathematical Computing*. Longman Cheshire, Melbourne.

Chapter 10

Edelstein-Keshet, L. (1988) *Mathematical Models in Biology*. Random House, New York.

Haldane, J.B.S. (1924). 'A mathematical theory of artificial and natural selection – I.' *Transactions of the Cambridge Philosophical Society*, **23**, 19–41. (Reprinted in the *Bulletin of Mathematical Biology*, **52**, 209–240. 1990.)

Hartl, D.L. (1980). *Principles of Population Genetics*. Sinauer Assoociates, Sunderland, Massachusetts.

Hexter,W. and Yost, H.T. (1976) *The Science of Genetics*. Prentice-Hall, Englewood Cliffs, New Jersey.

Maynard-Smith, J. (1968) *Mathematical Ideas in Biology*. Cambridge University Press.

Sandfur, J.T. (1968) 'Difference Equations in Genetics.' *The UMAP Journal* **10**, 257–274.

Chapter 11

Braun, M. (1975) *Differential Equations and their Applications: an Introduction to Applied Mathematics*. Springer, New York.

Emlen, J.M. (1984) *Population Biology: The Coevolution of Population Dynamics and Behaviour*. Macmillan, New York.

Emmel, T.C. (1976) *Population Biology*. Harper and Row, New York.

Giancoli, D.C. (1985) *Physics*. 2nd edition. Prentice Hall, Englewood Cliffs, New Jersey.

Hutchinson, G.E. (1971) *An Introduction to Population Ecology*. Yale University Press, Yale.

Keyfitz, N. (1977) *Introduction to the Mathematics of Populations*. Addison-Wesley, Reading, Massachusetts.

Kormondy, E.J. (1976) *Concepts of Ecology*. Prentice-Hall, Englewood Cliffs, New Jersey.

Marion, J.B. (1976) *Physics in the Modern World*. Academic Press, Reading, Massachusetts.

Rubinow, S.I. (1975) *An Introduction to Mathematical Biology*. Wiley, New York.

Chapter 12

Holman, J.P. (1981) *Heat Transfer*. McGraw-Hill, New York.
Modelling Heat. Open University Module, Open University Press, Milton
  Keynes.

Chapter 13

Rainey, R.H. (1967) 'Natural Displacement of Pollution from the Great Lakes.'
  Science, **155**, 1244.
Modelling Heat. Open University Module, Open University Press, Milton
  Keynes.

Chapter 11

Melissinos, A.C. (1966) *Experiments in Modern Physics*. Academic Press,
  Reading, Massachusetts.
Streeter, V.L. (1966) *Fluid Mechanics*. McGraw-Hill, New York.

Chapter 15

Braun, M. (1975) *Differential Equations and their Applications: an Introduction
  to Applied Mathematics*. Springer, New York.

Chapter 16

Cohen, I.B. (1987) *The Birth of a New Physics*. Penguin, Middlesex.
De Mestre, N. (1990) *The Mathematics of Projectiles in Sport*. Cambridge
  University Press.

Chapter 17

Toeplitz, O. (1963) *Calculus: A Genetic Approach*. University of Chicago Press,
  Chicago.

Chapter 18

Ackerman, E., Rosevar, J.W. and Mc Guckin, W.F. (1964) *Phys. Med. Biol.* **9**,
  203.
Bolie, V.W. (1960) 'Coefficients of normal glucose regulation.' *Journal of Applied
  Physiology*, **16**, 783.
Braun, M. (1975) *Differential Equations and their Applications: an Introduction
  to Applied Mathematics*. Springer, New York.
Burghes, D.N. and Borrie (1981) *Modelling with Differential Equations*. Ellis
  Horward Limited, Chichester.
Edelstein-Keshet, L. (1988) *Mathematical Models in Biology*. Random House,
  New York.
Middleman (1972) *Transport Phenomena in the Cardiovascular System*.
  Wiley-Interscience, New York.
Rodin, E,.Y. and Jacques, S. (1989) 'Countercurrent oxygen exchange in the
  swim bladders of deep sea fish: a mathematical model.' *Mathematical and
  Computer Modelling*, **12**, 389.

Chapter 19

Bailey, T.J. (1975) *The Mathematical Theory of Infectious Diseases and its Applications.* 2nd edition. Charles Griffin and Company Limited.

Braun, M. (1975) *Differential Equations and their Applications: an Introduction to Applied Mathematics.* Springer, New York.

Edelstein-Keshet, L. (1988) *Mathematical Models in Biology.* Random House, New York.

Giordano, F.R. & Wier, M.D. (1985) *A First Course in Mathematical Modeling.* Belmont California.

Lanchester: chapter in Newman (1956) *The World of Mathematics - Volume Two.* Simon and Schuster, New York.

Index

acceleration, 28, 322
acceleration, angular, 338, 339
air-resistance, 331
alleles, 178–185
amortization, 136
angular acceleration, 338, 339
angular velocity, 338, 343
Archimedes' principle, 282
Aristotle, 7
armies, 389
artillary, 394
artillary gun, 308
attractor, 172
Atwood's machine, 41
average acceleration, 29

battle, 389, 393
birth rate, 148
births, 147
block and tackle, 41
blood vessels, 374
breeding season, 145
buoyancy, 282

Carbon-14 dating, 230
carrying capacity, 152, 218
cell division, 147
chaotic growth, 159
characteristic equation, 296
circular motion, 343
cobweb diagrams, 118, 120, 129
cobweb model, 138
compartment, 257, 367
compartment diagram, 263, 270
compartment model, 258, 268
computer, 157–163, 166
concentration, 229, 258, 357
conduction, 241
conductivity, 243
constant solutions, 206

constant-coefficient, 127, 131, 295
convection, 239
convective heat transfer coefficient, 240
convergence, 120
corpse, cooling, 237
countercurrent, 377
coupled non-linear differential equations, 379
critically damped motion, 307

damping, 302, 304
damping constant, 302
dashpot, 302
death rate, 148
deaths, 147
diabetes, 366
difference equation, 107, 109
discrete logistic equation, 153
disease, 163, 395
displacement, 319
drag coefficient, 281
drag force, 277
drug absorption, 229
dynamics, 11

Earth, orbit of, 345
epidemic, 163, 394, 397
equilibrium states, 359
existence–uniqueness, 75
exponential decay, 227
exponential growth, 213, 216, 223

Fibonacci, 105, 108
fish populations, 379
fixed points, 114, 116
fluid mechanics, 277
forcing term, 312, 314
Fourier, 243
Fourier's law, 242, 250
friction, 60–70, 326, 328

friction, kinetic, 62
friction, static, 61

Galileo, 8, 9, 16, 326
genetics, 176
genotype, 178, 180
genotype proportions, 187
glucose, 366, 368
glucose-tolerance test, 368, 381
Great Lakes system, 265
guerilla combat, 390

Hardy-Weinberg law, 191
harmonic motion, 299, 304
harvesting fish, 222
heat and temperature, 237
heat flow, radial, 250
homogeneous, 78, 81, 127, 131, 204, 295, 298
Hooke's law, 86, 87
hot water tank, 270
housing loans, 136
human populations, 224
hydrometer, 284, 311

immune, 396
inclined plane, 326
infective, 164
infectives, 395
inhomogeneous, 205
injection, 229
insulation, 241, 242, 244, 249, 254, 274
insulin, 366, 368
interest, compound, 133, 134
interest, simple, 133, 134
intravenous infusion, 231
iteration, 111, 118

Kepler, 7, 9
Kepler's laws, 9, 348
kinematics, 10, 21–40, 318, 336

Lanchester, 393
latent period, 164
Leibniz's notation, 26
Leonardo of Pisa, 105
lethal recessive gene, 193–200
linear difference equations, 126–145
linear differential equation, 77, 81, 369
linear first-order difference equation, 129
linearization, 173, 348, 351
linearization about a fixed point, 171
logistic growth, 218
Lotka, 382
Lotka–Volterra equations, 382–384, 386–388

measles, 164–170, 394
Mendel, 178
military ballistics, 334
Millikan oil drop experiment, 288
mixing, 257, 356
mixture, 258, 356
mutation, 198

national income, 142
natural frequency, 314
natural length, 85
natural selection, 196
Newton, 13, 232
Newton's law of cooling, 232–234, 239, 252, 271
Newton's laws, 326
Newton's laws of motion, 13
normal vector, 340

overdamped motion, 305

particular solution, 297, 299, 313
pendulum, 349
period and amplitude, 82
phase-plane, 391, 398
phase-shift, 76
placenta, 373
planetary orbits, 9
pollution, 265
position vector, 319, 322
predator, 379–384
prey, 379–384
projectile, 322, 331, 334
pulley, 41

radial heat flow, 250
radioactive decay, 227
random mating, 185–192
repellor, 172
resonance, 314
resonant frequency, 314
Reynold's number, 278, 279, 280

shark populations, 379
small interval approximation, 259, 269
solute, 258
sport, mathematics in, 334
springs, 85–101, 302, 304
stable, 171
stable equilibrium point, 350
steady-state, 114, 154, 167, 171, 206, 369
steady-state conduction, 241
steady-state temperature, 235
stiffness, 87

stochastic models, 397
Stokes' law, 280
superposition, 78, 79, 81
superposition theorems, 127
supply and demand, 138–140
survival fraction, 190
susceptible, 164
susceptibles, 395
symmetry, 57, 58

tangent vector, 323, 340, 341
tension, 41
terminal velocity, 287, 288, 292
Toricelli's law, 269
trajectory, 331
transient, 311, 313
turbidity, 290

underdamped motion, 304
unstable, 172

vaccination, 169, 396
variables separable, 206
variables separable differential equations, 203
vector, tangent, 323
velocity, 322
velocity, angular, 338, 343
velocity-squared drag law, 281
viscosity, 277–279
Volterra, 379, 380, 382

wheel, 68

yeast cells, 212, 213, 220, 225

Made in the USA
Lexington, KY
31 August 2016